Singularities in Boundary Value Problems

NATO ADVANCED STUDY INSTITUTES SERIES

Proceedings of the Advanced Study Institute Programme, which aims at the dissemination of advanced knowledge and the formation of contacts among scientists from different countries

The series is published by an international board of publishers in conjunction with NATO Scientific Affairs Division

A	Life Sciences	Plenum Publishing Corporation
B	Physics	London and New York
C	Mathematical and Physical Sciences	D. Reidel Publishing Company Dordrecht, Boston and London
D	Behavioural and Social Sciences	Sijthoff & Noordhoff International Publishers
E	Applied Sciences	Alphen aan den Rijn and Germantown U.S.A.

Series C – Mathematical and Physical Sciences

Volume 65 – Singularities in Boundary Value Problems

Singularities in Boundary Value Problems

Proceedings of the NATO Advanced Study Institute held at Maratea, Italy, September 22 - October 3, 1980

edited by

H. G. GARNIR
University of Liège, Belgium

D. Reidel Publishing Company

Dordrecht : Holland / Boston : U.S.A. / London : England

Published in cooperation with NATO Scientific Affairs Division

Library of Congress Cataloging in Publication Data

CIP

Nato Advanced Study Institute, Maratea, Italy, 1980.
Singularities in boundary value problems.

(NATO advanced study institutes series: Series C, Mathematical and physical sciences; v. 65)
Sponsored by NATO Scientific Affairs Division.
Includes index.
1. Boundary value problems–Congresses. 2. Singularities (Mathematics)–Congresses. I. Garnir, Henri G. II. North Atlantic Treaty Organization. Division of Scientific Affairs. III. Title. IV. Series.
QA379.N37 1980 515.3'5 80-29685
ISBN 90-277-1240-9

Published by D. Reidel Publishing Company
P.O. Box 17, 3300 AA Dordrecht, Holland

Sold and distributed in the U.S.A. and Canada
by Kluwer Boston Inc.,
190 Old Derby Street, Hingham, MA 02043, U.S.A.

In all other countries, sold and distributed
by Kluwer Academic Publishers Group,
P.O. Box 322, 3300 AH Dordrecht, Holland

D. Reidel Publishing Company is a member of the Kluwer Group

Printed in The Netherlands

TABLE OF CONTENTS

PREFACE

The 1980 Maratea NATO Advanced Study Institute (= ASI) followed the lines of the 1976 Liège NATO ASI.

Indeed, the interest of boundary problems for linear evolution partial differential equations and systems is more and more acute because of the outstanding position of those problems in the mathematical description of the physical world, namely through sciences such as fluid dynamics, elastodynamics, electrodynamics, electromagnetism, plasma physics and so on.

In those problems the question of the propagation of singularities of the solution has boomed these last years.

Placed in its definitive mathematical frame in 1970 by L. Hörmander, this branch of the theory recorded a tremendous impetus in the last decade and is now eagerly studied by the most prominent research workers in the field of partial differential equations.

It describes the wave phenomena connected with the solution of boundary problems with very general boundaries, by replacing the (generallly impossible) computation of a precise solution by a convenient asymptotic approximation. For instance, it allows the description of progressive waves in a medium with obstacles of various shapes, meeting classical phenomena as reflexion, refraction, transmission, and even more complicated ones, called supersonic waves, head waves, creeping waves,......

The study of singularities uses involved new mathematical concepts (such as distributions, wave front sets, asymptotic developments, pseudo-differential operators, Fourier integral operators, microfunctions, ...) but emerges as the most sensible application to physical problems.

A complete exposition of the present state of this theory seemed to be still lacking.

H. G. Garnir (ed.), Singularities in Boundary Value Problems, vii–viii.

The Maratea ASI aimed to fill the gap by gathering most of the specialists in this field to present a unified, detailed, and up-to-date exposition of their recent results in an organized and pedagogical way.

The meeting had also some experimental character.

It was organized in the frame of an attempt by the NATO authorities to shift the ASIs to sites selected by them in specially convenient places.

That is why this meeting was located in Acquafredda di Maratea, a nice little village along the quiet and sunny Gulf of Policastro in Southern Italy.

The Institute was held from September 22 to October 3, 1980. It was attended by 77 participants; 59 from NATO countries (Belgium (15), Canada (1), France (11), West Germany (8), Greece (1), Italy (7), Turkey (7), U.K. (1), U.S.A. (8)) and 18 from non NATO countries (Algeria (1), Australia (2), Brazil (1), Egypt (1), Iran (1), Israel (2), Japan (7), Sweden (1), Switzerland (1), Venezuela (1)). During the session, 14 lecturers delivered 4 courses of 6 hours, 4 of 4 hours, 4 of 2 hours and 2 of 1 hour. Moreover, 20 advanced 20-minute seminars were organized by the participants to discuss the latest contributions to the field.

I wish to express my warmest thanks to NATO, which was the main sponsor of the Maratea meeting; I am specially grateful to the Scientific Affairs Division and more precisely to Dr. Di Lullo, NATO Scientific Officer in charge of the ASI programme.

It is also my pleasure to thank heartily all the institutions which contributed financially and specially the European Research Office of the U.S. Army, the Office of Naval Research of the U.S. Navy, the National Science Foundation of the U.S.A., Division of International Programs, and the University of Liège.

Many problems arising from the location of the meeting at Maratea were solved through the association "International Transfer of Science anf Technology", directed by Dr. T. Kester, for whose advice I am specially grateful.

For all the scientific and organizing tasks connected with this ASI, I have been helped efficiently by Prof. Léonard of the University of Liège, codirector of the Institute, who also deserves all my thanks.

H. G. GARNIR
Director of the Institute

LIST OF PARTICIPANTS

GARNIR H.G. : Inst. de Math. Univ. de Liège
15, avenue des Tilleuls / B-4000 LIEGE
BELGIUM

LEONARD P. : Inst. de Math. Univ. de Liège
15, avenue des Tilleuls / B-4000 LIEGE
BELGIUM

CHAZARAIN J : Institut de Mathématiques
Parc Valrose, 06034 Nice Cedex
FRANCE

ESKIN G. : Institute of Mathematics
The Hebrew University of Jerusalem
Jerusalem / ISRAEL

FRIEDLANDER F.G. : University of Cambridge
Department of Mathematics
Silver Street
Cambridge CB3 9EW / ENGLAND

HORMANDER L : University of Lund
Department of Mathematics
Box 725
220 07 Lund 7 / SWEDEN

IKAWA M : Osaka University
Department of Mathematics
Toyonaki, Osaka 560
JAPAN

KATAOKA K. : Department of Mathematics
Faculty of Science
University of Tokyo
Hongo, Tokyo / JAPAN

MELROSE R. : Department of Mathematics
Massachussetts Institute of Technology
(M.I.T.) Math. 2 - 171
Cambridge Mass. 02139 / U.S.A.

RALSTON J. : Department of Mathematics
University of California
Los Angeles, California 90024 / U.S.A.

SHIBATA Y. : University of Tsukuba
Department of Mathematics
Sakura-Mura, Niihari-Gun
Ibaraki 305 / JAPAN

SCHAPIRA P. : Université de Paris-Nord
Département de Mathématiques
Avenue J.B. Clément
Villetaneuse / FRANCE

SJOSTRAND J. : Département de Mathématiques
Université de Paris-Sud
Orsay / FRANCE

TAYLOR M. : Department of Mathematics
SUNY , Stony Brook
New York 11794 / U.S.A.

TSUJI M. : Kyoto University
Department of Mathematics
Kamigamo, Kita-Ku
Kyoto 603 / JAPAN

WAKABAYASHI S. : Institute of Mathematics
The University of Tsukuba
Sakura-Mura, Niihari-Gun
Ibaraki 305 / JAPAN

ALAYLIOGLU A. : Department of Mathematics
Middle Est Technical University
Ankara / TURKEY

ALBER H.D. : Institut für Angewandte Mathematik
Universität Bonn
Wegelerstr. 10
5300 Bonn 1 / WEST GERMANY

ALTIN A : Ankara Universitesi, Fen Fakültesi
Tatbiki Matematik Kürsüsü
Besevler
Ankara / TURKEY

AROSIO A. : Istituto di Matematica "Leonida Tonelli"
Via Buonarrotti, 2
56100 Pisa / ITALY

BASKAN T. : Hacettepe University
Fac. of Nat. Sciences and Engineering
Hacettepe, Ankara / TURKEY

BENAISSA L. : Université d'Alger
Département de Mathématiques
Alger / ALGERIE

BENGEL G. : Mathematisches Institut der Universität
Roxeler Strasse, 64
44 Münster / GERAMNY

BENNETT M. : University of Cambridge
D.P.M.M.S
16 Mill Lane
Cambridge CB2ISB / ENGLAND

CARDOSO F. : Universidade Federal de Pernambuco
Recife / BRAZIL

CATTABRIGA L. : Universita di Bologna
Istituto Matematico "Salvatore Pincerle"
Piazza di Porta S. Donato, 5
40127 Bologna / ITALY

CONSTANTIN P. : The Institute of Mathematics
The Hebrew University
Givat Ram Campus
Jerusalem / ISRAEL

COTTAFAVA G. : Istituto di Informatica e Sistemistica
dell Universita di Pavia
Strada Nuova 106/c
27100 Pavia / ITALY

DE JONGE J. : Institut Supérieur Industriel
6000 Charleroi / BELGIUM

DELANGHE R. : Seminar of Higher Analysis
State University of Ghent
Krijgslaan, 271
9000 Ghent / Belgium

DERMENJIAN Y. : Université de Paris-Nord
Centre Scientifique et Polytechnique
Avenue J.-B. Clément
93430 Villetaneuse / FRANCE

DIOMEDA L. : Universita degli Studi di Bari
Facolta di Scienze
Istituto di Analisi Matematica
Palazzo Ateneo
70121 Bari / ITALY

DUFF G. : University of Toronto
Department of Mathematics
Toronto, 181 / CANADA

ER U. : Academy Engineering and Architecture
State Academy of Konya / TURKEY

ETIENNE J. : Université de Liège
Institut de Mathématique
Avenue des Tilleuls, 15
4000 LIEGE / BELGIUM

FLORET K. : Mathematisches Seminar
Olshausenstrasse 40-60
Haus S 12
2300 Kiel 1 / GERMANY

GERARD P. : Université de Liège
Institut de Mathématique
Avenue des Tilleuls, 15
4000 LIEGE / BELGIUM

GERARD-HOUET C. : Université de Liège
Institut de Mathématique
Avenue des Tilleuls, 15
4000 LIEGE / BELGIUM

GEVECI T. : National Research Institute for Mathematical Sciences
P.O. Box 395
Pretoria 0001 / SOUTH AFRICA

GODIN P. : Département de Mathématiques
Université Libre de Bruxelles
Campus de la Plaine, C.P. 214
Boulevard du Triomphe
1050 Bruxelles / BELGIUM

GRIGIS A. : Ecole Polytechnique
Centre de Mathématique
91128 Palaiseau Cedex
FRANCE

von GRUDZINSKI O. : Mathematisches Seminar der Universität
Olshausenstr. 40-60
2300 Kiel 1 / WEST GERMANY

GUILLOT J.-C. : Université de Paris-Nord
Centre Scientifique et Polytechnique
Avenue J.-B. Clément
93430 Villetaneuse / FRANCE

HANGES N. : School of Mathematics
The Institute for Advanced Study
Princeton
New Jersey 08540 / U.S.A.

HANSEN S. : F.B. Mathematik - Informatik, Universität
Warburger Str. 100
4790 Paderborn / WEST GERMANY

HEINS A. : The University of Michigan
Department of Mathematics
347 West Engineering Building
Ann Arbor
Michigan 48109 / U.S.A.

HELFFER B. : Université de Nantes
Institut de Mathématiques et d'Informatique
2, chemin de la Houssinière
44072 Nantes Cedex
FRANCE

KANDIL M. : Department of Mathematics
Faculty of Science
University of Assiut
Assiut / EGYPT

KARACAY T. : Hacettepe University
Faculty of Natural Sciences and Engineering
Hacettepe
Ankara / TURKEY

KUO T.-C. : Department of Pure Mathematics
Sydney University
Sydney 2006 / AUSTRALIA

LANGE H. : Mathematisches Institut
Univ. Köln
Weyertal 86-90
5000 Köln 41 / GERMANY

LAUBIN P. : Université de Liège
Institut de Mathématique
15, avenue des Tilleuls
4000 Liège / BELGIUM

LEJEUNE Y. : Université de Liège
Institut de Mathématique
15, avenue des Tilleuls
4000 Liège / BELGIUM

LEJEUNE-RIFAUT E. : Université de Liège
Institut de Mathématique
15, avenue des Tilleuls
4000 Liège / BELGIUM

LIESS O. : Technische Hochschule Darmstadt
Fachbereich Mathematik
61 Darmstadt / GERMANY

LIEUTENANT J.-L. : Université de Liège
Institut de Mathématique
15, avenue des Tilleuls
4000 Liège / BELGIUM

LOUSBERG P. : Université de Liège
Institut de Mathématique
15, avenue des Tilleuls
4000 Liège / BELGIUM

MAIRE H.-M. : Section de Mathématiques
Case Postale 124
1211 Genève / SUISSE

MAMOURIAN A. : Department of Mathematics
Faculty of Sciences
University of Isfahan
Isfahan / IRAN

MATSUMOTO W. : Kyoto Sangyo University
Department of Mathematics
Kamigamo, Kita-Ku
Kyoto 603 / JAPAN

MEHMETI A. : Johannes Gutenberg Universität
Postfash 3980
Saarstrasse 21
6500 Mainz / GERMANY

MENDOZA G. : IVIC - Matematicas
Apartado 1827
Caracas / VENEZUELA

MORIMOTO Y. : Department of Engineering Mathematics
Faculty of Engineering
Nagoya University
464 Furo-Cho, Chikusa-Ku
Nagoya / JAPAN

MUNSTER M. : Université de Liège
Institut de Mathématique
15, avenue des Tilleuls
4000 Liège / BELGIUM

NOURRIGAT J. : Université de Rennes
UER Mathématiques et Informatique
Avenue du Général Leclerc
Rennes Beaulieu
35042 Rennes Cedex / FRANCE

PALMIERI G. : Universita degli Studi di Bari
Facolta di Scienze
Istituto di Analisi Matematica
Palazzo Ateneo
70121 Bari / ITALY

PIRIOU A. : Département de Mathématiques
Faculté des Sciences
Parc Valrose
06 Nice / FRANCE

RASSIAS J. : University of Athens
Department of Mathematics
Athens / GREECE

ROBERT D. : Université de Nantes
Institut de Mathématiques et d'Informatique
38, bd Michelet
BP 1044
44037 Nantes / FRANCE

RODINO L. : Istituto di Analisi Matematica
Universita di Torino
Via Carlo Alberto, 10
10123 Torino / ITALY

SILBERSTEIN P. : University of Western Australia
Department of Mathematics
Nedlands 6009 / WESTERN AUSTRALIA

TUNCAY H. : University of Egée
Izmir / TURKEY

VAILLANT J. : Académie de Paris
Université de Paris VI
Département de Mathématiques
4, place Jussieu
75230 Paris Cedex / FRANCE

WILLIAMS M. : Department of Mathematics
Massachussetts Institute of Technology (M.I.T.)
Math. 2 - 171
Cambridge Mass. 02139 / U.S.A.

WUIDAR J. : Université de Liège
Institut de Mathématique
15, avenue des Tilleuls
4000 Liège / BELGIUM

YINGST D. : Department of Mathematics
U.C.L.A.
Los Angeles
California / U.S.A.

ZANGHIRATI L. : Istituto Matematico
Universita
Via Machiavelli, 35
44100 Ferrara / ITALY

ZEMAN M. : Department of Mathematics
Southern Illinois University
Carbondale
Ill. 62901 / U.S.A.

SUR LE COMPORTEMENT SEMI CLASSIQUE DU SPECTRE ET DE L'AMPLITUDE DE DIFFUSION D'UN HAMILTONIEN QUANTIQUE.

J. CHAZARAIN

UNIVERSITE DE NICE

Résumé. Soit $\mathscr{H}_h = -\frac{1}{2} h^2 \Delta + V(x)$ un hamiltonien quantique qui dépend du paramètre $h = \frac{\hbar}{m} \in \left]0, h_o\right]$. Dans le cas où le potentiel V vérifie certaines conditions de croissance à l'infini, on relie l'ensemble de fréquence de la distribution $S_h(t) = \sum_j \exp(ih^{-1} \lambda_j t)$ à l'ensemble des périodes des trajectoires classiques.

Dans le cas où V est à support compact, on relie l'ensemble de fréquence de l'amplitude de diffusion à l'ensemble des durées de séjour des trajectoires classiques.

I - INTRODUCTION.

Un principe de la physique quantique demande de relier le comportement semi classique, c'est-à-dire quand $h \longrightarrow o$, des grandeurs associées à l'hamiltonien quantique

$$(1,1) \quad \mathscr{H}_h = -\frac{1}{2} h^2 \Delta + V(x)$$

H. G. Garnir (ed.), Singularities in Boundary Value Problems, 1–18.

à la mécanique classique de la particule ; voir par exemple : [BERRY & MOUNT] , [VOROS] .

Les trajectoires classiques dans $T^*\mathbb{R}^n$ sont les solutions des équations d'Hamilton

(1,2) $\dot{y}(t) = \partial_\eta H, \quad \dot{\eta}(t) = - \partial_y H$

correspondant à l'hamiltonien classique

(1,3) $H(y,\eta) = \frac{1}{2} |\eta|^2 + V(y),$

ou encore solutions des équations de Lagrange

(1,4) $\frac{d}{dt}(\partial_{\dot{y}} L) - \partial_y L = 0$

correspondant au lagrangien

(1,5) $L = \dot{y} \cdot \eta - H = \frac{1}{2} |\dot{y}|^2 - V(y).$

Lorsque l'on s'interesse au comportement asymptotique, quand $h \longrightarrow 0$, d'une fonction (ou d'une distribution) T_h définie sur un ouvert X et paramétrée par $h \in]0,h_0]$, il faut préciser en particulier s'il y a un comportement oscillant ou non. Dans ce but [GUILLEMIN & STERNBERG] on introduit le concept d'"ensemble de fréquence" (frequency set) pour localiser l'ensemble des fréquences d'oscillation quand $h \longrightarrow 0$. Cet ensemble, noté $F[T_h]$, semble jouer un rôle analogue à celui du spectre singulier (wave front set) pour les questions de comportement asymptotique. Rappelons la

<u>DEFINITION</u>. Un point (x^0,ξ^0) de T^*X n'appartient pas à l'ensemble de fréquence $F[T_h]$ de T_h si et seulement si il existe $\rho \in C_0^\infty(X)$ non nulle en x^0 et un voisinage $\mathcal{V}$ de ξ^0 (non conique en général !) tel que

(1,6) $$< \rho(x)\, e^{-i h^{-1} x.\xi} , T_h > \; = O(h^\infty)$$

uniformément pour ξ dans $\mathcal{V}$.

Ainsi $F[T_h]$ est un fermé de T^*X. Par exemple, si $T_h = a(x)\, e^{ih^{-1}\psi(x)}$ avec a et ψ dans $C^\infty(X)$ et ψ réelle, on trouve immédiatement que

$$F[T_h] \subset \{(x, d\psi(x)) \mid x \in X\} = \Lambda_\psi .$$

Plus généralement, si T_h est une fonction oscillante de Maslov associée à une variété lagrangienne Λ , on trouve que $F[T_h] \subset \Lambda$.

Rappelons également la définition, selon [DUISTERMAAT] d'une fonction oscillante de degré μ associée à une variété lagrangienne Λ . C'est une fonction C^∞ paramétrée par h et qui est somme localement finie en x de fonctions du type suivant

$$(1,7) \qquad \left(\frac{1}{2\pi h}\right)^{N/2} \int e^{-i h^{-1} \varphi(x,\zeta)} a(x,\zeta,h)d\zeta$$

où φ est une phase locale pour Λ et a est un symbole d'ordre μ en h^{-1} et à support compact en la variable ζ dans un espace $\mathbb{R}^N$.

Dans le premier exposé, on suppose que le potentiel V a une certaine croissance à l'infini et on étudie l'ensemble de fréquence de la distribution $S_h(t) = \sum_j \exp(i h^{-1} \lambda_j t)$ associée aux valeurs propres λ_j de l'hamiltonien $\mathcal{H}_h$.

Dans le deuxième exposé, on suppose au contraire que $\mathcal{V} \in C_o^\infty(\mathbb{R}^n)$ et l'on s'intèresse à l'ensemble de fréquence de l'amplitude de diffusion $A_h(k,\omega,\theta)$ pour ω et θ fixés.

II - CAS OU LE POTENTIEL TEND VERS L'∞.

On suppose que V est une fonction C^∞ à valeurs réelles qui vérifie pour $|x| \longrightarrow +\infty$ les estimations suivantes :

$$(2,1) \qquad \partial_x^\alpha V(x) = O\left(|x|^{(2-|\alpha|)_+}\right) \text{ et } V(x) \geq c\,|x|^2 \text{ avec}$$

$c > o$.

L'exemple le plus simple est l'oscillateur harmonique. Dans ces conditions on sait que $\mathcal{H}_h$ est un opérateur auto-adjoint dans $L^2(\mathbb{R}^n)$ et son spectre est constitué de valeurs propres

$$\lambda_1(h) \leq \lambda_2(h) \leq \ldots\ldots \text{ avec } \lambda_j(h) \underset{j}{\longrightarrow} +\infty.$$

La mesure spectrale de $\mathscr{H}_h$ est définie par

(2,2) $\sigma_h(\lambda) = \sum_j \delta(\lambda - \lambda_j)$,

et par Fourier en λ, on obtient la distribution $S_h(t)$ paramétrée par h.

Notons $\mathscr{L}_\tau$ l'ensemble des périodes des solutions périodiques de (1,2) d'énergie $\frac{1}{2}|\eta|^2 - V(y) = -\tau$.

On définit l'ensemble $\mathscr{L}$ comme étant la fermeture dans $T^*\mathbb{R}$ de $\bigcup_{\tau \in \mathbb{R}} (\mathscr{L}_\tau \times \{\tau\})$. Alors, on a le

THEOREME 1. Avec les hypothèses précédentes sur V, on a l'inclusion $F[S_h] \subset \mathscr{L}$.

Ce type d'inclusion est à rapprocher des travaux de [BALIAN & BLOCH]. Un résultat analogue a été annoncé simultanément par [ALBEVERIO - BLANCHARD - HØEGH KROHN] qui utilisent la technique des intégrales de Feynman.

Il y a également une ressemblance avec la relation de Poisson relative à un opérateur elliptique sur une variété compacte, mais elle est, à notre avis, purement formelle.

Nous allons indiquer seulement le principe de la démonstration de ce théorème ; les détails sont publiés dans [CHAZARAIN] où l'on trouvera aussi d'autres résultats dans cette direction.

Soit $U_h(t) = \exp(-ih^{-i}\, t\, \mathscr{H}_h)$ le groupe solution de l'équation de Shrödinger.

(2,3) $(ih\,\partial_t - \mathscr{H}_h)U = 0, \quad U(0) = I.$

Ce groupe est lié à la distribution S_h par l'égalité

(2,4) $< S_h(t), \Theta(t) > = \text{trace} \left(\int U_h(t)\, \Theta(t)\, dt \right)$

pour toute $\Theta \in \mathscr{S}(\mathbb{R})$. Aussi, pour étudier le comportement asymptotique de la quantité

$$I_h = < e^{-ih^{-1}\tau t} \rho(t), S_h(t) >$$

on est conduit à chercher une approximation $E_h(t,x,y)$ du noyau distribution de l'opérateur $U_h(t)$ et ensuite à montrer que le comportement de

$$(2,5) \qquad J_h = \int e^{-ih^{-1}\tau t} \rho(t) E_h(t,x,x)\,dx\,dt$$

est égal à celui de I_h modulo $O(h^\infty)$.

On commence par supposer que $\rho(t)$ est à support voisin de 0 , de façon à pouvoir chercher l'opérateur $E_h(t)$ sous la forme

$$(2,6) \qquad (E_h(t)u)(x) =$$

$$= (2\pi h^{-1})^{-n} \int e^{ih^{-1}(S(t,x,\eta)-y.\eta)} a(t,x,\eta,h)\, u(x)\,d\eta\,dx$$

quand t est petit .

La phase S doit être solution de

$$(2,7) \qquad \partial_t S + H(x, \partial_x S) = 0 \quad , \quad S|_{t=0} = x.\eta \quad ;$$

et l'amplitude $\qquad a \sim \sum_{j\geq 0} h^j a_j(t,x,\eta)$

vérifie les équations de transport correspondantes.
On renvoie pour les détails à [CHAZARAIN] . Une construction voisine a été faite également par [FUJIWARA] et [KITADA] .

La difficulté vient du fait que l'on a besoin de contrôler simultanément le comportement à l'infini en x et η et le comportement vers 0 en h des fonctions $S(t,x,\eta)$ et $a(t,x,\eta;h)$ pour pouvoir donner un sens et étudier les intégrales (2,5) et (2,6) ; c'est la partie technique de ce travail. Ensuite, on obtient facilement le comportement de J_h en appliquant le théorème de la phase non stationnaire. Enfin, pour déduire le comportement de I_h de celui de J_h , on s'appuie sur des résultats de [ASADA & FUJIWARA] concernant la continuité L^2 d'intégrales du type (2,6) .

Terminons ce paragraphe par quelques remarques.

Remarque 1. Pour construire la solution asymptotique $E_h(t)$ de l'équation de Schrödinger, on peut penser se ramener, après une transformation de Fourier par rapport à h^{-1}, à chercher une solution modulo C^∞ de l'équation transformée.

De façon plus précise, si on pose

$$W(s,t) = \int e^{-ish^{-1}} U_h(t)dh^{-1}$$

on trouve que W doit résoudre un problème de Cauchy caractéristique pour l'opérateur

$$\partial_s\partial_t + \tfrac{1}{2}\Delta_x + V(x)\,\partial_s^2$$

relativement à l'hyperplan $t = 0$.

Cette méthode ne semble donc pas très fructueuse !

Remarque 2. On trouvera dans [HELFFER & ROBERT] une extension de nos résultats au cas où $\mathcal{H}_h$ est dans certaines classes d'opérateurs pseudo-différentiels elliptiques.

III – CAS OU LE POTENTIEL V EST A SUPPORT COMPACT.

Dans ce paragraphe, on suppose $V \in C_0^\infty(\mathbb{R}^n)$, et pour simplifier certaines expressions on prend $\mathcal{H}_h = -h^2\Delta + V(x)$. Alors le spectre de $\mathcal{H}_h$ n'est plus discret et on a une théorie de la diffusion (scattering) pour le spectre continu.

On démontre que pour tout réel $k > 0$ et tout vecteur unitaire $\omega \in S_{n-1}$ il existe une fonction propre généralisée $\Phi_h(k, x, \omega)$ définie par les conditions :

$$(3,1) \qquad (-h^2\Delta + V(x) - k^2)\,\Phi_h = 0$$

$$(3,2) \qquad \Phi_h - \exp(i\,h^{-1}k\langle x,\omega\rangle) = O(|x|^{\frac{1}{2}(1-n)} \cdot e^{ih^{-1}k|x|})$$

De plus, le premier membre de (3,2) multiplié par $|x|^{\frac{1}{2}(n-1)} \cdot e^{-ih^{-1}k|x|}$ admet une limite quand x tend vers l'infini dans la direction du vecteur unitaire $\theta \in S_{n-1}$.

Cette limite est notée $A_h(k, \theta, \omega)$ et s'appelle l'amplitude de diffusion. Pour la démonstration de ces résultats de la théorie de la diffusion, on renvoie par exemple à l'article de [SHENK & THOE] .

Dans ce paragraphe, on se propose de relier l'ensemble de fréquence de l'application $k \to A_h(k, \theta, \omega)$ pour θ et ω fixés, à l'ensemble des durées de séjour des trajectoires classiques arrivant à la vitesse $2k\omega$ et sortant à la vitesse $2k\theta$. Pour définir cette notion, on se donne un réel R assez grand pour que le support de V soit inclus à l'intérieur de la sphère $|X| = R$, sphère notée Σ.

Soient A et B les points de Σ définis par $\vec{OA} = -R\omega$ et $\vec{OB} = R\theta$; on note Π_ω et Π_θ les plans tangents à Σ en A et B.

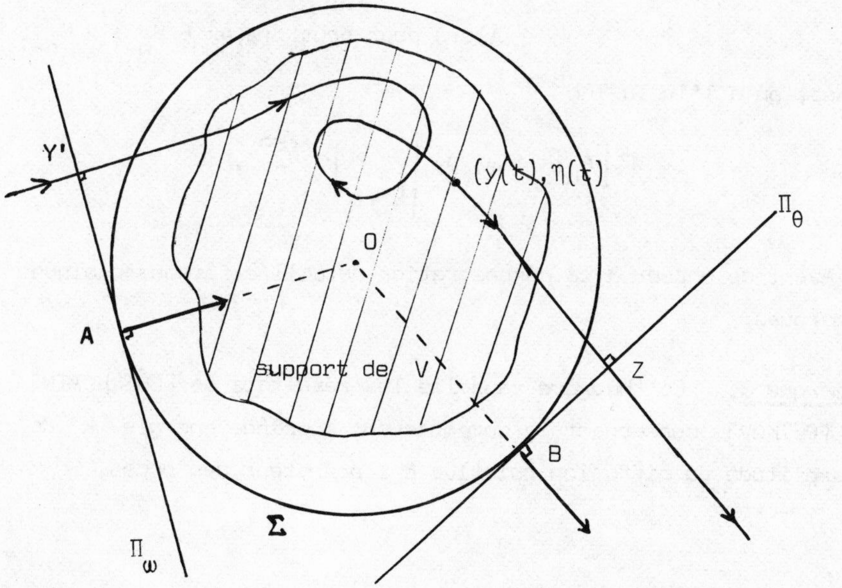

(3,3) On note $(y(t,y'), \eta(t,y'))$ la trajectoire classique solution de (1,2) issue du point $y' \in \Pi_\omega$ avec la vitesse $2k\omega$. Bien entendu, il faut prendre $H = |\eta|^2 + V(y)$ et $L = \left(\frac{|\dot{y}|}{2}\right)^2 - V(y)$. On suppose que pour t assez grand elle sort de Σ à la vitesse $2k\theta$, alors elle coupe le plan Π_θ en un point z à l'instant $t_{y'}$. A la durée de séjour $t_{y'}$ de cette trajectoire, on associe la longueur $T_{y'} = 2k\, t_{y'} - 2R$. Soit $\mathcal{C}_k$ l'ensemble des nombres $T_{y'}$ correspondant aux trajectoires issues des points $y' \in \Pi_\omega$ à la vitesse $2k\omega$ et quittant la sphère Σ à la vitesse $2k\theta$. On note $\mathcal{C}$, la fermeture dans $T^*\mathbb{R}$ de l'ensemble

$$\bigcup_{k>0} (\{k\} \times \mathcal{C}_k) .$$

Alors, on a le

THEOREME 2. On fait l'hypothèse suivante sur les trajectoires classiques :

(H) Toute trajectoire (3,3) ne reste qu'un temps fini dans supp V.

Alors, pour tout $(\omega,\theta) \in S_{n-1} \times S_{n-1}$ fixés, on a l'inclusion

$$F\left[A_h(k\,;\,\omega,\theta)\Big|_{k>0}\right] \subset \mathcal{C} .$$

Avant de passer à la démonstration détaillée faisons quelques remarques.

Remarque 3. Ce théorème rappelle les résultats de [GUILLEMIN] et [PETKOV] concernant le comportement à grande energie k de l'amplitude de diffusion relative à l'opérateur des ondes.

Remarque 4. L'amplitude de diffusion est aussi égale à un facteur près, au noyau de l'opérateur de transmission $T_h(k) = I - \hat{S}_h(k)$, où $\hat{S}_h(k)$ désigne la matrice de diffusion en variable de Fourier. On peut donc étudier le comportement semi-classique de $\hat{S}_h(k)$ à partir de celui de $A_h(k, \omega, \theta)$; ce comportement est étudié par une autre méthode par [YAJIMA] .

De plus, la distribution paramétrée

$$\mathcal{S}_h(k) = \text{Trace}\ \frac{d}{dk}\,(\text{Log}\ \hat{S}_h(k))$$

s'explicite facilement en fonction de $A_h(k, \omega, \theta)$; les travaux de [MAJDA & RALSTON] montrent que $\mathcal{S}_h(k)$ joue un rôle analogue à celui de la densité spectrale.

Revenons à la démonstration du théorème 2 .

Elle s'appuie sur des résultats de [VAINBERG] qui étudie le comportement semi-classique de $A_h(k, \omega, \theta)$ dans le cas où la section différentielle efficace est non nulle. Cet auteur montre que, à une constante multiplicative près, $A_h(k, \omega, \theta)$ est égale à

$$(3,4)\qquad (2\pi k h^{-1})^{\frac{1}{2}(3-n)} \int f(r) \left[\frac{\partial \Phi_h}{\partial r} + ikh^{-1} < \theta, \frac{x}{r} > \Phi_h \right] e^{-ikh^{-1}<\theta,x>} dx$$

où $f \in C_0^\infty(]R, R+1[)$, $\int f(r)dr = 1$ et $r = |x|$.

Cette expression a l'avantage de ne nécessiter la connaissance de Φ_h que sur le compact $R \leq |x| \leq R+1$. Pour obtenir une approximation de A_h (quand $h \to 0$), on va remplacer dans (3,4) Φ_h par une approximation ψ_h . On construit ψ_h en cherchant une solution du problème de Cauchy asymptotique suivant :

$$(3,5)\qquad (-h^2\Delta + V(x) - k^2)\,\psi_h = O(h^\infty)$$

$$(3,6)\qquad \psi_h\Big|_{\Sigma_\omega} = e^{ih^{-1}k\omega . x} = e^{-ih^{-1}kR}$$

dans la classe des fonctions oscillantes de Maslov de degré 0 relative à la variété lagrangienne Λ engendrée par les trajectoires classiques. Pour définir Λ, on montre que l'application

$$\mathbb{R} \times \Sigma_{\omega} \ni (t,y') \xrightarrow{\ j\ } (y(t,y'), \eta(t,y')) \in T^{*}\mathbb{R}^{n}$$

est une immersion lagrangienne injective.

Pour cela, il suffit de vérifier que $j^{*}(d\eta \wedge dy) = 0$.

On a $j^{*}(d\eta \wedge dy) = (\dot{\eta}\, dt + \partial_{y'}\eta\, dy') \wedge (\dot{y}\, dt + \partial_{y'}y\, dy') =$
$= (dV(y) \cdot \partial_{y'}y + 2\eta \cdot \partial_{y'}\eta)(dy' \wedge dt) = 0$
car l'énergie $V(y) + |\eta|^{2} = k^{2}$ est indépendante de y'.

Pour montrer que l'image de j est une sous-variété lagrangienne Λ de $T^{*}\mathbb{R}^{n}$, on utilise l'hypothèse (H) du théorème 2 qui entraîne que j est un plongement.

Alors , on sait (voir [DUISTERMAAT] Théorème 1.4.1 page 225) qu'il existe une solution ψ_h, ici unique, modulo $O(h^{\infty})$, à ce problème asymptotique. Le lien entre Φ_h et ψ_h est donné par le résultat suivant de [VAINBERG] :

Pour tout R' et tout N, il existe C tel que

$$(3,7) \qquad \sup_{\substack{|\alpha| \leq N \\ |x| \leq R'}} |\partial_x^{\alpha}(\psi_h - \Phi_h)| \leq C\, h^{N}$$

On peut donc étudier le comportement modulo $O(h^{\infty})$ de A_h en remplaçant Φ_h par ψ_h dans (3,4).

Pour démontrer le théorème 2 , il est inutile d'expliciter complètement ψ_h, il suffit de savoir que pour $R \leq |x| \leq R+1$, elle est somme finie de fonctions du type (1,7) où φ est une phase locale pour Λ et l'amplitude a est de degré 0 . En reportant dans (3,4) , on est conduit à représenter $A_h(k, \omega, \theta)$ modulo $O(h^{\infty})$ par une somme finie d'expressions du type

$$(2\pi h)^{-N/2}\int e^{ih^{-1}(\varphi(x,t,\omega,\zeta)-k\theta.x)}b(x,\ k,\ \zeta,\ \omega,\ h)d\zeta\ dx$$

où b est une amplitude de degré $\frac{1}{2}(5-n)$ par rapport à h^{-1}.

Pour démontrer l'inclusion du théorème 2, on se donne $(k_0,\ \tau_0)\notin\mathcal{C}$ et il s'agit de vérifier que la quantité $I_h = \langle \rho(k)\ e^{-ih^{-1}\tau k},\ A_h(k,\ \omega,\ \theta)\rangle$ est d'ordre $O(h^{\infty})$, quand la fonction ρ est à support voisin de k_0 et quand τ est voisin de τ_0. On est donc amené à étudier le comportement asymptotique d'intégrales du type

$$(3,9)\quad \int e^{ih^{-1}(\varphi(x,k,\omega,\zeta)-k\theta.x\ -k\tau)}\ \rho(k)\ b(x,k,\zeta,\omega,h)dx\ dk$$

Les points critiques de la phase de (3,9) sont les (x,k) du support d'intégration qui vérifient :

$$(3,10)\qquad \partial_\zeta\varphi(x,\ k,\ \zeta) = 0$$

$$(3,11)\qquad \partial_x\varphi(x,\ k,\ \zeta) = k\theta$$

$$(3,12)\qquad \partial_k\varphi(x,\ k,\ \zeta) - \theta x = \tau$$

Les deux premières équations siginifient qu'il existe une trajectoire $(y(t,y'),\ \eta(t,y'))$ qui sort de la sphère Σ à la vitesse $k\theta$ et passe par le point x.

La troisième équation est plus délicate à interpréter. Pour le faire, nous allons expliciter un moyen de construire une phase $\varphi(x,\ k,\ \zeta)$ pour la variété Λ. Rappelons que l'on peut recouvrir Λ par des cartes du type :

$$\Lambda\supset\Lambda_j\ni(x,\xi)\longrightarrow(x_\alpha,\ \xi_\beta)\in U_j\subset\mathbb{R}^n$$

où (α,β) est une partition de l'ensemble $\{1,2,\dots,n\}$; on note $x_\beta(x_\alpha,\ \xi_\beta)$, $\xi_\alpha(x_\alpha,\ \xi_\beta)$ le diffeomorphisme inverse.
En inversant le diffeomorphisme

$$(t,y') \longrightarrow (y(t,y'),\ \eta(t,y')) \in \Lambda_j \longrightarrow (y_\alpha(t,y'),\ \eta_\beta(t,y')) \in U_j$$

on obtient un diffeomorphisme $t = t(x_\alpha, \xi_\beta)$, $y' = y'(x_\alpha, \xi_\beta)$. On note $y(t, x_\alpha, \xi_\beta)$, $\eta(t, x_\alpha, \xi_\beta)$ la trajectoire (3,3) correspondant à $y'(x_\alpha, \xi_\beta)$.

L'expression de la dérivée d'une phase le long d'une trajectoire conduit à définir une phase par

$$(3,13)\quad \varphi(x, \xi_\beta) = \varphi_0(y'(x_\alpha, \xi_\beta)) + \int_0^{t(x_\alpha,\xi_\beta)} \eta(t, x_\alpha, \xi_\beta).\dot{y}(t, x_\alpha, \xi_\beta)dt$$
$$+ < x_\beta - x_\beta(x_\alpha, \xi_\beta), \xi_\beta >$$

où ξ_β joue le rôle de la variable de fréquence ζ .
Ici, φ_0 est donnée sur Σ_ω par (3,6) et tout est fonction de la variable k ; il vient

$$(3,14)\quad \varphi(x, k, \xi_\beta) = -kR + \int_0^{t(x_\alpha,\xi_\beta)} 2|\eta(t, x_\alpha, \xi_\beta, k)|^2\, dt$$
$$+ < x_\beta - x_\beta(x_\alpha, \xi_\beta), \xi_\beta > \quad .$$

Il reste à vérifier que ceci définit bien une phase pour Λ . Ce calcul, bien que fastidieux, est à notre avis assez instructif, et de toute façon on en aura besoin pour interpréter $\partial_k\varphi$.
Pour calculer $\partial_{x_\alpha}\varphi$, $\partial_{\xi_\beta}\varphi$, $\partial_k\varphi$, il est commode de calculer $\partial_u\varphi$ en supposant que x_α et ξ_β dépendent du paramètre u . On obtient ainsi des fonctions $t(u) = t(x_\alpha(u), \xi_\beta(u))$, $y(t,u)$ et $\eta(t,u)$. On commence par calculer la dérivée de l'action

$$S(u) = \int_0^{t(u)} 2|\eta(t,u)|^2\, dt \quad .$$

En écrivant que $2|\eta|^2 = (|\eta|^2 + V(y)) + (|\eta|^2 - V(y))$, il vient

$$S(u) = k^2\, t(u) + \int_0^{t(u)} L\, dt \quad .$$

D'où, l'on obtient

$$\partial_u S(u) = k^2\, \partial_u t(u) + L\big|_{t=t(u)} \cdot \partial_u t(u) + \int_0^{t(u)} (\partial_u L)\, dt \quad .$$

Or

$$\int_0^{t(u)} (\partial_u L) dt = \int_0^{t(u)} (\partial_y L - \frac{d}{dt}\partial_{\dot y} L)\partial_u y\ dt + \Big[\partial_{\dot y} L \cdot \partial_u y\Big]_{t=0}^{t=t(u)}$$

$$= \partial_{\dot y} L \cdot \partial_u y\big|_{t=t(u)} = \xi \cdot \partial_u y\big|_{t=t(u)} \quad ,$$

car, pour $t = 0$ on a $y(0,u) \in \Sigma_\omega$ pour tout u .
Enfin, si on suppose x en dehors de supp V , il vient

(3,15) $$\partial_u S(u) = 2k^2 \partial_u t(u) + \xi \cdot \partial_u y\big|_{t=t(u)} \quad .$$

D'autre part, on a

$$\partial_u(< x_\beta - y_\beta(t(u),u),\ \xi_\beta(u) >)$$

$$= < x_\beta - y_\beta(t(u),u),\ \partial_u \xi_\beta > - 2\xi_\beta^2 \partial_u t(u) - \partial_u y_\beta \cdot \xi_\beta$$

Finalement, en écrivant $\xi \cdot \partial_u y\big|_{t=t(u)} = \xi_\alpha \cdot \partial_u y_\beta + \xi_\beta \cdot \partial_u y_\beta$,
il vient :

(3,16) $$\partial_u \varphi(x_\alpha(u),\ x_\beta,\ \xi_\beta(u),\ k) = 2k^2 \partial_u t(u) - 2\xi_\beta^2 \partial_u t(u) + \xi_\alpha \cdot \partial_u y_\alpha + < x_\beta - y_\beta(t(u),\ u), \partial_u \xi_\beta > \quad .$$

La dérivée $\partial_{\xi_\beta} \varphi(x,\ k,\ \xi_\beta)$ correspond au cas où $u = \xi_\beta$ et
x est indépendant de u . On trouve, en notant que
$\partial_u y_\alpha = -2\xi_\alpha \cdot \partial_u\ t(u)$ car $x_\alpha = y_\alpha(t(u),\ u)$, que

$$\partial_u \varphi = 2k^2 \partial_u t(u) - 2(\xi_\beta^2 + \xi_\alpha^2)\partial_u t(u) + x_\beta - x_\beta(x_\alpha,\ \xi_\beta)$$

(3,17) $$\partial_u \varphi = x_\beta - x_\beta(x_\alpha, \xi_\beta) \quad .$$

Donc, la condition $\partial_{\xi_\beta} \varphi = 0$ signifie que $x_\beta = x_\beta(x_\alpha, \xi_\beta)$.

La dérivée $\partial_{x_\alpha} \varphi(x, k, \xi_\beta)$ correspond au cas où $u = x_\alpha$ et ξ_β sont indépendants de u . On trouve, en notant que $\xi_\alpha \cdot \partial_u y_\alpha = \xi_\alpha - 2\xi_\alpha^2 \cdot \partial_u t(u)$ car $y_\alpha(t(x_\alpha), x_\alpha) = x_\alpha$,

$\partial_u \varphi = 2k^2 \partial_u t(u) - 2(\xi_\rho^2 + \xi_\alpha^2) \partial_u t(u) + \xi_\alpha$; c'est-à-dire

(3,18) $$\partial_{x_\alpha} \varphi = \xi_\alpha(x_\alpha , \xi_\beta) \quad .$$

Enfin, le calcul de $\partial_{x_\beta} \varphi = \xi_\beta$ est immédiat et montre que φ est bien une phase pour Λ . Il reste à calculer $\partial_k \varphi(x, k, \xi_\beta)$, pour cela, il faut modifier un peu les calculs précédents qui supposaient k fixe. On trouve

(3,19) $$\partial_k \varphi = -R + 2k\, t(k) + 2k^2 \partial_k t(u) - 2\xi_\beta^2 \partial_k t(k) + \xi_\alpha \cdot \partial_k y_\alpha ,$$

or l'égalité $x_\alpha = y_\alpha(t(k), k)$ montre que

$$\partial_k y_\alpha = -2\xi_\alpha \partial_k t(k) ,$$ et en reportant

dans (3,19), il vient :

(3,20) $$\partial_k \varphi = -R + 2k\, t(k) \quad .$$

Pour interpréter (3,12), on remarque que $\theta \cdot x = R + \overline{zy}$, et comme la trajectoire de z à y est en dehors du support de V , on a $\overline{zy} = (2k) \times$ (temps pour aller de z à y) , car la vitesse $\dot{y}(t)$ est égale à $2k\theta$. De sorte que

(3,21) $$\partial_k \varphi - \theta \cdot x = (2k) \times (\text{temps pour aller de } y' \text{ à } z) - 2R = T_{y'}$$

Alors la condition $\tau = T_{y'}$ n'est pas possible pour τ voisin de τ_0 , ce qui montre qu'il n'y a pas de point critique pour l'intégrale (3,9) et termine la démonstration du théorème 2 .

REFERENCES

S. ALBEVERIO - PH. BLANCHARD - R.HØEGH KROHN ; Feynman path integrals, the Poisson formula and the function for the Scrödinger operators, à paraître.

K. ASADA - D. FUJIWARA ; Jap. J. Math 4 (1978), 299-361

R. BALIAN - C. BLOCH ; Ann. Phys. 85(1974) 514-545 .

M.V. BERRY - K.E. MOUNT ; Rep prog. Phys. 35 (1972), 315-397 .

J. CHAZARAIN ; Comm. Partial Diff. Equat. 5(6) , 595-644(1980) .

J.J. DUISTERMAAT ; Comm. P. Appl. Math. 27 (1974), 207-281 .

D. FUJIWARA ; A construction of the fundamental solution for the Schrödinger equation, à paraître .

V. GUILLEMIN ; Publ. RIMS, Kyoto Univ. 12 Suppl(1977), 69-88 .

V. GUILLEMIN - S. STERNBERG ; Geometric asymptotics , A.M.S. (1977) .

B. HELFFER - D. ROBERT ; Comportement semi-classique du spectre des hamiltoniens quantiques elliptiques, à paraître

H. KITADA ; J. Fac of Sc. Tokyo Univ. 27(1980), 193-226 .

A. MAJDA - J. RALSTON ; Duke Math. J. 45(1978), 513-536 .

V.P. MASLOV ; Theorie des perturbations et méthodes asymptotiques Trad. Dunod, Paris (1972) .

V. PETKOV ; High frequency asymptotics of the scattering amplitude for non convex bodies, à paraître .

N. SHENK - D. THOE ; Rocky Mountain J. of Math. 1 (1971), 89-125 .

B.R. VAINBERG ; Functional Analysis and Appl. 11 (1977), 247-257 .

A. VOROS ; Développements semi-classiques, thèse ORSAY (1977)

K. YAJIMA ; Comm. Math. Phys. 69 (1979), 101-129 .

English summary

Let

$$\mathcal{H}_h = -\frac{1}{2} h^2 \Delta + V(x)$$

be a quantum hamiltonian depending on the real parameter $h = \frac{\hbar}{2m} \in]0, h_o]$; V is a real C^∞ potential.

We are interested in the asymptotic behaviour of parametric distributions T_h ($h \in]0, h_o]$) associated to the hamiltonian $\mathcal{H}_h$ as $h \to 0$.

Let us first recall some definitions.

Definition 1

The classical trajectories in T^*R^n are the solutions of the hamiltonian equations :

$$\begin{cases} \dot{y}(t) = \partial_\eta H = \eta, \\ \dot{\eta}(t) = -\partial_y H = -dV(y), \end{cases}$$

where H is the classical hamiltonian :

$$H(y, \eta) = \frac{1}{2}|\eta|^2 + V(y)$$

Definition 2

A point $(x^o, \xi^o) \in T^*X$ is not in the frequency set $F[T_h]$ of the parametric distribution T_h if and only if there exist $\rho \in C_o^\infty(x)$ $(\rho(x^o) \neq 0)$ and a (not necessarily conic) neighborhood V of ξ^o such that

$$\langle \rho(x) e^{ih^{-1}x.\xi}, T_h \rangle = O(h^\infty)$$

uniformly with respect to ξ in V when $h \to 0$.

For example, if $T_h = a(x)e^{ih^{-1}\psi(x)}$, with a and ψ in $C^\infty(x)$, ψ real, one has

$$F[T_h] \subset \{(x, d_x\psi(x)) : x \in [a]\}.$$

In the same way, if

$$T_h(x) = \int_{R^n} e^{ih^{-1}\psi(x,\xi)} a(x,\xi, h)d\xi,$$

where ψ is a real C^∞ phase and a is a symbol with respect to h^{-1} with compact support in ξ, we have

$$F[T_h] \subset \{(x,\partial_x\psi(x,\xi)) : \partial_\xi\psi(x,\xi) = 0\} \equiv \Lambda\psi.$$

We shall examine the following two cases :

a) $V(x)\to\infty$ as $|x|\to+\infty$;

b) V has compact support.

Case a)

Let us suppose that, for $|x|\to+\infty$,

$$\partial_x^\alpha V(x) = O(|x|^{(2-|\alpha|)+}) \text{ and } V(x) \geq c|x|^2 \quad (c>0).$$

Then, we know that $\mathcal{H}_h$ is self-adjoint in $L^2(R^n)$ and its spectrum is a sequence of eigenvalues

$$\lambda_1(h) \leq \lambda_2(h) \leq \ldots, \lambda_j(h)\to+\infty.$$

The spectral measure of $\mathcal{H}_h$ is defined by

$$\sigma_h(\lambda) = \sum_j \delta(\lambda-\lambda_j(h)).$$

Its Fourier transform is the tempered distribution

$$S_h(t) = \sum_j e^{ih^{-1}\lambda_j(h)t}.$$

Let us denote by $\mathcal{L}_\tau$ the set of periods of the periodic classical trajectories with energy $\frac{1}{2}|\eta|^2 - V(y) = -\tau$.

We define $\mathcal{L}$ as the closure in T^*R of $\bigcup_{\tau\in R} (\mathcal{L}_\tau \times \{\tau\})$.

Then, we have the

Theorem 1

With these assumptions, we have $F[S_h] \subset \mathcal{L}$.

Case b)

Let us now suppose that $V \in C_o^\infty(R^n)$. To simplify some expressions, we put

$$\mathcal{H}_h = -h^2\Delta + V(x).$$

Then, the spectrum is no longer discrete and we have a scattering theory for the continuous spectrum.

It is known that for any $k > 0$ and $\omega \in S_{n-1}$, there exists a generalised eigen-function $\phi_h(k,x,\omega)$ defined by the conditions

$$\begin{cases} (-h^2\Delta + V(x) - k^2)\phi_h = 0, \\ \phi_h - e^{ikh^{-1}x.\omega} = O(|x|^{\frac{n-1}{2}} e^{ikh^{-1}|x|}. \end{cases}$$

Moreover,

$$(\phi_h - e^{ikh^{-1}x.\omega})|x|^{\frac{n-1}{2}} e^{ikh^{-1}|x|}$$

has a limit as $x \to \infty$ in any direction $\theta \in S_{n-1}$. This limit is denoted by $A_h(k,\theta,\omega)$ and is called the <u>amplitude of diffusion</u>.

Let Σ be a sphere of radius R including supp V, A and B the points of Σ such that $\vec{OA} = -R\omega$ and $\vec{OB} = R\theta$, Π_ω and Π_θ the tangent planes to Σ through A and B.

Let $(y(t,y'),\eta(t,y'))$ be the classical trajectory issued from $y' \in \Pi_\omega$ with speed $2k\omega$ (here $H = |\eta|^2+V(y)$). We suppose that for t sufficiently large, it leaves Σ with speed $2k\theta$; then it cuts Π_θ at a point z at time $t_{y'}$. We put $T_{y'} = 2kt_{y'}-2R$. Let $\mathcal{T}_k$ be the set of those $T_{y'}$ for all $y' \in \Pi_\omega$ and $\mathcal{T}$ be the closure in T^*R of $\bigcup_{k>0} (\{k\}x\mathcal{T}_k)$.

Then, we have the following

<u>Theorem</u> 2

If any classical trajectory leaves supp V after a finite time, then, for every $(\omega,\theta) \in S_{n-1}xS_{n-1}$, one has

$$F[A_h(k;\omega,\theta)|_{k>0}] \subset \mathcal{T}.$$

GENERAL INITIAL-BOUNDARY PROBLEMS FOR SECOND ORDER HYPERBOLIC EQUATIONS

Gregory Eskin

Hebrew University of Jerusalem, Israel

Abstract

The general initial-boundary problem for a second order hyperbolic equation is considered for two classes of domains: domains with boundaries which are strictly concave with respect to the bicharacteristics of the hyperbolic operator and domains with boundaries strictly convex with respect to these bicharacteristics. The exposition is given for the model equations although all results can be extended to the general case. Propagation of singularities of solutions is also studied.

§1. FORMULATION OF THE PROBLEM

Let $A(x,D)$ be a hyperbolic operator of second order in $\mathbb{R}^{n+1}$, where $x = (x_0,x_1,\dots,x_n)$, $D = (i\partial/\partial x_0, i\partial/\partial x_1,\dots,i\partial/\partial x_n)$. Consider the initial-boundary problem with zero initial condition:

$$A(x,D)u = 0, \qquad x \in \Omega, \tag{1.1}$$

$$u = 0, \qquad x_0 < 0, \ x \in \Omega, \tag{1.2}$$

$$B(x,D)u|_{\partial\Omega} = h(x'), \qquad x' \in \partial\Omega, \tag{1.3}$$

where Ω is a cylindrical domain in $\mathbb{R}^{n+1}$, $\Omega = (-\infty,+\infty) \times G$, $x_0 \in (-\infty,+\infty)$, $(x_1,\dots,x_n) \in G \subset \mathbb{R}^n$, and $B(x,D)$ is a differential operator of order r.

We assume that $h(x')$ is given distribution on $\partial\Omega$ and $h(x') = 0$ for $x_0 < 0$. The problem is to find what are conditions on $B(x,D)$ such that the initial-boundary problem (1.1), (1.2), (1.3) is well-posed. The initial-boundary problem is called well-

H. G. Garnir (ed.), Singularities in Boundary Value Problems, 19–54.

posed if for any $h \in \mathcal{D}'(\partial\Omega)$ there exists a unique solution $u \in \mathcal{D}'(\Omega)$. We note that the boundary $\partial\Omega$ is noncharacteristic with respect to the operator $A(x,D)$; therefore the restriction $B(x,D)u|_{\partial\Omega}$ exists for any distribution such that $A(x,D)u = 0$, since $A(x,D)$ is partially hypoelliptic with respect to the normal to $\partial\Omega$. The second problem, which we shall study, is the propagation of singularities of the solution $u(x)$, assuming that problem (1.1), (1.2), (1.3) is well-posed and the singularities of $h(x')$ are given.

Since $u = 0$ and $h = 0$ for $x_0 < 0$ the functions $u_\tau = e^{-x_0\tau}$ and $h_\tau = e^{-x_0\tau}h$ are well-defined for any $\tau \geq 0$. If the problem (1.1), (1.2), (1.3) is well-posed then the following problem is well-posed for any $\tau \geq 0$:

$$A(x,D_0+i\tau,D_1,\ldots,D_n)u_\tau = 0, \qquad x \in \Omega, \tag{1.4}$$

$$u_\tau = 0 \quad \text{for} \quad x_0 < 0, \qquad x \in \Omega, \tag{1.5}$$

$$B(x',D_0+i\tau,D_1,\ldots,D_n)u_\tau = h_\tau(x'), \qquad x' \in \partial\Omega. \tag{1.6}$$

Consider the initial-boundary problem in Ω with the Dirichlet boundary condition:

$$A(x,D_0+i\tau,D_1,\ldots,D_n)u_\tau = 0, \qquad x \in \Omega, \tag{1.7}$$

$$u_\tau = 0 \quad \text{for} \quad x_0 < 0, \qquad x \in \Omega, \tag{1.8}$$

$$u_\tau|_{\partial\Omega} = v(x'). \tag{1.9}$$

It is well-known that the problem (1.7), (1.8), (1.9) has a unique solution. Denote by N^τ the following operator in $\mathcal{D}'(\partial\Omega)$:

$$N^\tau v_\tau = -i\left.\frac{\partial u_\tau}{\partial\nu}\right|_{\partial\Omega},$$

where u_τ is the solution of the initial boundary Dirichlet problem (1.7), (1.8), (1.9) and $\partial/\partial\nu$ is the normal derivative with respect to $\partial\Omega$. The operator N^τ is called the Neumann operator. Let $\Omega_\delta \approx \partial\Omega \times [0,\delta)$ be the δ-neighbourhood of $\partial\Omega$. We introduce in Ω_δ the coordinates $(x',\nu) = (x_0,x'',\nu)$ where $x' = (x_0,x'') \in \partial\Omega$ and $\nu \in [0,\delta)$, δ is small. Let $A(x',\nu,D',D_\nu)$ and $B(x',D',D_\nu)$ be the operators $A(x,D)$ and $B(x',D)$ written in new system of coordinates in Ω. Since $A(x',\nu,D',D_\nu)$ is a differential operator of second order, the boundary condition (1.6) can be rewritten in the following form

$$B_1(x',D_0+i\tau,D'')(-i)\frac{\partial u_\tau}{\partial\nu} + B_2(x',D_0+i\tau,D'')u_\tau|_{\partial\Omega} = h_\tau(x') \tag{1.10}$$

where $B_k(x',D_0+i\tau,D'')$ are differential operators on $\partial\Omega$ of order $r-2+k$, $k = 1,2$.

The well-posedness of the initial boundary problem (1.4), (1.5), (1.6) (or (1.1), (1.2), (1.3)) is equivalent to the existence and uniqueness of the solution of the following equation on the boundary $\partial\Omega$:

$$B_1(x',D_0+i\tau,D'')N^{\tau}v_{\tau} + B_2(x',D_0+i\tau)v_{\tau} = h_{\tau}(x') . \tag{1.11}$$

To study the equation (1.11) we need to describe more explicitly the operator N^{τ}. The symbol of the principal part of the operator $A(x',\nu,D_0+i\tau,D'',D_{\nu})$ has the following form:

$$A^{(0)}(x',\nu,\xi_0+i\tau,\xi'',\sigma) = (\sigma-\lambda(x',\nu,\xi_0+i\tau,\xi''))^2 - \\ - \mu(x',\nu,\xi_0+i\tau,\xi'') , \tag{1.12}$$

where $\lambda(x',\nu,\xi_0,\xi'')$ and $\mu(x',\nu,\xi_0,\xi'')$ are real polynomials, $\deg_{\xi'}\lambda = 1$, $\deg_{\xi'}\mu = 2$.

By choosing a suitable system of coordinates in $\partial\Omega \times [0,\delta)$ we can make

$$\lambda(x',\nu,\xi_0,\xi'') \equiv 0 . \tag{1.13}$$

Denote by $T_0^*(\partial\Omega)$ the cotangent bundle on $\partial\Omega$. Let N_0 be the part of $T_0^*(\partial\Omega)$, where $\mu(x',0,\xi_0,\xi'') = 0$, $N_+ \subset T_0^*(\partial\Omega)$, where $\mu(x',0,\xi_0,\xi'') > 0$, and $N \subseteq T_0^*(\partial\Omega)$ is such that $\mu(x',0,\xi_0,\xi'') < 0$. For $(x',\xi') \in N_-$ the equation

$$\sigma^2 - \mu(x',0,\xi_0,\xi'') = 0 \tag{1.14}$$

has no real zeros in σ. Therefore the region N_- is called the "elliptic" region. The region N_+ where the equation (1.14) has two real zeros $\sigma_1 = +\sqrt{\mu(x',0,\xi_0,\xi'')}$ and $\sigma_2 = -\sqrt{\mu(x',0,\xi_0,\xi'')}$ is called the "hyperbolic" region.

The region N_0 where equation (1.14) has a double real root, is called the diffraction region. We shall study the operator N^{τ} microlocally for each $(x',\xi',\tau) \in T_0^*(\partial\Omega) \times \mathbb{R}_+^1$.

We shall say that two operators $C_{1\tau}$ and $C_{2\tau}$, depending on the parameter τ, are equal microlocally in an open conic set $\Sigma \subset T_0^*(\partial\Omega) \times \mathbb{R}_+^1$ if

$$C_{1\tau}\chi(x',D',\tau)v = C_{2\tau}\chi(x',D',\tau)v \ (\mathrm{mod}\ C^{\infty}) \tag{1.15}$$

for any $v \in \mathcal{D}'(\partial\Omega)$ with a compact support and for any $\chi(x',\xi',\tau)$ which is homogeneous of order zero with respect to (ξ',τ) and such that $\mathrm{supp}\,\chi(x',\xi',\tau) \subset \Sigma$. Since the operator $A(x',\nu,D_0+i\tau,D'',D_{\nu})$ is an elliptic operator microlocally when $(x',\xi') \in N_-$ or when $\tau > c|\xi'|$ we can prove by using the pseudodifferential technique that N^{τ} is a pseudodifferential operator in this region and the principal symbol of N^τ has the following form:

$$N_0^\tau(x',\xi') = \sqrt{\mu(x',0,\xi_0+i\tau,\xi'')}\ , \tag{1.16}$$

where we take the branch of the square root which has positive imaginary part for $\tau > 0$. We note that $\mu(x',0,\xi_0+i\tau,\xi'')$ is not real for $\tau > 0$ because of the hyperbolicity with respect to x_0. The sign in (1.16) is related to our definition of the Fourier transform: We denote by $\tilde{v}(\xi') = \int_{-\infty}^{\infty} v(x')e^{i(x',\xi')}dx'$ the Fourier transform of $v(x')$ and then the inverse Fourier transform has the following form

$$v(x') = \frac{1}{(2\pi)^n}\int_{-\infty}^{\infty}\tilde{v}(\xi')e^{-i(x',\xi')}d\xi'\ . \tag{1.17}$$

In the case when $A(D_0+i\tau,D'',D_\nu)$ has constant coefficients and $\partial\Omega = \mathbb{R}^n$ we have the following formula for the solution of the initial-boundary problem with the Dirichlet boundary condition

$$u_\tau(x',\nu) = \frac{1}{(2\pi)^n}\int_{-\infty}^{\infty} e^{+i\nu\sqrt{\mu(\xi_0+i\tau,\xi'')}-i(x',\xi')}\tilde{v}_\tau(\xi')d\xi'. \tag{1.18}$$

Since $\operatorname{Im}\sqrt{\mu(\xi_0+i\tau,\xi'')} > 0$ for $\tau > 0$ the exponential in (1.18) is bounded.

It follows from (1.18) that

$$N^\tau v_\tau(x') = -i\,\frac{\partial u_\tau(x',0)}{\partial\nu} = \frac{1}{(2\pi)^n}\int_{-\infty}^{\infty}\sqrt{\mu(\xi_0+i\tau,\xi'')}\ e^{-i(x',\xi')}\tilde{v}_\tau(\xi')d\xi'\ .$$

Therefore the formula (1.16) is proved for the constant coefficient case. By constructing a parametrix one can also prove (1.16) in the case of variable coefficients.

In the region N_+ the operator N^τ is equal also to a pseudodifferential operator with the principal symbol (1.16). To prove this we used the eiconal method for the construction of a parametrix and then take the almost analytic extension with respect to τ. (See [7] for details.)

The most complicated form the operator N^τ has in the neighbourhood of N_0 and this form depends on the geometry of the boundary. We shall construct N^τ only in two cases: when the boundary $\partial\Omega$ is strictly convex with respect to bicharacteristics of the operator $A(x,D)$ and when the boundary is strictly concave with respect to these bicharacteristics.

Let $x' = x'(t)$, $\xi' = \xi'(t)$, $\nu = \nu(t)$, $\sigma = \sigma(t)$ be the null-bicharacteristics of the operator $A(x,D)$ passing through the point $(x_0',0,\xi_0',0)$, where $(x_0',\xi_0') \in N_0$, i.e., $\mu(x_0',0,\xi_0') = 0$. Therefore

$$\frac{dx'}{dt} = -\frac{\partial\mu(x',\nu,\xi')}{\partial\xi'}, \quad \frac{d\xi'}{dt} = \frac{\partial\mu(x',\nu,\xi')}{\partial x'}, \tag{1.19}$$
$$\frac{d\nu}{dt} = 2\sigma, \quad \frac{d\sigma}{dt} = -\frac{\partial\mu(x',\nu,\xi')}{\partial\nu}.$$

Since $\sigma(0) = 0$ we have $d\nu(0)/dt = 2\sigma(0) = 0$. So that the bicharacteristics, passing through the point $(x_0',0,\xi_0',0) \in \partial\Omega$ is tangent to $\partial\Omega$. The boundary is called convex with respect to the tangent bicharacteristics of $A(x,D)$ if $\nu(0) = 0$, $d\nu(0)/dt = 0$ and $d^2\nu(0)/dt^2 < 0$. Since $d^2\nu/dt^2 = 2\, d\sigma/dt = 2\,\partial\mu/\partial\nu$ the convexity means that

$$\frac{\partial\mu(x',0,\xi')}{\partial\nu} < 0 . \tag{1.21}$$

Analogously the boundary $\partial\Omega$ is called concave with respect to the tangent bicharacteristics passing through the point $(x_0',0,\xi_0',0)$ if

$$\frac{\partial\mu(x',0,\xi')}{\partial\nu} > 0 \tag{1.22}$$

at this point.

We shall start with the construction of the Neumann operator and the investigation of the whole problem in the case of model equations.

We note that $\partial\mu/\partial\xi_0 \neq 0$ for $\mu = 0$ since the operator $A(x,D)$ is hyperbolic with respect to x_0. Also $\partial\mu/\partial\nu > 0$ for $\mu = 0$ in the case of strictly concave boundary. Therefore the simplest form of the symbol $A^{(0)}(x',\nu,\xi_0+i\tau,\xi'',\sigma)$ in the neighbourhood of N_0 for the case of strictly concave boundary, will be the following:

$$a_+(\nu,\xi_0+i\tau,\xi'',\sigma) = \sigma^2 - (\xi_0 + i\tau + \nu|\xi''|)|\xi''| . \tag{1.23}$$

For the case of strictly convex boundary the simplest form of the symbol will be the following:

$$a_-(\nu,\xi_0+i\tau,\xi'',\sigma) = \sigma^2 - (\xi_0 + i\tau - \nu|\xi''|)|\xi''| . \tag{1.24}$$

R. Melrose proved in [24] that there exists an operator which transforms microlocally the second order hyperbolic operator to the operator with the symbol (1.23) when the boundary is strictly concave and to the operator with the symbol (1.24) when the boundary is strictly convex.

The operator with the symbol (1.23) is a model for the case of a concave boundary in the neighbourhood of the region of glancing rays, i.e. when $\nu = 0$, $\xi_0 = 0$, $\xi'' \neq 0$. If we study such a model in the half-space $\mathbb{R}_+^{n+1}$ it is convenient to replace $|\xi''|$ by $|\hat{\xi}''| = (1 + |\xi''|^2)^{\frac{1}{2}}$ and to consider the following equation

$$\hat{a}_+(x_n, D_0 + i\tau, D'', D_n)u_\tau = 0, \quad x \in \mathbb{R}_+^{n+1}, \tag{1.25}$$

where $x = (x_0, x'', x_n)$, $x'' = (x_1, \ldots, x_{n-1})$, $x_n \geq 0$,

$$\begin{aligned} &\hat{a}_+(x_n, \xi_0 + i\tau, \xi'', \xi_n) = \xi_n^2 - (\xi_0 + i\tau + x_n|\hat{\xi}''|)|\hat{\xi}''|, \\ &|\hat{\xi}''| = (1 + |\xi''|^2)^{\frac{1}{2}}. \end{aligned} \tag{1.26}$$

Analogously, the model equation for the case of strictly convex boundary will be the following

$$\hat{a}_-(x_n, D_0 + i\tau, D'', D_n)u_\tau = 0, \quad x \in \mathbb{R}_+^{n+1}, \tag{1.27}$$

where

$$\hat{a}_-(x_n, \xi_0 + i\tau, \xi'', \xi_n) = \xi_n^2 - (\xi_0 + i\tau - x_n|\hat{\xi}''|)|\hat{\xi}''|. \tag{1.28}$$

Equation (1.25) is similar to the model equation

$$\sum_{k=1}^{n} \frac{\partial^2 u}{\partial x_k^2} - x_n \frac{\partial^2 u}{\partial x_0^2} = 0,$$

considered by G. Friedlander [9].

§2. SOLUTION OF THE MODEL INITIAL-BOUNDARY PROBLEM FOR THE CASE OF STRICTLY CONCAVE BOUNDARY

Consider the following initial-boundary problem in $\mathbb{R}_+^{n+1}$ with zero initial conditions:

$$a_-(x_n, D_0 + i\tau, D'', D_n)u_\tau = 0, \quad x_n > 0, \tag{2.1}$$

$$u_\tau = 0 \quad \text{for} \quad x_0 < 0, \quad x_n > 0, \tag{2.2}$$

$$B(x', D_0 + i\tau, D'', D_n)u_\tau\big|_{x_n = 0} = h_\tau(x'), \tag{2.3}$$

where $h_\tau(x') = 0$ for $x_0 < 0$. $a_-(x_n, \xi_0 + i\tau, \xi'', \xi_n)$ is the same as in (1.26) and

$$B(x', \xi_0+i\tau, \xi'', \xi_n) = -\xi_n + \lambda_1(x', \xi''). \tag{2.4}$$

To find the condition of well-posedness of the problem (2.1), (2.2), (2.3) we shall reduce it to the solution of an equivalent equation on the boundary $x_n = 0$:

$$N^\tau v_\tau + \lambda_1(x', D'')v_\tau = h_\tau , \tag{2.5}$$

where $v_\tau(x') = u_\tau(x',0)$ and N^τ is the Neumann operator, i.e., $N^\tau v_\tau = -i\, \partial u_\tau(x',0)/\partial x_n$.

To solve equation (2.1) we take the Fourier transform with respect to $x' = (x_0, x_1, \ldots, x_{n-1})$. We note that in fact we take the Laplace transform with respect to x_0, since $u(x', x_n)$ and $h(x')$ are equal to zero for $x_0 < 0$.

After the Fourier transform equation (2.1) has the following form:

$$\frac{\partial^2 \tilde{u}(\xi_0+i\tau, \xi'', x_n)}{\partial x_n^2} + (\xi_0+i\tau+x_n|\hat{\xi}|)|\xi|\tilde{u}(\xi_0+i\tau, \xi'', x_n) = 0, \tag{2.6}$$

where

$$\begin{aligned} \tilde{u}(\xi_0+i\tau, \xi'', x_n) &= \int_{-\infty}^{\infty} u_\tau(x', x_n)\, e^{i(x',\xi')}\, dx' \\ &= \int_{-\infty}^{\infty} u(x', x_n) e^{i(\xi_0+i\tau)+i(x'',\xi'')}\, dx' . \end{aligned} \tag{2.7}$$

Equation (2.6) can be easily reduced to the Airy equation, so that the general solution of (2.6) can be expressed by the Airy function. We look for the solution of (2.6) which is tempered distribution for $x_n > 0$, $\xi' \in \mathbb{R}^n$, $\tau > 0$, and which is analytic in $\xi_0+i\tau$ for $\tau > 0$.

It follows by the WKB-method that $\tilde{u}(\xi_0+i\tau, \xi'', x_n)$ must have the following asymptotic behavior for $x_n > 0$, $\tau > 0$, $|\xi_0+i\tau+x_n|\hat{\xi}||\,|\xi|^{-1/3} \gg 1$:

$$\tilde{u}(\xi_0+i\tau, \xi'', x_n) \sim C\, \rho^{-1/4} \exp i\, \frac{2}{3}\, \rho^{3/2} , \tag{2.8}$$

where $\rho(x_n, \xi_0+i\tau, \xi'') = (\xi_0+i\tau+x_n|\hat{\xi}|)|\xi|^{-1/3}$. We take in (2.8) such branch of $\rho^{3/2}$ which is positive for $\rho > 0$, so that

$$\operatorname{Im}(\rho(x_n, \xi_0+i\tau, \xi''))^{3/2} \geq \operatorname{Im}(\rho(0, \xi_0+i\tau, \xi''))^{3/2} \tag{2.9}$$

for $x_n \geq 0$. Since $\operatorname{Im}(\rho(x_n, \xi_0+i\tau, \xi''))^{3/2}$ takes both negative and positive values for $x_n > 0$, $\tau > 0$, $(\xi_0, \xi'') \in \mathbb{R}^n$, the only choice to obtain a solution of (2.6), which has a tempered growth, is to take the solution having the asymptotic behavior (2.8) with $C = \exp(-i(2/3)(\rho(0, \xi_0+i\tau, \xi''))^{3/2})$. Denote by $A(z)$

the Airy function which has the asymptotic behavior (2.8), i.e.,

$$A(x) = \frac{C}{z^{1/4}} \exp i \frac{2}{3} z^{3/2}(1 + O(1/z^{3/2})) \tag{2.10}$$

for $z \to \infty$, $\operatorname{Im} z \geq 0$. We note that $A(z)$ is oscillating on the real positive semi-axis and $A(z)$ is exponentially increasing on real negative semi-axis when $z \to -\infty$. The solution of the initial-boundary problem in $\mathbb{R}^{n+1}_+$ for the equation (2.1) with zero initial condition and with the Dirichlet boundary condition

$$u_\tau(x',0) = v_\tau(x') , \tag{2.11}$$

has the following form

$$u(x',x_n) = \frac{1}{(2\pi)^n} \int_{-\infty}^{\infty} \frac{A((\xi_0+i\tau+x_n|\hat{\xi}|)|\hat{\xi}|^{-1/3})}{A((\xi_0+i\tau)|\hat{\xi}|^{-1/3})} \cdot$$
$$\cdot\, e^{-i(\xi_0+i\tau)x_0-i(\xi'',x'')}\tilde{v}(\xi_0+i\tau,\xi'')\, d\xi_0\, d\xi'' . \tag{2.12}$$

Therefore for the Neumann operator N^τ we obtain:

$$N^\tau v_\tau = \frac{1}{(2\pi)^n} \int_{-\infty}^{\infty} -i \frac{A'((\xi_0+i\tau)|\hat{\xi}|^{-1/3})}{A((\xi_0+i\tau)|\hat{\xi}|^{-1/3})}|\hat{\xi}|^{2/3} \cdot$$
$$\cdot\, e^{-i(x',\xi')}\tilde{v}(\xi_0+i\tau,\xi'')\, d\xi_0\, d\xi'' , \tag{2.13}$$

i.e. N^τ is a pseudodifferential operator with the symbol

$$N(\xi_0+i\tau,\xi'') = -i|\hat{\xi}|^{2/3} \frac{A'((\xi_0+i\tau)|\hat{\xi}|^{-1/3})}{A((\xi_0+i\tau)|\hat{\xi}|^{-1/3})} . \tag{2.14}$$

We shall use some properties of the Airy function $A(z)$.

We shall show that for $\operatorname{Im} z \geq 0$

$$\operatorname{Re} i \frac{A'(z)}{A(z)} < 0 , \qquad \operatorname{Im} i \frac{A'(z)}{A(z)} < 0 . \tag{2.15}$$

To prove (2.15) we denote

$$A_1(z) = \overline{A(\bar{z})} .$$

It is clear that $A_1(z)$ is a solution of the Airy equation, which is linearly independent of $A(z)$.

For real t we have

$$\begin{aligned} \operatorname{Re} i \frac{A'(t)}{A(t)} &= \frac{1}{2}\left(i \frac{A'(t)}{A(t)} - i \overline{\frac{A'(t)}{A(t)}}\right) \\ &= \frac{i}{2} \frac{A'(t)A_1(t) - A(t)A_1'(t)}{|A(t)|^2} = \frac{c_0}{|A(t)|^2}, \end{aligned} \tag{2.16}$$

since $A'(t)A_1(t) - A(t)A_1'(t)$ is a constant (it is the Wronskian of the Airy equation).

By taking $t \to +\infty$ and using the asymptotics of $A(t)$ one can see that c_0 is negative. It follows from the asymptotics of $A(z)$ that $\operatorname{Re} i(A'(z)/A(z)) < 0$ for $|z|$ large and $\operatorname{Im} z \geq 0$. Therefore $\operatorname{Re} i(A'(z)/A(z)) < 0$ for all z such that $\operatorname{Im} z \geq 0$.

Analogously we have for real t:

$$\operatorname{Im} i \frac{A'(t)}{A(t)} = \frac{1}{2i}\left(i \frac{A'(t)}{A(t)} + i \overline{\frac{A'(t)}{A(t)}}\right) = \frac{1}{2} \frac{A'(t)\overline{A(t)} + A(t)\overline{A'(t)}}{|A(t)|^2}$$

$$= \frac{1}{2|A(t)|^2} \frac{d}{dt} |A(t)|^2 < 0, \tag{2.17}$$

since $|A(t)|^2$ is decreasing (see [27]). For large $|z|$, with $\operatorname{Im} z \geq 0$ the inequality $\operatorname{Im} i(A'(z)/A(z)) < 0$ follows from the asymptotics of $A(z)$. Therefore $\operatorname{Im} i(A'(z)/A(z)) < 0$ for all z such that $\operatorname{Im} z \geq 0$.

Using the asymptotics of $A(z)$ and the inequalities (2.15) we obtain the following estimates for the symbol $N(\xi_0+i\tau,\xi'')$ of the Neumann operator N^τ:

$$\operatorname{Re} N(\xi_0+i\tau,\xi'') > C\tau K(\xi',\tau), \tag{2.18}$$

$$\operatorname{Im} N(\xi_0+i\tau,\xi'') > C\tau K(\xi',\tau), \tag{2.19}$$

where we denote

$$\begin{aligned} K(\xi',\tau) &= \frac{|\hat{\xi}''|^{1/3}}{(1 + |\zeta|)^{1/2}}, \\ \zeta &= \frac{\xi_0 + i\tau}{|\hat{\xi}''|^{1/3}}. \end{aligned} \tag{2.20}$$

Denote by $H_s^\tau(\mathbb{R}^n)$ the Sobolev space with the following norm

$$\|v\|_{s,\tau}^2 = \int_{-\infty}^{\infty} (1 + \xi_0^2 + |\xi''|^2 + \tau^2)^s |\tilde{v}(\xi_0+i\tau,\xi'')|^2 d\xi_0 d\xi'' \tag{2.21}$$

Assume that $\lambda_1(x',\xi'')$ is such that

$$\sqrt{\xi_0 + i\tau}\ |\xi''|^{1/2} + \lambda_1(x',\xi'') \neq 0 \quad \text{for} \quad \xi'' \neq 0,\ \tau > 0\ . \tag{2.22}$$

The condition (2.22) is called the weak Lopatinskii condition.

The following theorem holds:

Theorem 2.1. Let the assumption (2.22) be satisfied and $\tau \geq \tau_0$ where τ_0 is sufficiently large. Then for each $h \in H_s^\tau(\mathbb{R}^n)$, $h(x') = 0$ for $x_0 < 0$, there exists a unique solution $v(x') \in \mathcal{D}'(\mathbb{R}^n)$ of the equation (2.5) such that $v = 0$ for $x_0 < 0$. The following estimate holds:

$$\tau(\|v\|_s^2 + \|K^{1/2}v\|_s^2) \leq C(\|K^{-1/2}h\|_s^2 + \|K^{-1}h\|_s^2)\ . \tag{2.23}$$

The well-posedness of the initial-boundary problem (2.1), (2.2), (2.3) follows immediately from Theorem 2.1. We note that the condition (2.22) is also necessary for the well-posedness of the mixed problem. The proof of Theorem 2.1 and its extension for the general initial-boundary problem (1.1), (1.2), (1.3) for the case of convex boundary is given in [7]. We shall consider here only two particular cases of $\lambda_1(x',\xi'')$ which satisfies the condition (2.22):

a) Assume that

$$\operatorname{Im} \lambda_1(x',\xi'') \geq 0\ . \tag{2.24}$$

It is obvious that (2.24) implies (2.22). Taking the imaginary part of the scalar product of (2.5) with $K^{-1}(D',\tau)v_\tau$ and with v_τ and applying the sharp Garding inequality we obtain (2.23). The case when $\operatorname{Re} \lambda_1(x',\xi'') \geq 0$ is considered analogously.

b) Assume that for some point $(\hat{x}',\hat{\xi}'')$ we have

$$\lambda_1(\hat{x}',\hat{\xi}'') = 0, \tag{2.25}$$

$$\frac{\partial}{\partial x_0} \operatorname{Re} \lambda_1(\hat{x}',\xi'') \neq 0, \tag{2.26}$$

$$\frac{\partial}{\partial x_0} \operatorname{Im} \lambda_1(\hat{x}',\xi'') \neq 0. \tag{2.27}$$

We assume that values of $\lambda_1(x',\xi'')$ and $\partial\lambda_1(x',\xi'')/\partial x_0$ are close to $\lambda_1(\hat{x}',\hat{\xi}'')$, $\partial\lambda_1(\hat{x}',\hat{\xi}'')/\partial x_0$. Therefore by using the implicit function theorem we obtain

$$\operatorname{Re} \lambda_1(x',\xi'') = \lambda_2(x',\xi'')(x_0 - \lambda_3(x'',\xi'')), \tag{2.28}$$

where

$$\lambda_2(x',\xi'') \neq 0.$$

It follows from the Taylor formula that

$$\begin{aligned}\operatorname{Im} \lambda_1(x_0,x'',\xi'') &= \operatorname{Im} \lambda_1(\lambda_3(x'',\xi''),x'',\xi'') + \\ &\quad + \lambda_4(x',\xi'')(x_0 - \lambda_3(x'',\xi'')) = \\ &= \lambda_5(x'',\xi'') + \lambda_6(x',\xi'') \operatorname{Re} \lambda_1(x',\xi''),\end{aligned} \tag{2.29}$$

where

$$\lambda_5(x'',\xi'') = \operatorname{Im} \lambda_1(\lambda_3(x'',\xi''),x'',\xi''),$$

$$\lambda_6(x',\xi'') = \lambda_4(x',\xi'')/\lambda_2(x',\xi'').$$

Assume that (2.22) is satisfied. Then we have

$$\lambda_5(x'',\xi'') \geq 0, \tag{2.30}$$

$$\lambda_6(x',\xi'') < 0. \tag{2.31}$$

Indeed (2.22) is equivalent to the following assumption:

$$\operatorname{Re} \lambda_1(x',\xi'') < 0 \quad \text{implies} \quad \operatorname{Im} \lambda_1(x',\xi'') \geq 0$$

and (2.32)

$$\operatorname{Im} \lambda_1(x',\xi'') < 0 \quad \text{implies} \quad \operatorname{Re} \lambda_1(x',\xi'') \geq 0.$$

When $x_0 - \lambda_3 < 0$ and small we have $\operatorname{Re} \lambda_1(x',\xi'') < 0$ and so that $\operatorname{Im} \lambda_1(x',\xi'') \geq 0$. Taking a limit for $x_0 - \lambda_3 < 0$, $x_0 - \lambda_3 \to 0$, we obtain (2.30).

It follows from (2.26), (2.27) that $\lambda_6 \neq 0$. Therefore we have

$$\operatorname{Re} \lambda_1 = \frac{\operatorname{Im} \lambda_1 - \lambda_5}{\lambda_6}.$$

Since $\lambda_5 \geq 0$ and since $\operatorname{Im} \lambda_1 < 0$ implies $\operatorname{Re} \lambda_1 \geq 0$ we must have (2.31).

We note that

$$\begin{aligned}&\operatorname{Re}(-\lambda_6(x',\xi'') - i)(N(\xi_0+i\tau,\xi'') + \lambda_1(x',\xi'')) = \\ &\quad = -\lambda_6 \operatorname{Re} N + \operatorname{Im} N - \lambda_6 \operatorname{Re} \lambda_1 + \operatorname{Im} \lambda_1 = \\ &\quad = -\lambda_6 \operatorname{Re} N + \operatorname{Im} N + \lambda_5(x'',\xi'') \geq -\lambda_6 \operatorname{Re} N + \operatorname{Im} N.\end{aligned} \tag{2.33}$$

Therefore taking the real part of the scalar product of $(-\lambda_6(x',D'') - i)(N(D_0+i\tau,D'') + \lambda_1(x',D''))$ with $v_\tau(x')$ and applying the sharp Garding inequality we shall prove the estimate (2.23).

We note that when the assumptions (2.22), (2.25), (2.26), (2.27) are satisfied we can obtain an estimate which is better than (2.22).

It was shown by M. Taylor and R. Melrose that for the case of strictly concave boundary the estimate of the solution of the initial boundary value problem with the Neumann boundary condition is sharper than in the flat case.

Since $-\lambda_6(x',\xi'') \geq c > 0$ and $-\lambda_6 \operatorname{Re} N(\xi_0+i\tau,\xi'') + \operatorname{Im} N(\xi_0+i\tau,\xi'') \geq CK^{-1}(\xi',\tau)|\hat{\xi}|$ we can obtain by using the Garding inequality the same estimate as for the initial-boundary Neumann problem.

A detailed proof of the estimate of the form (2.23) without additional assumptions (2.25), (2.26), (2.27) is given in [7].

§3. PROPAGATION OF SINGULARITIES FOR THE MODEL PROBLEM WITH STRICTLY CONCAVE BOUNDARY

Consider the initial-boundary problem (2.1), (2.2), (2.3), and assume that $\lambda_1(x',\xi'')$ in (2.4) is real. If $\lambda_1(x',\xi'') \geq 0$ for all (x',ξ''), then multiplying equation (2.5) by $(1-i)$ and applying the sharp Garding inequality we obtain the following estimate:

$$\| |\hat{D}''| K^{-1} v_\tau \|_s \leq C \|h_\tau\|_s, \tag{3.1}$$

where $|\hat{D}''|$ is the pseudodifferential operator with the symbol $|\hat{\xi}''|$.

To obtain (3.1) we use that $\operatorname{Re}(1-i)(N+\lambda_1) = N_1 + \lambda_1 + N_2 \geq N_1 + N_2 \geq C|\hat{\xi}''|K^{-1}$, where $N_1 = \operatorname{Re} N$, $N_2 = \operatorname{Im} N$.

Let $\varphi(x',\xi')$ be a C^∞-function for $\xi' \neq 0$ which is homogeneous in ξ' of order 0. Since the symbol of the commutator of $N+\lambda$ and φ has order $K(\xi',\tau)$ we obtain that $WF(v) \subset WF(h)$.

Therefore it remains to describe the wave front set of v when $\lambda_1(x',\xi'')$ is not a nonnegative function.

We shall describe the wave front set of the solution $u(x',x_n)$ assuming that the wave front set of $WF(h)$ is given. We note that it is enough to find $WF(v)$ since then $WF(u)$ can be found easily. Since N^0 is a pseudodifferential operator with the symbol $N(\xi_0,\xi'')$ belonging to the class $S^1_{1/3,0}$ we have $WF(N^0 v) \subset WF(v)$.

For $\xi_0|\xi''|^{-1/3}$ large the symbol $N(\xi_0+i\tau,\xi'')$ has the following asymptotics (see (2.14)):

$$N(\xi_0+i\tau,\xi'') \approx \sqrt{\xi_0+i\tau}\ |\hat{\xi}''|^{1/2},$$

i.e. it is close to the form (1.16).

Consider the equation

$$\sqrt{\xi_0}\ |\xi''|^{1/2} + \lambda_1(x',\xi'') = 0. \tag{3.2}$$

The equation of null-bicharacteristics for the symbol (3.1) has the following form, when we take x_0 as a parameter:

$$\begin{aligned} \frac{d\xi_k}{dx_0} &= -2\sqrt{\xi_0}\ |\xi''|^{-1/2}\frac{\partial\lambda_1}{\partial x_k}, \quad 0 \le k \le n-1, \\ \frac{dx_k}{dx_0} &= 2\sqrt{\xi_0}\ |\xi''|^{-1/2}\left(\frac{1}{2}\sqrt{\xi_0}\ |\xi''|^{-3/2}\ \xi_k + \frac{\partial\lambda_1}{\partial\xi_k}\right), \\ & \qquad 0 \le k \le n-1. \end{aligned} \tag{3.3}$$

We note that for the symbol (3.2) the "elliptic" ("hyperbolic") region is given by the inequality $\xi_0 < 0$ ($\xi_0 > 0$) and the diffraction region (the region of gliding rays) is the surface $\xi_0 = 0$. It is well known that in the "elliptic" region $WF(v) = WF(h)$ and in the "hyperbolic" region the singularities propagate along the null-bicharacteristics of (3.1). We shall call by the outgoing bicharacteristics starting at some point $(\hat{x}',\hat{\xi}')$ the part of bicharacteristic curve passing through $(\hat{x}',\hat{\xi}')$, where $x_0 \ge \hat{x}_0$.

The only open question is what will be the wave front set of $v(x')$ for $\xi_0 = 0$ and for ougtoing bicharacteristics starting at the points where $\xi_0 = 0$. It is convenient to parametrize the outgoing bicharacteristics in the following way:

Take the following change of variables

$$\begin{aligned} y' &= x', \\ \zeta_0 &= \sqrt{\xi_0}\ |\xi''|^{1/2}, \\ \zeta'' &= \xi''. \end{aligned} \tag{3.4}$$

In new variables the equations (3.3) have the form

$$\frac{d\zeta_0}{dy_0} = -|\zeta''|^{-1}\,\frac{\partial\lambda_1(y',\zeta'')}{\partial y_0},$$
$$\frac{d\zeta''}{dy_0} = -2\zeta_0|\zeta''|^{-1}\,\frac{\partial\lambda_1(y',\zeta'')}{\partial y''}, \tag{3.5}$$
$$\frac{dy_k}{dy_0} = 2\zeta_0|\zeta''|^{-1}\left(\frac{1}{2}\,\zeta_0\,\frac{\zeta_k}{|\zeta''|^2} + \frac{\partial\lambda_1(y',\zeta'')}{\partial\zeta_k}\right), \quad 1 \le k \le n-1 .$$

The right hand sides of (3.5) are homogeneous in ζ' and smooth for $\zeta' \neq 0$ functions of (y',ζ') and therefore for any $(\tilde{y}',\tilde{\zeta}')$, $\tilde{\zeta}' \neq 0$, there exists a unique solution of (3.5) for all $-\infty < y_0 < +\infty$.

Let $(\tilde{x}',\tilde{\xi}')$ be arbitrary such that $\tilde{\xi}_0 \geq 0$ and

$$\sqrt{\tilde{\xi}_0}\;|\tilde{\xi}''| + \lambda_1(\tilde{x}',\tilde{\xi}'') = 0. \tag{3.6}$$

We shall denote by $\tilde{L}(\tilde{y}',\tilde{\zeta}')$ the outgoing solution of the equations (3.5) with the initial conditions

$$y'(y_0) = \tilde{y}', \qquad \zeta'(y_0) = \tilde{\zeta}', \tag{3.7}$$

where

$$\tilde{y}' = \tilde{x}', \qquad \tilde{\zeta}_0 = \sqrt{\tilde{\xi}_0}\;|\tilde{\xi}''|^{1/2}, \qquad \tilde{\zeta}'' = \tilde{\xi}''. \tag{3.8}$$

Let $\tilde{L}^+(\tilde{y}',\tilde{\zeta}')$ be the connected part of $\tilde{L}(\tilde{y}',\tilde{\zeta}')$ such that $\zeta_0(y_0) \geq 0$. It may happen that $\zeta_0(y_0) \geq 0$ for all $y_0 \geq \tilde{y}_0$. Otherwise there exists y_0^* which has the following properties: $\zeta_0(y_0) \geq 0$ for any y_0 such that $\tilde{y}_0 \leq y_0 \leq y_0^*$ and there is a sequence $\{y_0^{(n)}\}$ such that $\zeta_0(y_0^{(n)}) < 0$ and $y_0^{(n)} \to y_{0+}^*$, $y_0^{(n)} > y_0^*$. In the case when $\tilde{y}_0 = y_0^*$ the curve $\tilde{L}^+(\tilde{y}',\tilde{\zeta}')$ reduces to a single point.

Denote by $L^+(\tilde{x}',\tilde{\xi}')$ the pre-image of $L^+(\tilde{y}',\tilde{\zeta}')$ under the transformation (3.4). It is obvious that for $\xi_0 > 0$ $L^+(\tilde{x}',\tilde{\xi}')$ is an outgoing bicharacteristics for the symbol (3.1). When $(\tilde{x}',\tilde{\xi}')$ is such that $\tilde{\xi}_0 < 0$ or $\tilde{\xi}_0 \geq 0$ but $\sqrt{\tilde{\xi}_0}\;|\tilde{\xi}''|^{1/2} + \lambda_1(\tilde{x}',\tilde{\xi}'') \neq 0$ we also denote by $L^+(\tilde{x}',\tilde{\xi}')$ the one point set $\{(\tilde{x}',\tilde{\xi}')\}$.

<u>Theorem 3.1</u>. Assume that $\lambda_1(x',\xi'')$ is real and is not nonnegative. Let $v(x')$ be the solution of the equation

$$N(D_0,D'')v + \lambda_1(x',D'')v = h, \tag{3.9}$$

where $h \in \mathcal{D}'(\mathbb{R}^n)$, $h = 0$ for $x_0 < 0$. Denote by $L^+(h)$ the union of all $L^+(\tilde{x}',\tilde{\xi}')$, where $(\tilde{x}',\tilde{\xi}')$ belongs to $WF(h)$. Then

$$WF(v) \subset L^+(h) . \tag{3.10}$$

The proof of the Theorem 3.1 is given in [7]. The Theorem 3.1 proves the inclusion of $WF(v)$ in $L^+(h)$. We shall show in some examples that for some $h(x')$ there is an equality $WF(v) = L^+(h)$.

Example 3.1. Consider the initial-boundary problem (2.1), (2.2),(2.3) with the boundary operator B of the following form:

$$B(D',D_n) = -i \frac{\partial}{\partial x_n} + i \frac{\partial}{\partial x_1} , \tag{3.11}$$

i.e. $\lambda_1(x',\xi'') = \xi_1$.

Therefore equation (2.5) after the Fourier transform with respect to (x_0,x'') will take the following form:

$$b(\xi_0+i\tau,\xi'')\tilde{v}(\xi_0+i\tau,\xi'') = \tilde{h}(\xi_0+i\tau,\xi''), \tag{3.12}$$

where

$$b(\xi_0+i\tau,\xi'') = -i \frac{A'((\xi_0+i\tau)|\hat{\xi}|^{-1/3})}{A((\xi_0+i\tau)|\hat{\xi}|^{-1/3})} |\hat{\xi}|^{2/3} + \xi_1 . \tag{3.13}$$

So that

$$\tilde{v}(\xi_0+i\tau,\xi'') = \frac{\tilde{h}(\xi_0+i\tau,\xi'')}{b(\xi_0+i\tau,\xi'')} . \tag{3.14}$$

We note that $b(\xi_0+i\tau,\xi'') \neq 0$ for $\tau \geq 0$. Let $\xi_1/|\xi''|^{1/2}$ be large and $\xi_1/|\xi''|$ be small. Then $b(\xi_0+i\tau,\xi'')$ for $\xi_0|\xi''|^{1/3}$ large has the following asymptotics (c.f. (2.10)):

$$\begin{aligned} b(\xi_0+i\tau,\xi'') &= \sqrt{\xi_0+i\tau}\ |\hat{\xi}''|^{1/2} + \frac{i|\hat{\xi}''|}{4(\xi_0+i\tau)} + \xi_1 + \cdots = \\ &= \sqrt{\xi_0}\ |\hat{\xi}''|^{1/2} + \frac{i\tau}{2\sqrt{\xi_0}} |\hat{\xi}''|^{1/2} + \frac{i|\hat{\xi}''|}{4\xi_0} + \xi_1 + \cdots \end{aligned} \tag{3.15}$$

Therefore the equation $b(\xi_0+i\tau,\xi'') = 0$ has a root $\xi_0+i\tau = Z(\xi'')$ with the following asymptotics:

$$Z(\xi'') = \frac{\xi_1^2}{|\hat{\xi}''|} + \frac{i|\xi''|}{2\xi_1} + \cdots , \tag{3.16}$$

where $\xi_1 < 0$.

Let $g(t) \in C_0^\infty(\mathbb{R}^1_+)$, $\int_{-\infty}^{\infty} g(t)\, dt = 1$, and let

$$\tilde{h}_1(\eta_0+i\tau,\eta'') = \tilde{g}(\xi_0|\hat{\xi}''|^{-1} \log^2|\xi''|)g(-\xi_1|\hat{\xi}''|^{-1} \log|\hat{\xi}''|). \tag{3.17}$$

Then $h_1(x') = 0$ for $x_0 < 0$ and the wave front set of $h_1(x')$ consists of all points (x',ξ') such that $x' = 0$, $\xi_0 = \xi_1 = 0$

and ξ'' are arbitrary. Now we shall check what are $L^+(h_1)$ and $WF(v_1)$, where $\tilde{v}_1 = \tilde{h}_1/b$.

The solutions of the equations (3.3) for $\lambda_1(x',\xi'') = \xi_1$ will be the straight lines

$$\begin{aligned} &\xi' = \tilde{\xi}', \\ &x_1 = (\tilde{\xi}_0|\tilde{\xi}''|^{-2}\tilde{\xi}_1 + 2\sqrt{\tilde{\xi}_0}\ |\tilde{\xi}''|^{-1/2})x_0 + \tilde{x}_1 \qquad (3.18) \\ &x_k = \tilde{\xi}_0|\tilde{\xi}''|^{-2}\tilde{\xi}_k x_0 + \tilde{x}_k \quad \text{for} \quad 2 \le k \le n-1 , \end{aligned}$$

where $(\tilde{x}',\tilde{\xi}')$ is the initial condition for $x_0 = \tilde{x}_0$. For $\tilde{\xi}_0 = 0$ we have from (3.18)

$$x' = \tilde{x}' , \qquad \xi' = \tilde{\xi}' . \qquad (3.19)$$

Therefore $L^+(h_1)$ consists of all rays $0 \le x_0 < +\infty$, $x'' = 0$, $\xi_0 = \xi_1 = 0$, ξ'' are arbitrary

We note that

$$-C_2|\hat{\xi}''|(\log|\hat{\xi}''|)^{-1} \le \xi_1 \le -C_1|\hat{\xi}''|(\log|\hat{\xi}''|)^{-1}$$

on the support of $\tilde{h}_1(\xi_0+i\tau,\xi'')$. Therefore

$$-C_3|\hat{\xi}''|(\log|\hat{\xi}''|)^{-2} \le \operatorname{Re} Z(\xi'') \le C_4|\hat{\xi}''|(\log|\hat{\xi}''|)^{-2} ,$$

$$-C_5\log|\hat{\xi}''| \le \operatorname{Im} Z(\xi'') \le -C_6\log|\hat{\xi}''| .$$

Since $Z(\xi'')$ is the only root of $b(\xi_0+i\tau,\xi'')$ for $-C|\hat{\xi}''|^{1/3} < \tau < 0$ we obtain by using the Cauchy integral formula:

$$\begin{aligned} \tilde{v}_1(x_0,\xi'') &= \frac{1}{2\pi}\int_{-\infty}^{\infty} \frac{\tilde{h}_1(\xi_0,\xi'')}{b(\xi_0,\xi'')} \exp(-ix_0\xi_0)\, d\xi_0 = \\ &= -i\left(\frac{\partial b(Z(\xi''),\xi'')}{\partial\xi_0}\right)^{-1} \tilde{g}(Z(\xi'')|\hat{\xi}''|(\log|\xi''|)^2) \cdot \\ &\quad \cdot g(\xi_1|\hat{\xi}''|^{-1}\log|\hat{\xi}''|)\exp(-ix_0Z(\xi'')) + \\ &\quad + O(\exp(-cx_0|\hat{\xi}''|^{1/3}) , \end{aligned} \qquad (3.20)$$

where $\tilde{v}_1(x_0,\xi'')$ is the Fourier transform of $v_1(x_0,x'')$ with respect to x''. It follows from (3.20) that

$$v_1 \notin C^\infty \quad \text{since} \quad |\exp(-ix_0Z(\xi''))| \ge |\xi''|^{-C_5x_0} .$$

Therefore $WF(v_1)$ consists of all rays $0 \le x_0 < +\infty$ such that

$\xi_1 = \xi_0 = 0$, $x'' = 0$ and ξ'' are arbitrary, so that $WF(v_1) = L^+(h_1)$.

Consider now $\tilde{h}_2(\xi_0+i\tau,\xi'')$ given by the following formula:

$$\tilde{h}_2(\xi_0+i\tau,\xi'') = \tilde{g}\left(\frac{\xi_0+i\tau}{|\hat{\xi}''|^{\varepsilon}}\right) g\left(\frac{\xi_1}{|\hat{\xi}''|^{1-\varepsilon}}\right) \tag{3.21}$$

where $0 < \varepsilon < 1/4$. Let $\tilde{v}_2(\xi_0+i\tau,\xi'') = \tilde{h}_2(\xi_0+i\tau,\xi'')/b(\xi_0+i\tau,\xi'')$. Since $|b(\xi_0+i\tau,\xi'')| \geq C\xi_1 \geq C|\hat{\xi}''|^{1-\varepsilon}$ for $\xi_1 \geq C|\hat{\xi}''|^{1-\varepsilon}$, $|\xi_0+i\tau| \leq C|\hat{\xi}''|^{2\varepsilon}$ we have that $b^{-1}(\xi_0+i\tau,\xi'')$ belongs to the class $S^{-1+\varepsilon}_{(1/3)-\varepsilon,0}$ in this region.

Therefore $WF(v_2)$ is equal to $WF(h_2)$ and consists of all (x',ξ') such that $\tilde{x}' = 0$, $\xi_0 = \xi_1 = 0$, ξ_2 are arbitrary, so that $L^+(h_2) \neq WF(v_2)$. Analogously we can consider a more general case when $\lambda_1(\tilde{\xi}'')$ is independent of x' and such that there exists a sequence $\{\xi''_n\}$ where $\lambda_1(\xi''_n) < 0$, $\lambda_1(\xi''_n) \to 0$.

Example 3.2. Consider the initial-boundary problem (2.1), (2.2), (2.3) with the boundary operator B of the form

$$B(x_0,D'',D_n) = -i\frac{\partial}{\partial x_n} - \gamma x_0 |\hat{D}''|, \tag{3.22}$$

where $\gamma > 0$ is a constant. Taking the Fourier transform with respect to (x_0,x'') we obtain from (2.5) the following ordinary differential equation

$$-i|\hat{\xi}''|^{\frac{2}{3}} \frac{A'((\xi_0+i\tau)|\hat{\xi}''|^{-1/3})}{A((\xi_0+i\tau)|\hat{\xi}''|^{-1/3})} \tilde{v}(\xi_0+i\tau,\xi'') +$$

$$+ i\gamma|\hat{\xi}''| \frac{\partial\tilde{v}(\xi_0+i\tau,\xi'')}{\partial\xi_0} = \tilde{h}(\xi_0+i\tau,\xi'') . \tag{3.23}$$

The solution of a homogeneous differential equation (3.23) is the following

$$\tilde{w}(\xi_0+i\tau,\xi'') = C\, A^{1/\gamma}((\xi_0+i\tau)|\hat{\xi}''|^{-1/3}) ,$$

where we choose the branch of $A^{1/\gamma}(\zeta)$ which has the following asymptotics: $A^{1/\gamma}(\zeta) = C\zeta^{-1/4\gamma} \exp i(2/3\gamma)\zeta^{3/2}$ for $\zeta \to \infty$, $\operatorname{Im}\zeta \geq 0$. The unique solution of the equation (3.23) which is analytic in $\xi_0+i\tau$ for $\tau > 0$ and which is a tempered distribution for $\tau > 0$, $(\xi_0,\xi'') \in \mathbb{R}^n$, is given by the following formula

$$\tilde{v}(\xi_0+i\tau,\xi'') =$$

$$= A^{1/\gamma}((\xi_0+i\tau)|\hat{\xi}''|^{-\frac{1}{3}}) \int_{-\infty}^{\xi_0+i\tau} \frac{\tilde{h}(t,\xi'')}{i\gamma|\hat{\xi}''|} A^{-1/\gamma}(t|\hat{\xi}''|^{-\frac{1}{3}})\,dt. \tag{3.24}$$

We take $\tilde{h}(\xi_0,\xi'') = \tilde{g}(\xi_0|\hat{\xi}''|^{-1/3})$, where $g(t)$ is the same as in Example 3.1, i.e. $g(t) \in C_0^\infty(\mathbb{R}^1_+)$, $\int_{-\infty}^{\infty} g(t)dt = 1$. One can consider $\tilde{g}(\xi_0|\hat{\xi}''|^{-1/3})$ as a symbol of the class $S^0_{1/3,0}$. Therefore $WF(h) = WF(g(D_0|\hat{D}''|^{-1/3})\delta(x))$ consists of all points (x',ξ') such that $x' = 0$, $\xi_0 = 0$ and ξ'' are arbitrary, $\xi'' \neq 0$.

The solutions of equations (3.3) with initial conditions $x'' = 0$, $\xi_0 = 0$ for $x_0 = 0$ will be the parabolas

$$\sqrt{\xi_0} - \gamma x_0 |\xi''|^{1/2} = 0, \tag{3 25}$$

where $\xi'' \neq 0$ are arbitrary.

Therefore $L^+(h)$ is a union of such parabolas for all $\xi'' \neq 0$. Suppose that $g(t)$ is such that

$$\int_{-\infty}^{\infty} g(t)A^{-1/\gamma}(t)\,dt = 0. \tag{3.26}$$

Taking in (3.24) $t|\hat{\xi}''|^{-1/3} = \tau$, we obtain

$$\tilde{v}(\xi_0+i\tau,\xi'') =$$

$$= A^{1/\gamma}((\xi_0+i\tau)|\hat{\xi}''|^{-1/3}) \int_{-\infty}^{(\xi_0+i\tau)|\hat{\xi}''|^{-1/3}} \frac{g(\tau)}{i\gamma|\hat{\xi}''|^{2/3}} A^{-1/\gamma}(\tau)\,d\tau. \tag{3.27}$$

If (3.26) holds then $\tilde{v}(\xi_0+i\tau,\xi'')$ rapidly tends to zero, when $\xi_0 \to -\infty$ and $\xi_0 \to +\infty$. Therefore $WF(v) = WF(g) \neq L^+(h)$.

If (3.26) is not satisfied then $\tilde{v}(\xi_0+i\tau,\xi'')$ tends rapidly to zero when $\xi_0 \to -\infty$ and oscillates when $\xi_0 \to +\infty$. Using the asymptotics of $A((\xi_0+i\tau)|\hat{\xi}''|^{-1/3})$ for $\xi_0 \to +\infty$ one can find that $WF(v) = L^+(h)$ when (3.26) is not satisfied.

§4. MODEL INITIAL-BOUNDARY PROBLEM FOR THE CASE OF STRICTLY CONVEX BOUNDARY

Consider the following initial-boundary problem

$$\hat{a}_-(x_n,D_0+i\tau,D'',D_n)u_\tau = 0, \quad x \in \mathbb{R}^{n+1}_+, \tag{4.1}$$

$$u_\tau = 0 \quad \text{for} \quad x_0 < 0, \quad x \in \mathbb{R}^{n+1}_+, \tag{4.2}$$

$$B(D'',D_n)u_\tau|_{x_n=0} = h_\tau(x'), \quad x' \in \mathbb{R}^n, \tag{4.3}$$

where $h(x') = 0$ for $x_0 < 0$,

$$\hat{a}_-(x_n,\xi_0+i\tau,\xi'',\xi_n) = \xi_n^2 - (\xi_0 + i\tau - x_n|\hat{\xi}''|)|\hat{\xi}''|, \tag{4.4}$$

$$B(\xi'',\xi_n) = -\xi_n + \lambda_1(\xi'') . \tag{4.5}$$

To solve the equation (4.1) we take the Fourier transform with respect to (x_0,x''). As in §2 we obtain the following ordinary differential equation

$$\frac{\partial^2 u(\xi_0+i\tau,\xi'',x_n)}{\partial x_n^2} + (\xi_0+i\tau-x_n|\hat{\xi}''|)|\hat{\xi}''|\tilde{u}(\xi_0+i\tau,\xi'',x_n) = 0, \tag{4.6}$$

where $\tilde{u}(\xi_0+i\tau,\xi'',x_n) = \int_{-\infty}^{\infty} u_\tau(x_0,x'',x_n)e^{i(x',\xi')}\,d\xi'$.

Denote $\rho_-(x_n,\xi_0+i\tau,\xi'') = (\xi_0 + i\tau - x_n|\hat{\xi}''|)|\hat{\xi}''|^{-1/3}$. When $|\rho_-|$ is large the equation (4.6) has two solutions $\tilde{u}_+$ and $\tilde{u}_-$ with the following asymptotics:

$$\tilde{u}_\pm(\xi_0+i\tau,\xi'',x_n) \approx \frac{C_\pm}{\rho_-^{1/4}} \exp \pm \frac{2}{3}\, i\rho_-^{3/2} ,$$

where the branch of $\rho_-^{3/2}$ is taken which has a negative imaginary part for the negative ρ_-. Since for $x_n > 0$ we have

$$\mathrm{Im}(\rho_-(x_n,\xi_0+i\tau,\xi''))^{3/2} < \mathrm{Im}(\rho_-(0,\xi_0+i\tau,\xi'')^{3/2} \tag{4.7}$$

and $\mathrm{Im}(\rho_-(x_n,\xi_0+i\tau,\xi''))^{3/2} \to -\infty$ when $x_n \to +\infty$, therefore the only solution of (4.6), which is analytic in $\xi_0+i\tau$ and a tempered distribution, has the following asymptotics:

$$\tilde{u}(\xi_0+i\tau,\xi'',x_n) \sim \frac{C}{\rho_-^{1/4}} \exp(-\frac{2}{3}\, i)\, \rho_-^{3/2} . \tag{4.8}$$

Denote by $A_0(z)$ the Airy function which has asymptotics (4.8), i.e.,

$$A_0(t) \sim \frac{C}{|t|^{1/4}} \exp - \frac{2}{3}\, |t|^{3/2} \quad \text{for} \quad t \to -\infty . \tag{4.9}$$

We note that $A_0(z)$ has the following properties (see for example [27]):

$$A_0(t) \sim \frac{C}{t^{1/4}} \sin\left(\frac{2}{3}\, t^{3/2} + \frac{\pi}{4}\right) \quad \text{for} \quad t \to +\infty, \tag{4.10}$$

$$A_0(z) \sim \frac{C}{z^{1/4}} \exp - \frac{2}{3}\, iz^{3/2} \quad \text{for} \quad |z| \to \infty, \qquad 2\pi - \varepsilon < \arg z < \varepsilon, \quad \forall\, \varepsilon > 0, \tag{4.11}$$

$$A_0(z) = A(z) - A_1(z), \tag{4.12}$$

where $A(z)$ is the same Airy function as in (2.10) and $A_1(z) = \overline{A(\overline{z})}$.

The solution of the initial-boundary problem for equation (4.1) with zero initial condition (4.2) and with the Dirichlet boundary condition

$$u_\tau(x',0) = v_\tau(x') \tag{4.13}$$

has the following form:

$$u(x',x_n) = \frac{1}{(2\pi)^n} \int_{-\infty}^{\infty} \frac{A_0((\xi_0+i\tau-x_n|\hat{\xi}''|)|\hat{\xi}''|^{-1/3})}{A_0((\xi_0+i\tau)|\hat{\xi}''|^{-1/3})} \cdot e^{-i(\xi_0+i\tau)x_0-i(x'',\xi'')} \tilde{v}(\xi_0+i\tau,\xi'')\, d\xi_0\, d\xi''. \tag{4.14}$$

Therefore the Neumann operator for the initial-boundary problem (4.1), (4.2), (4.3) has the following form

$$N^\tau v_\tau = -i\frac{\partial u_\tau(x',0)}{\partial x_n} = \frac{1}{(2\pi)^n} \int_{-\infty}^{\infty} i|\hat{\xi}''|^{2/3} \frac{A_0'((\xi_0+i\tau)|\hat{\xi}''|^{-1/3})}{A_0((\xi_0+i\tau)|\hat{\xi}''|^{-1/3})} \cdot e^{-i(x',\xi')} \tilde{v}(\xi_0+i\tau,\xi'')\, d\xi_0\, d\xi''. \tag{4.15}$$

We note that in the "elliptic" region, i.e., for $\xi_0 < 0$, the operator N^τ is equal microlocally to a pseudodifferential operator with a symbol of the class $S^1_{1/3,0}$. In the "hyperbolic" region for $\xi_0 > 0$, $\xi_0|\hat{\xi}''|^{-1/3} >> 1$ the behavior of N^τ is much more complicated. (In the case of a strictly concave boundary N^τ is a pseudodifferential operator with a symbol of the class $S^1_{1/3,0}$ for all ξ' (cf. §2).)

Let $\chi(t) \in C^\infty(\mathbb{R}^1)$, $\chi(t) = 1$ for $t > -1$, $\chi(t) = 0$ for $t < -2$. Denote by B the following operator

$$Bv_\tau = \frac{1}{(2\pi)^n} \int_{-\infty}^{\infty} \chi(\xi_0|\hat{\xi}''|^{-1/3}) \frac{A((\xi_0+i\tau)|\hat{\xi}''|^{-1/3})}{A_1((\xi_0+i\tau)|\hat{\xi}''|^{-1/3})} \cdot \tilde{v}(\xi_0+i\tau,\xi'')e^{-i(x',\xi')}\, d\xi_0\, d\xi''. \tag{4.16}$$

It follows from the asymptotics of the Airy function $A(z)$ that

$$\frac{A(z)}{A_1(z)} \sim \exp\left(\frac{4}{3} iz^{3/2}\right) \quad \text{for} \quad 0 \le \operatorname{Im} z \le C_0, \quad \operatorname{Re} z \to +\infty. \tag{4.17}$$

Therefore for large $\xi_0|\hat{\xi}''|^{-1/3}$ the operator B is a Fourier integral operator with the phase function $L(x',\xi')$, where

$$L(x',\xi') = (x',\xi') - \frac{4}{3}\xi_0^{3/2}|\xi''|^{-1/2}. \tag{4.18}$$

The basic fact of the theory of Fourier integral operators (see [12]) is that the wave front set of Bv is contained in the image of $WF(v)$ under the canonical transformation $(y',\zeta')\to(x',\xi')$ with the generating function $L(x',\zeta')$. This canonical transformation is defined by the following relations:

$$\begin{aligned}
\xi' &= L_{x'}(x',\zeta') = \zeta', \\
y_0 &= L_{\zeta_0}(x',\zeta') = x_0 - 2\zeta_0^{1/2}|\zeta''|^{-1/2}, \qquad\qquad (4.19)\\
y_k &= L_{\zeta_k}(x',\zeta') = x_k + \frac{2}{3}\zeta_0^{3/2}|\zeta''|^{-5/2}\zeta_k, \quad 1 \le k \le n-1.
\end{aligned}$$

The canonical transformation (4.19) has the following geometrical meaning:

$(x',0,\xi',\xi_0^{1/2}|\xi''|^{1/2})$ is the endpoint in the half-space $x_n > 0$ of the outgoing bicharacteristics of the operator $a_-(x_n,D)$ which starts at the point $(y',0,\zeta',-\zeta_0^{1/2}|\zeta''|^{1/2})$. Here $\xi_0 > 0$ and $\zeta_0 > 0$. Indeed the equations for the bicharacteristics have the form: (cf. (1.19), (1.20))

$$\begin{aligned}
&\frac{dx_n}{dt} = 2\xi_n, \qquad \frac{d\xi_n}{dt} = -|\xi''|^2, \\
&\frac{dx_0}{dt} = -|\xi''|, \qquad \frac{dx''}{dt} = -\xi_0\frac{\xi''}{|\xi''|} + 2x_n\xi'', \qquad\qquad (4.20)\\
&\frac{d\xi'}{dt} = 0.
\end{aligned}$$

The solution of (4.20) with initial conditions

$$x'(0) = y', \quad \xi'(0) = \zeta', \quad x_n(0) = 0, \quad \xi_n(0) = -\zeta_0^{1/2}|\zeta''|^{1/2} \tag{4.21}$$

has the following form

$$\begin{aligned}
&\xi'(t) \equiv \zeta', \quad x_0(t) = -|\xi''|t + y_0, \qquad\qquad (4.22)\\
&\xi_n(t) = -|\zeta''|^2 t - \zeta_0^{1/2}|\zeta''|^{1/2}, \quad x_n(t) = -|\zeta''|^2 t^2 - 2\zeta_0^{1/2}|\zeta''|^{1/2} t, \\
&x''(t) = -\zeta_0\zeta''|\zeta''|^{-1}t - \frac{2}{3}|\zeta''|^2\zeta'' t^3 - 2\zeta_0^{1/2}\zeta''|\zeta''|^{1/2}t^2 + y''.
\end{aligned}$$

The endpoint of (4.22) will be reached for $t_0 = -2\zeta_0^{1/2}|\zeta''|^{-3/2}$, since $x_n(t_0) = 0$. It is easy to see that for $t = t_0$ we obtain the transformation (4.19).

We note that B^n is a Fourier integral operator with the phase function which is the generating function for the following canonical transformation:

$$(t',\zeta') \to (x',\xi') :$$

$(x',0,\xi',\xi_0^{1/2}|\xi''|^{1/2})$ is the endpoint of the broken outgoing bicharacteristics of the operator $a_-(x_n,D)$ which starts at the point $(y',0,\zeta',-\zeta_0^{1/2}|\zeta''|^{1/2})$ and reaches the endpoint $(x',0,\xi',\xi_0^{1/2}|\xi''|^{1/2})$ after $n-1$ reflections from the boundary $x_n = 0$. Denote $N(\xi_0+i\tau,\xi'') = A_0'(\zeta)/A_0(\zeta)$ where $\zeta = (\xi_0+i\tau)|\xi''|^{-1/3}$. We have

$$\chi(\tfrac{1}{2}\xi_0|\xi''|^{-1/3})\, N(\xi_0+i\tau,\xi'') = \tag{4.23}$$
$$= \chi(\tfrac{1}{2}\xi_0|\xi''|^{-1/3})\, i|\hat{\xi}''|^{2/3}\left(\frac{A_1'(\zeta)}{A_1(\zeta)} - \frac{A'(\zeta)}{A(\zeta)}\, B\right)(1-B)^{-1}.$$

We note that, for $\zeta \to \infty$, $\xi_0 > 0$ we have

$$i|\hat{\xi}''|^{2/3}\frac{A_1'(\zeta)}{A_1(\zeta)} = \sqrt{\xi_0+i\tau}\;|\hat{\xi}''|^{1/2} + O(1)\ . \tag{4.24}$$

Expanding $(1-B)^{-1}$ into the power series we obtain a representation of $\chi((1/2)\xi_0|\xi''|^{-1/3})\, N(\xi_0+i\tau,\xi'')$ as a series of Fourier operators. This representation corresponds to the multiple reflections of the outgoing bicharacteristics from the boundary. The solution of the initial-boundary problem (4.1), (4.2), (4.3) is equivalent to the solution of the following equation:

$$N^\tau v_\tau + \lambda_1(D')v_\tau = h_\tau\ . \tag{4.25}$$

Taking the Fourier transform with respect to (x_0,x'') we obtain from (4.25)

$$b(\xi_0+i\tau,\xi'')\tilde{v}(\xi_0+i\tau,\xi'') = \tilde{h}(\xi_0+i\tau,\xi'')\ , \tag{4.26}$$

where

$$b(\xi_0+i\tau,\xi'') = i|\hat{\xi}''|^{2/3}\frac{A_0'(\zeta)}{A_0(\zeta)} + \lambda_1(\xi'')\ , \tag{4.27}$$
$$\zeta = (\xi_0+i\tau)|\hat{\xi}''|^{-1/3}\ .$$

We assume that $h(x') \in H_s^\tau(\mathbb{R}^n)$ for some $\tau > 0$ and some s and that $h(x') = 0$ for $x_0 < 0$.

The Paley-Wiener-Schwartz theorem gives a necessary and sufficient condition for the existence of a unique solution of equation (4.26) which is equal to zero for $x_0 < 0$ and which belongs to $H_{s_1}((-\infty,T_1) \times \mathbb{R}^{n-1})$ for any finite T_1, where $x_0 \in (-\infty,T_1)$, $x'' \in \mathbb{R}^{n-1}$ and s_1 may depend on T_1.

This condition is the following: There exist constants C_1,C_2,C_3,C_4 such that

$$|b(\xi_0+i\tau,\xi'')| \geq C_1(1+|\xi_0|+\tau+|\xi''|)^{C_2} \tag{4.28}$$

$$\text{for} \quad \tau > C_3 \log|\hat{\xi}''| + C_4 . \tag{4.29}$$

Theorem 4.1. The necessary and sufficient conditions for the well-posedness of the initial-boundary problem (4.1), (4.2), (4.3) are the following:

$$\text{a)} \quad \sqrt{\xi_0+i\tau}\,|\xi''|^{1/2} + \lambda_1(\xi'') \neq 0 \quad \text{for} \quad \tau>0, \quad \xi''\neq 0, \tag{4.30}$$

i.e. the weak Lopatinskii condition is satisfied,

b) For any ω_0 such that $|\omega_0| = 1$, $\lambda_1(\omega_0) = 0$, there exists a constant C such that

$$-\text{Re}\,\lambda_1(\omega)\left(\log\frac{1}{|\text{Re}\,\lambda_1(\omega)|}\right)^{-1} \leq C(\text{Im}\,\lambda_1(\omega))^2 \tag{4.31}$$

for any ω such that $|\omega| = 1$, belonging to some neighbourhood of ω_0.

At first we shall prove the necessity of the condition (4.31). If condition (4.31) is not satisfied then there exists a sequence $\{\omega_m\}$ such that $|\omega_m| = 1$, $\omega_m \to \omega_0$,

$$\text{Re}\,\lambda_1(\omega_m) < 0, \quad \text{Re}\,\lambda_1(\omega_0) = \text{Im}\,\lambda_1(\omega_0) = 0, \tag{4.32}$$

and

$$\begin{aligned}|\text{Re}\,\lambda_1(\omega_m)|\left((\text{Re}\,\lambda_1(\omega_m))^2 + (\text{Im}\,\lambda_1(\omega_m))^2\right)^{-1} &\cdot \\ \cdot\,|\log|\text{Re}\,\lambda_1(\omega_m)||^{-1} &\to +\infty ,\end{aligned} \tag{4.33}$$

when $m \to \infty$. It follows from (4.33) that

$$\begin{aligned}&\text{Re}\,\frac{1}{\lambda_1(\omega_m)} = \frac{\text{Re}\,\lambda_1(\omega_m)}{(\text{Re}\,\lambda_1(\omega_m))^2 + (\text{Im}\,\lambda_1(\omega_m))^2} < 0 \quad \text{and} \\ &\left|\text{Re}\,\frac{1}{\lambda_1(\omega_m)}\right|\,|\log|\text{Re}\,\lambda_1(\omega_m)||^{-1} \to +\infty .\end{aligned} \tag{4.34}$$

We define ξ''_m in the following way:

$$\xi_m'' = |\xi_m''|\omega_m, \quad |\xi_m''| = |\lambda_1(\omega_m)|^{-3-\varepsilon}, \quad \text{where} \quad \varepsilon > 0.$$

Then the equation

$$\frac{A_0(\zeta)}{A_0'(\zeta)} = \frac{-i}{|\xi''|^{1/3}\ \lambda_1(\omega)}, \qquad \omega = \frac{\xi''}{|\xi''|}, \tag{4.35}$$

(which is equivalent to the equation $b(\xi_0+i\tau,\xi'') = 0$) has a root $\xi_0^{(m)} + i\tau_m$ with the following asymptotics:

$$(\xi_0^{(m)} + i\tau_m)|\xi_m''|^{-1/3} = k_0 - \frac{i}{\lambda_1(\omega_m)|\xi_m''|^{1/3}}(1 + O(|\xi_m''|^{-\varepsilon_1})),$$

$$\varepsilon_1 > 0, \tag{4.36}$$

where k_0 is an arbitrary root of the Airy function $A_0(z)$. In particular,

$$\tau_m = -\mathrm{Re}\,\frac{1}{\lambda_1(\omega_m)}(1 + O(|\xi_m''|^{-\varepsilon_1})).$$

We have $|\mathrm{Re}\,\lambda_1(\omega_m)| \le |\lambda_1(\omega_m)| = |\xi_m''|^{-1/(3+\varepsilon)}$. It follows from (4.33) that

$$|\mathrm{Re}\,\lambda_1(\omega_m)|^{1-\varepsilon} \ge C|\mathrm{Re}\,\lambda_1(\omega_m)|\,|\log|\mathrm{Re}\,\lambda_1(\omega_m)||^{-1} \ge$$

$$\ge C_1|\lambda_1(\omega_m)|^2 = C_1|\xi_m''|^{-2/(3+\varepsilon)}.$$

Therefore

$$C|\xi_m''|^{-1/(3+\varepsilon)(1-\varepsilon)} \le |\mathrm{Re}\,\lambda_1(\omega_m)| \le |\xi_m''|^{-2/(3+\varepsilon)}$$

and so that

$$C_2 \log|\xi_m''| \le |\log|\mathrm{Re}\,\lambda_1(\omega_m)|| \le C_1 \log|\xi_m''|. \tag{4.37}$$

It follows from (4.34) and (4.37) that

$$\frac{\tau_m}{\log|\xi_m''|} \to +\infty.$$

Therefore the condition (4.28), (4.29) is not satisfied, so that the condition (4.31) is necessary for the well-posedness of the problem (4.1), (4.2), (4.3). The necessity of condition (4.30) is known (see [20]). To prove that conditions (4.30), (4.30) are sufficient we need the following proposition:

<u>Proposition 4.1</u>. The following inequalities hold for $\xi_0 \ge 0$, $\tau|\hat{\xi}''|^{-1/3}(1 + |\zeta|^{1/2}) \le C$:

$$\operatorname{Im} \frac{A_0'(\zeta)}{A_0(\zeta)} \le -C\tau(1 + |\zeta|)|\hat{\xi}''|^{-1/3} , \tag{4.38}$$

$$\operatorname{Im} \frac{A_0(\zeta)}{A_0'(\zeta)} \ge C\tau|\hat{\xi}''|^{-1/3} . \tag{4.39}$$

Proof. Taking the imaginary part of the scalar product of (4.6) with $\tilde{u}(\xi_0+i\tau,\xi'',x_n)$ and integrating by parts, we obtain

$$-\operatorname{Im} \frac{\partial\tilde{u}(\xi_0+i\tau,\xi'',0)}{\partial x_n} \overline{\tilde{u}(\xi_0+i\tau,\xi'',0)} + \tau \int_0^\infty |\hat{\xi}''||\tilde{u}(\xi_0+i\tau,\xi'',x_n)|^2 \, dx_n = 0 . \tag{4.40}$$

If we choose

$$\tilde{u}(\xi_0+i\tau,\xi'',x_n) = \frac{A_0((\xi_0 + i\tau - x_n|\hat{\xi}''|)|\hat{\xi}''|^{-1/3})}{A_0'((\xi_0+i\tau)|\hat{\xi}''|^{-1/3})} ,$$

we obtain the following equality

$$\operatorname{Im} \frac{A_0(\zeta)}{A_0'(\zeta)} = \tau \int_0^\infty |\hat{\xi}''|^{1/3} \frac{|A_0(\zeta - x_n|\hat{\xi}''|^{2/3})^2}{|A_0'(\zeta)|^2} dx_n , \tag{4.41}$$

where $\zeta = (\xi_0+i\tau)|\hat{\xi}''|^{-1/3}$.

Since $|A_0'(\zeta)|^2 \le C(1+|\zeta|)^{1/2}$ for $\tau < C|\hat{\xi}''|^{1/3}(1+|\zeta|)^{-1/2}$ and

$$\int_0^\infty |A(\zeta - x_n|\hat{\xi}''|^{2/3})|^2 \, dx_n = \int_0^\infty |A_0(\zeta - y_n)|^2 \, dy_n \, |\hat{\xi}''|^{-2/3} \ge$$

$$\ge C(1+|\zeta|)^{1/2}|\hat{\xi}''|^{-2/3} \tag{4.42}$$

$$\text{for } \xi_0 > 0, \quad \tau < C|\hat{\xi}''|^{1/3}(1+|\zeta|)^{-1/2} ,$$

we obtain the inequality (4.39).

To prove the estimate (4.38) we choose in (4.40)

$$\tilde{u}(\xi_0+i\tau,\xi'',x_n) = A_0(\zeta - x_n|\hat{\xi}''|^{2/3})/A_0(\zeta) .$$

Then we have analogously to (4.41):

$$\operatorname{Im} \frac{A_0'(\zeta)}{A_0(\zeta)} = -\tau \int_0^\infty |\hat{\xi}''|^{1/3} \frac{|A_0(\zeta - x_n|\hat{\xi}''|^{2/3})|^2}{|A_0(\zeta)|^2} dx_n . \tag{4.43}$$

We note that $|A_0(\zeta)|^2 \le C(1+|\zeta|)^{-1/2}$ for $\xi_0 > 0$,

$\tau < C|\hat{\xi}''|^{1/3}(1+|\zeta|)^{-1/2}$. Therefore using (4.42) we obtain the inequality (4.38).

Denote by Γ_0 the region where $\xi_0 \geq 0$, $\tau < C_0|\hat{\xi}''|^{1/3}(1+|\zeta|)^{-1/2}$ and by Γ_0' the complement to Γ_0. We note that in Γ_0' the asymptotics (4.11) is also valid.

Now we are able to prove that the conditions (4.30) and (4.31) are sufficient for the well-posedness of the problem (4.1), (4.2), (4.3). Assume that the constant C_0 defining the region Γ_0 (and Γ_0') is sufficiently large. In the region Γ_0' assuming that $\tau \geq \tau_0$ for $\xi_0 \leq 0$ where τ_0 is sufficiently large, we have

$$|b(\xi_0+i\tau,\xi'')| \geq C ,$$

since the condition (4.30) is satisfied and the asymptotics (4.11) is valid (cf. with the proof of the Theorem 2.1). Now consider the region Γ_0. We have

$$\begin{aligned} b(\xi_0+i\tau,\xi'') &= N(\xi_0+i\tau,\xi'') + \lambda_1(\xi'') = \\ &= N(\xi_0+i\tau,\xi'')\lambda_1(\xi'')\left(\frac{1}{N}+\frac{1}{\lambda_1}\right) , \end{aligned} \tag{4.44}$$

where $N(\xi_0+i\tau,\xi'') = i|\xi''|^{2/3} A_0'(\zeta)/A_0(\zeta)$.

It follows from (4.39) that

$$\operatorname{Re}\frac{1}{N} = \operatorname{Im}|\hat{\xi}''|^{-2/3}\frac{A_0(\zeta)}{A_0'(\zeta)} \geq \frac{C\tau}{|\hat{\xi}''|} \quad \text{in } \Gamma_0 . \tag{4.45}$$

The condition (4.31) gives

$$-\operatorname{Re}\frac{1}{\lambda_1(\xi'')} \leq C|\hat{\xi}''|^{-1}|\log|\operatorname{Re}\lambda_1(\omega)|| ,$$

where $\omega = \xi''/|\hat{\xi}''|$. Consider the part of Γ_0 where $|\operatorname{Re}\lambda_1(\omega)| \geq |\hat{\xi}''|^{-2/3}$. Then $-\operatorname{Re} 1/\lambda_1(\xi'') \leq C_1|\hat{\xi}''|^{-1}\log|\hat{\xi}''|$ and therefore

$$\operatorname{Re}\left(\frac{1}{N}+\frac{1}{\lambda_1}\right) \geq \frac{C\tau}{|\hat{\xi}''|} - C_1\frac{\log|\hat{\xi}''|}{|\hat{\xi}''|} \geq C_2\frac{\log|\hat{\xi}''|}{|\hat{\xi}''|} , \tag{4.46}$$

when $\tau > C_3\log|\hat{\xi}''|$.

We note that $|A_0'(\zeta)/A_0(\zeta)| \geq |\operatorname{Im}(A_0'(\zeta)/A_0(\zeta))| \geq C\tau|\hat{\xi}''|^{-1/3}$ (see (4.38)). Therefore for $\tau > C_3\log|\hat{\xi}''|$, $(\xi_0+i\tau,\xi'') \in \Gamma_0$ and $|\operatorname{Re}\lambda_1(\omega)| \geq |\hat{\xi}''|^{-2/3}$ we have

$$|b(\xi_0+i\tau,\xi'')| \geq |N||\lambda_1|\left|\frac{1}{N}+\frac{1}{\lambda_1}\right| \geq C\frac{\log^2|\hat{\xi}''|}{|\hat{\xi}''|^{1/3}} ,$$

i.e., the estimate (4.28) holds.

It remains to consider the region where $(\xi_0+i\tau,\xi'') \in \Gamma_0$ and $|\mathrm{Re}\,\lambda_1(\omega)| \le |\xi''|^{-2/3}$. In this region we have, using (4.38),

$$\mathrm{Re}\, b(\xi_0+i\tau,\xi'') = -|\xi''|^{2/3}\ \mathrm{Im}\frac{A_0'(\zeta)}{A_0(\zeta)} + \mathrm{Re}\,\lambda_1(\xi'') \ge$$

$$\ge C\tau|\xi''|^{1/3} - |\xi''|^{1/3} \ge C_4|\xi''|^{1/3}\log|\xi''|,$$

when $\tau > C_3\log|\xi''|$. Therefore the condition (4.28), (4.29) is satisfied and the Theorem 4.1 is proved.

Remark 4.1. Consider now the case when the condition (4.31) is not satisfied. Then acoording to the Theorem 4.1 the initial boundary problem (4.1), (4.2), (4.3) is ill-posed in the space of distributions. We shall show that this problem is well-posed in some space of ultra-distributions, i.e. linear continuous functionals on the Gevrey classes of functions.

The general initial-boundary problems for hyperbolic equations in the spaces of ultra-distributions was studied by Chazarian [2] and Beals [1]. We note that G.E. Shilov was the first who began to study the spaces of ultradistribution and who applied it to the Cauchy problem for partial differential equations with constant coefficients.

Since in the region Γ_0' the asymptotics of the form (4.11) is valid we obtain, assuming that the weak Lopatinskii condition (4.30) is satisfied,

$$|b(\xi_0+i\tau,\xi'')| = |N(\xi_0+i\tau,\xi'') + \lambda_1(\xi'')| \ge C \qquad (4.47)$$

for $(\xi_0+i\tau,\xi'') \in \Gamma_0'$ and τ sufficiently large.

In particular, (4.47) holds when $\tau > C_1|\xi''|^{1/3}$, where C_1 is sufficiently large.

Consider a Banach space $\hat{G}$ of measurable functions $\tilde{g}(x_0,\xi'')$, equal to zero for $x_0 < 0$ and with the following finite norm:

$$\|\tilde{g}\|^2 = \int_0^\infty |\tilde{g}(x_0,\xi'')|^2 \exp 2x_0C_1|\xi''|^{1/3}\, dx_0\, d\xi'' . \qquad (4.48)$$

Since $\tilde{g}(x_0,\xi'')$ decreases exponentially in ξ'' for $x_0 > 0$ its inverse Fourier transform $g(x_0,x'')$ belongs to a Gevrey class for $x_0 > 0$. Denote by G the inverse Fourier transform of the space $\hat{G}$. The space G' dual to G is a some space of the ultradistributions. We shall describe now the Fourier transform of the space G'. We note that, as usual, the Fourier transform of a space of generalized functions is defined by using the Parseval formula (see [11]).

Let $\hat{H}$ be the space dual to $\hat{G}$ with respect to the L_2 scalar product. The space $\hat{H}$ consists of functions $\tilde{h}(x_0, \xi'')$ equal to zero for $x_0 < 0$ and such that

$$\int_0^\infty |\tilde{h}(x_0,\xi'')|^2 \exp - 2x_0 C_1 |\hat{\xi}''|^{1/3} \, dx_0 \, d\xi'' < +\infty \tag{4.49}$$

Denote by $\tilde{H}$ the Fourier transform of $\hat{H}$ with respect to x_0. It follows from (4.49) that $\tilde{H}$ consists of functions $\tilde{h}(\xi_0+i\tau, \xi'')$ analytic in $\xi_0+i\tau$ for $\tau > C_1|\hat{\xi}''|^{1/3}$. The norm in the space $\tilde{H}$ is given by the formula

$$\|\tilde{h}\|^2 = \int_{-\infty}^{\infty} |\tilde{h}(\xi_0 + iC_1|\hat{\xi}''|^{1/3}, \xi'')|^2 \, d\xi_0 \, d\xi'' .$$

According to the definition of the Fourier transform of the generalized functions the space $\tilde{H}$ is the Fourier transform of the space G'.

It follows from (4.47) that for any $\tilde{h}(\xi_0+i\tau,\xi'') \in \tilde{H}$ there exists a unique solution

$$\tilde{v}(\xi_0+i\tau,\xi'') = \frac{\tilde{h}(\xi_0+i\tau,\xi'')}{b(\xi_0+i\tau,\xi'')} \in \tilde{H}$$

of the equation (4.26). Therefore taking the inverse Fourier transform we obtain that the equation (4.25) has a unique solution in the space G'.

§5. MODEL PROBLEM FOR A BOUNDARY OPERATOR WITH THE SYMBOL DEPENDING ON x'

Consider the initial-boundary problem (4.1), (4.2), (4.3) where

$$B(x',D'',D_n) = - \xi_n + \lambda_1(x',\xi'') . \tag{5.1}$$

Let for simplicity $\lambda_1(x',\xi'')$ be real. In the case when λ_1 is independent of x' it follows from the Theorem 4.1 that the problem (4.1), (4.2), (4.3) is well-posed when $\lambda_1 \geq 0$ and it is ill-posed when there exists a sequence $\{\omega_m\}$, $|\omega_m| = 1$, such that $\lambda_1(\omega_m) < 0$, $\lambda_1(\omega_m) \to 0$. Consider now the case when $\lambda_1(x',\xi'')$ depends on x'. If $\lambda_1(x',\xi'') \geq 0$ for all (x',ξ'') then the problem (4.1), (4.2), (4.3) is well-posed. This can be proved by solving an equation of the form (4.25) but it is easier in this case to use the energy estimates (cf. the equality (4.40)).

Consider the case when $\lambda_1(x',\xi'')$ is not nonnegative, so that there exists a sequence $\{(x'_m,\omega_m)\}$, $|\omega_m| = 1$, such that

$$(x'_m,\omega_m) \to (x'_0,\omega_0), \quad \lambda_1(x'_m,\omega_m) < 0, \quad \lambda_1(x'_0,\omega_0) = 0. \tag{5.2}$$

It was shown in §4 that there exists a root $\xi_0^{(m)} + i\tau_m$ of the equation

$$i|\xi''_m|^{2/3} \frac{A'_0(\zeta_m)}{A_0(\zeta_m)} + \lambda_1(x'_m, \xi''_m) = 0 \ , \tag{5.3}$$

where $\zeta_m = (\xi_0^{(m)} + i\tau_m)|\xi''_m|^{-1/3}$, $\xi''_m = |\xi''_m|\omega_m$, which has the asymptotic behavior of the form (4.30).

We shall look for an asymptotic solution of the equation

$$(N^\tau + \lambda_1(x', D''))v = 0 \tag{5.4}$$

in the following form

$$v_m = \sum_{k=0}^{N_m} d_{mk}(x')\exp\{-i(x_0 - x_0^{(m)})\zeta_m|\xi''_m|^{1/3} - |x'' - x''_m|^2|\xi''_m|^{\gamma_1} - \\ - i(x'' - x''_m, \omega_m)|\xi''_m|\}\ \theta(x_0^{(m)} - x_0) \ , \tag{5.5}$$

where $\theta(t) = 1$ for $t > 0$, $\theta(t) = 0$ for $t < 0$. We shall choose v_m such that

$$(N^\tau + \lambda_1)v_m = O\left(\frac{1}{|\xi''_m|^{N_1}}\right) \qquad \text{for } x_0 < x_0^{(m)} \ , \tag{5.6}$$

where $N_1 \to \infty$ when $N \to \infty$, $d_{mk}(x') \in C^\infty$, $d_{m0}(x'_m) = 1$, $d_{mk}(x') \equiv 0$ for $x_0 < 0$. The construction of an asymptotic solution of the form (5 4) is possible when the following additional condition is satisfied:

$$\frac{\partial\lambda_1(x'_0, \xi''_0)}{\partial x_0} = 0 \ . \tag{5.7}$$

The existence of an asymptotic solution of the form (5.5) proves the ill-posedness of the initial-boundary problem (4.1), (4.2), (4.3) on the time interval $(0, x_0^{(0)} + \varepsilon)$ for any $\varepsilon > 0$. Indeed, if this initial-boundary problem is well-posed, then there exists an estimate of some Sobolev's norm of $v(x')$ on $(0, x_0^{(0)} + \varepsilon)$ by some Sobolev's norm of $h(x')$ on the same time interval for any ε. Since there is no such estimate for v_m, the initial-boundary problem (4.1), (4.2), (4.3) is ill-posed, when the conditions (5.2) and (5.7) hold.

It can be proven that when condition (5.7) is not satisfied the initial-boundary problem (4.1), (4.2), (4.3) is still well-posed. We shall consider here the following model problem: Find a solution of equation (4.1) with initial condition (4.2), satisfying the boundary condition

$$B(x_0,D'',D_n)u_\tau|_{x_n=0} = h_\tau(x') , \tag{5.8}$$

where

$$B(x_0,\xi'',\xi_n) = -\xi_n + \gamma(x_0 - a)|\hat{\xi}''| , \tag{5.9}$$

$a > 0$ and γ are real constants.

As in §4 the solution of the initial-boundary problem (4.1), (4.2), (5.8) is equivalent to the solution of the equation

$$N^\tau v_\tau + \gamma(x_0 - a) \ |\hat{D}''|v_\tau = h_\tau , \tag{5.10}$$

where N^τ is the same as in (4.15). Taking the Fourier transform of (5.10) with respect to (x_0,x'') we obtain the following ordinary differential equation:

$$i|\hat{\xi}''|^{2/3} \frac{A_0'(\zeta)}{A_0(\zeta)} \tilde{v}(\xi_0+i\tau,\xi'') - i\gamma|\hat{\xi}''| \ \frac{\partial\tilde{v}(\xi_0+i\tau,\xi'')}{\partial\xi_0} -$$

$$- \gamma a|\hat{\xi}''|\tilde{v}(\xi_0+i\tau,\xi'') = \tilde{h}(\xi_0+i\tau,\xi'') . \tag{5.11}$$

The solution of the homogeneous equation (5.11) (i.e. when $\tilde{h} = 0$) is given by the following formula

$$\tilde{w}(\xi_0+i\tau,\xi'') = C(\xi'')A_0^{1/\gamma}((\xi_0+i\tau)|\hat{\xi}''|^{-1/3})\exp ia(\xi_0+i\tau), \tag{5.12}$$

where we take the branch of $A_0^{1/\gamma}(\zeta)$ which has the following asymptotics for $\pi \leq \arg\zeta < \varepsilon, \quad \forall\, \varepsilon > 0$:

$$A_0^{1/\gamma}(\zeta) \approx C\zeta^{-1/4\gamma} \exp -\frac{2}{3\gamma}\, i\zeta^{3/2} , \tag{5.13}$$

where $\operatorname{Im}\zeta^{3/2} < 0$ for $\operatorname{Im}\zeta = 0, \ \ \zeta < 0$.

We note for any $\gamma \neq 0$ the function (5.12) is not a Fourier transform of a tempered distribution which is equal to zero for $x_0 < 0$. Indeed when $\gamma < 0$ then $\tilde{w}(\xi_0+i\tau,\xi'')$ grows exponentially for $\zeta \to -\infty$. When $\gamma > 0$ then $\tilde{w}(\xi_0+i\tau,\xi'')$ is a tempered distribution but it grows exponentially in the half-plane $\tau > 0$. Therefore its inverse Fourier transform does not vanish for $x_0 < 0$.

Let $\gamma < 0$ and let $h(x') \in C_0^\infty(\mathbb{R}_+^n)$. Then the unique solution of the equation (5.11), which is a tempered distribution, is given by the following formula:

$$\tilde{v}(\xi_0+i\tau,\xi'') = A_0^{1/\gamma}((\xi_0+i\tau)|\hat{\xi}''|^{-1/3})e^{ia(\xi_0+i\tau)} \ \cdot \tag{5.14}$$

$$\cdot \int_{-\infty}^{\xi_0} A_0^{-1/\gamma}((\zeta_0+i\tau)|\hat{\xi}''|^{-1/3})e^{-ia(\zeta_0+i\tau)}(-\gamma^{-1}i|\hat{\xi}''|^{-1})\tilde{h}(\zeta_0+i\tau,\xi'')d\zeta_0 .$$

For $\gamma > 0$ the unique solution of (5.11) is given by the formula

$$\tilde{v}(\xi_0+i\tau,\xi'') = A_0^{1/\gamma}((\xi_0+i\tau)|\xi''|^{-1/3}) \cdot$$
$$\cdot \int_{\xi_0}^{\infty} e^{ia(\xi_0+i\tau)-ia(\zeta_0+i\tau)} A_0^{-1/\gamma}((\zeta_0+i\tau)|\hat{\xi}''|^{-1/3}) \cdot \tag{5.15}$$
$$\cdot (-\gamma^{-1} i|\hat{\xi}''|^{-1}) \tilde{h}(\zeta_0+i\tau,\xi'')\, d\zeta_0 .$$

Repeating the proof of the Proposition 2.1 from [5] we obtain the following estimate

$$1 - \left|\frac{A(\zeta)}{A_1(\zeta)}\right| \geq C\tau|\hat{\xi}''|^{-1/3}(1 + |\zeta|)^{1/2} \tag{5.16}$$

for $\xi_0 \geq 0$, $(\xi_0+i\tau,\xi'') \in \Gamma_0$, i.e. for $\tau < C|\hat{\xi}''|^{1/3}(1 + |\zeta|)^{-1/2}$. It follows from (5.16) that for $(\xi_0+i\tau,\xi'') \in \Gamma_0$ we have

$$|A_0(\zeta)| = |A_1(\zeta)|\left|1 - \left|\frac{A(\zeta)}{A_1(\zeta)}\right|\right| \geq C\tau|\hat{\xi}''|^{-1/3}(1+|\zeta|)^{1/4}. \tag{5.17}$$

Using (5.17) in the region Γ_0 and the asymptotics (4.11) in the region Γ_0' we can estimate the solution, given by the formula (5.14) for $\gamma < 0$ or by the formula (5.15) for $\gamma > 0$, and therefore prove the well-posedness of the initial-boundary value problem (4.1), (4.2), (5.8).

Consider now the propagation of singularities of the solution of the problem (4.1), (4.2), (5.8).

Let $\tau > 0$ be fixed. In the region $\xi_0 > -C_0|\xi''|^{1/3}$, where $C_0 > 0$ is small, we can factorize the operator

$$\tilde{L} = i|\hat{\xi}''|^{2/3}\frac{A_0'(\zeta)}{A_0(\zeta)} - i\gamma|\hat{\xi}''|\frac{\partial}{\partial\xi_0} - \gamma a|\hat{\xi}''|$$

in the following way:

$$\tilde{L} = (1 - B)^{1/\gamma}\tilde{L}_0(1 - B)^{-1/\gamma} , \tag{5.18}$$

where

$$\tilde{L}_0 = i|\hat{\xi}''|^{2/3}\frac{A_1'(\zeta)}{A_1(\zeta)} - i\gamma|\hat{\xi}''|\frac{\partial}{\partial\xi_0} - \gamma a|\hat{\xi}''| \tag{5.19}$$

and $B = A(\zeta)/A_1(\zeta)$ for $\xi_0 > -C_0|\hat{\xi}''|^{1/3}$.

To prove (5.18) we note that $\chi(\xi_0|\hat{\xi}''|^{-1/3}) \equiv 1$ for $\xi_0 > -C_0|\hat{\xi}''|^{1/3}$ and

$$\frac{A_0'(\zeta)}{A_0(\zeta)} = \frac{A'(\zeta) - A_1'(\zeta)}{A_1(B-1)} = \frac{A_1'(\zeta)}{A_1(\zeta)} - \frac{B'}{1-B}, \tag{5.20}$$

where

$$B' = \frac{d}{d\zeta}\frac{A(\zeta)}{A_1(\zeta)} = \frac{A'}{A_1} - \frac{A_1'}{A_1}, \qquad \xi_0 > -C_0|\hat{\xi}''|^{1/3}.$$

Since the commutator $[(1-B)^{1/\gamma}, \partial/\partial\xi_0]$ is equal to $(1/\gamma)B'(1-B)^{(1/\gamma)-1}\,|\hat{\xi}''|^{-1/3}$ for $\xi_0 > -C_0|\hat{\xi}''|^{-1/3}$, we obtain the factorization (5.18). We note that the factorization (5.18) can be also obtained from the formulas (5.14) or (5.15).

Therefore microlocally for $\xi_0 > -C_0|\hat{\xi}''|^{1/3}$ the inverse L^{-1} of the operator $L = N^\tau + \gamma(x_0 - a)|\hat{D}''|$ has the following form

$$L^{-1} = (I-B)^{1/\gamma}\, L_0^{-1}(I-B)^{-1/\gamma}, \tag{5.21}$$

where L_0 is a pseudodifferential operator with the symbol $i|\hat{\xi}''|^{2/3}A_1'(\zeta)/A_1(\zeta) + \gamma(x_0 - a)|\hat{\xi}''|$ and $(I-B)^\alpha$ is the following operator:

$$(I-B)^\alpha v_\tau = \frac{1}{(2\pi)^n}\int_{-\infty}^{\infty}\left(1 - \chi(\xi_0|\xi''|^{-1/3})\,\frac{A(\zeta)}{A_1(\zeta)}\right)\cdot \tag{5.22}$$
$$\cdot\ \tilde{v}(\xi_0 + i\tau, \xi'')\, e^{-i(x',\xi')}\, d\xi_0\, d\xi''.$$

Since $(I-B)^\alpha$ is an infinite series of the Fourier integral operators it can be proven (see [4] where the case $\alpha = -1$ was considered) that the wave front set of $(I-B)^\alpha v$ is the billiard ball map of the wave front set of v, i.e. it is the union of $WF(B^k v)$ for all $k \geq 0$ (see §4) and the union of all outgoing gliding rays which begin at the points of $WF(v)$. We note that the gliding rays are the null-bicharacteristics of $\mu(x', 0, \xi_0, \xi'')$ (see (1.14)), so that for the operator $a_-(x_n, D)$ the gliding rays are straight lines $\xi_0 = 0$, $\xi'' = \text{const}$, $x'' = \text{const}$, $-\infty < x_0 < +\infty$ (see for example [4]). We have assumed above that α is not a positive integer.

When $\alpha = m$ is a positive integer then it is obvious that

$$WF((I-B)^m v) \subset \bigcup_{k=0}^{m} WF(B^k v). \tag{5.23}$$

The wave front set of $L_0^{-1}h$ for $\xi_0 \geq 0$ is the same as the wave front set of the solution of the equation of the Example 3.2, i.e. $WF(L_0^{-1}h_0)$ for $\xi_0 \geq 0$ is contained in $L^+(h_0)$, where $L^+(h_0)$ is the union of all parabolas

$$\sqrt{\xi_0} + \gamma(x_0 - a)|\xi_0''|^{1/2} = 0 \quad \text{for} \quad x_0 \geq x_0^{(0)}, \tag{5.24}$$

where

$$\sqrt{\xi_0^{(0)}} + \gamma(x_0^{(0)} - a)|\xi_0''|^{1/2} = 0 \quad \text{and}$$

$$(x_0^{(0)}, x_0'', \xi_0^{(0)}, \xi_0'') \text{ belongs to } WF(h_0) .$$

It can be shown also that microlocally for $\xi_0 < -C|\xi''|^{1/3}$ we have $WF(v) \subset WF(h)$, where v is the solution of the equation (5.10).

Therefore we obtain a full description of the wave front set of the solution $v(x')$ of the equation (5.10) and consequently the description of the wave front set of $u(x',x_n)$, where $u(x',x_n)$ is the solution of the initial-boundary problem (4.1), (4.2), (5.8). Hence the singularities of $u(x',x_n)$ propagate along the gliding rays and the boundary bicharacteristics (5.24) which in its turn propagate inside the domain and overgo multiple reflections from the boundary.

There is one exception: if $\gamma = 1/m$ where m is a positive integer, then there are only m reflections of the singularities which propagate along the boundary bicharacteristics (see (5.23)).

§6. GENERALIZATIONS AND COMMENTARIES

The results of §§2-5 are valid for the initial-boundary value problem (1.1), (1.2), (1.3), where $A(x,D)$ is an arbitrary hyperbolic operator of second order. Moreover these results can be extended on strictly hyperbolic equations or systems of equations of an arbitrary order, assuming that each component of the characteristic cone of the hyperbolic operator is strictly convex and the boundary is strictly convex or strictly concave with respect to the bicharacteristics (cf. [3]). With these assumptions the initial-boundary problem can be reduced near the diffraction region to the solution of a second order equation which is pseudo-differential in the tangential variables. The boundary conditions will be also pseudodifferential in the tangential variables. In fact, we need only that the boundary will be strictly convex or strictly concave only for the intersection of the diffraction region with the region, where the Lopatinskii determinant is equal to zero.

To study the initial-boundary problem we construct the Neumann operator N^τ for the general hyperbolic equation and then we solve the equation on the boundary of the form (1.11). We don't use the Melrose transformation of the general hyperbolic equation to the model equation near the diffraction region, because it seems that the presence of the parameter τ doesn't allow us to use his results. Our constructions of the Neumann operator is similar to the constructions in [30], [23], [3], where the Neumann operator was constructed for the case $\tau = 0$. The case $\tau > 0$ is more difficult.

Initial-boundary problem for the wave equation with the oblique derivative boundary condition in the exterior of a convex domain was studied by M. Ikawa [15], [16].

M. Ikawa also found independently a particular case of the ill-posed interior problem for two space dimensions [17], [1]. Our results are more general and are obtained in a much simpler way.

General initial-boundary problems for the case when the uniform Lopatinskii condition (or Agmon-Kreiss-Sakamoto condition) is not satisfied, were studied by S. Miyatake [28], L. Garding [10], and R. Melrose and J. Sjostrand [25], [26]. They considered general smooth domains without restrictions on the curvature. But for the case of strongly convex and strongly concave domains our results are stronger.

REFERENCES

[1] Beals, R., Mixed boundary value problems for nonstrict hyperbolic equations, Bull. AMS (1972), pp. 520-521.

[2] Chazarain, J., Problemes de Cauchy abstraits et application a quelques problemes mixtes, J. Funct. Analysis 7 (1971), pp. 386-446.

[3] Eskin, G., A parametrix for mixed problems for strictly hyperbolic equations of an arbitrary order, Comm. P.D.E. 1 (1976), pp. 521-560.

[4] Eskin, G., Propagation of singularities for the interior mixed hyperbolic problem, Sem. Goulaouic-Schwartz, 1976-77 expose no. XII.

[5] Eskin, G., Parametrix and propagation of singularities for the interior mixed hyperbolic problem, Journ. d'Analyse Math. 32 (1977), pp. 17-62.

[6] Eskin, G., Well-posedness and propagation of singularities for initial-boundary value problem for second order hyperbolic equation with general boundary condition, Sem. Goulaouic-Schwartz, 1979-80, expose no. 2.

[7] Eskin, G., Initial-boundary value problem for second order hyperbolic equation with general boundary condition, I (to appear in Journ. d'Analyse).

[8] Eskin, G., Initial-boundary value problem for second order hyperbolic equation with general boundary condition, II (in preparation).

[9] Friedlander, F.G., The wave front set of the solution of a simple initial boundary value problem with glancing rays, Math. Proc. Camb. Phil. Soc. 79, part I (1976), pp. 145-160.

[10] Garding, L., Le probleme de la derivee oblique pour l'equation des ondes, C.R. Acad. Sci. Paris t. 285 (1977), pp. 773-775, et C.R. Acad. Sci. Paris t. 286 (1978), p. 1199.

[11] Gelfand, I.M., and Shilov, G.E., Generalized functions, v. 3, Fizmatgiz, Moscos, 1958.

[12] Hormander, L., Fourier integral operators I, Acta Math. 127 (1971), pp. 79-183.

[13] Hormander, L., Linear Partial Differential Operators, Springer Verlag, Berlin, 1963.

[14] Hormander, L., Spectral analysis of singularities, pp. 3-49, Seminar on singularities, Annals of Math. Studies, No. 91, Princeton, 1979.

[15] Ikawa, M., Problemes Mixtes pour l'Equations des Ondes II, Publ. R.I.M.S. Kyoto Univ. V. 13 No. 1 (1987), pp. 61-106.

[16] Ikawa, M., Mixed problems for the wave equation IV, The existence and exponential decay of solutions, Journ. of Math. Kyoto Univ. Vol. 19, no. 3, 1979, pp. 375-411.

[17] Ikawa, M., On the mixed problems for the wave equation in an interior domain, Comm. P.D.E. 3 (3), pp. 249-295 (1979).

[18] Ikawa, M., Preprint (1979).

[19] Ivrii, V. Ja., The nonclassical propagation of singularities of a solution of a wave equation near a boundary, Soviet Math. Sokl. 19 (1978) no. 4, pp. 947-949

[20] Kajitani, K., A necessary condition for the well-posed hyperbolic mixed problem with variable coefficients, J. Math. Kyoto Univ. 14 (1974).

[21] Kreiss, H.O., Initial-boundary value problems for hyperbolic systems, Comm. Pure Appl. Math. 23 (1970), pp. 277-298.

[22] Melrose, R., Microlocal parametrices for diffractive boundary value problems, Duke Math. J. 42 (1975), pp. 605-635.

[23] Melrose, R., Airy operators, Comm. in P.D.E. 3 (1978), pp. 1-76.

[24] Melrose, R., Singularities of solutions to boundary value problems, Proc. of Int Congress of Math., pp. 785-790, Acad. Sci. Fennica, Helsinki, 1980.

[25] Melrose, R. and Sjostrand, J., Singularities of boundary value problems I, Comm. Pure Appl Math. 31 (1978), pp. 593-617.

[26] Melrose, R. and Sjostrand, J., Singularities of boundary value problems II (to appear).

[27] Miller, J.C.P., Airy integral, Cambridge, 1946.

[28] Miyatake, S., A sharp form of the existence theorem for hyperbolic mixed problems of second order, J. Math. Kyoto Univ. 17 (2), 1977, pp. 199-223.

[29] Sakamoto, R., Mixed problems for hyperbolic equations, I, J. Math. Kyoto Univ. 10 (1970), pp. 349-373, II 10 (1970), pp. 403-417.

[30] Taylor, M.E., Grazing rays and reflection of singularities of solutions to wave equations, Comm. Pure Appl. Math. 29 (1976), pp. 1-37.

NOTE ON A SINGULAR INITIAL-BOUNDARY VALUE PROBLEM

F.G. Friedlander

Department of Applied Mathematics and Theoretical Physics,
University of Cambridge

1. The classical Euler-Poisson-Darboux equation in $\mathbb{R}^2$ is

$$\frac{\partial^2 u}{\partial y^2} - \frac{\partial^2 u}{\partial x^2} - \frac{2\mu}{x}\frac{\partial u}{\partial x} = 0 , \tag{1.1}$$

where μ is a complex number. A higher dimensional analogue of this, known under the same name, is

$$\frac{\partial^2 u}{\partial t^2} + \frac{2\mu}{t}\frac{\partial u}{\partial t} - \sum_{i=1}^{n-1}\frac{\partial^2 u}{\partial x_i^2} = 0 , \tag{1.2}$$

where $n > 2$. The literature on this is extensive; for a recent survey, see [1]. The equation which will be discussed here is a different generalisation of (1.1),

$$Pu \equiv \frac{\partial^2 u}{\partial t^2} - \sum_{i=1}^{n-1}\frac{\partial^2 u}{\partial x_i^2} - \frac{2\mu}{x_1}\frac{\partial u}{\partial x_1} = f(x,t) , \tag{1.3}$$

defined on the half-space

$$X = \{(x,t) \in \mathbb{R}^{n-1} \times \mathbb{R} : x_1 > 0\} . \tag{1.4}$$

For $\mu = \frac{1}{2}$, the equation (1.3) is satisfied by axisymmetric solutions of the standard wave equation in $\mathbb{R}^{n+1}$, so one can look upon (1.3) as a generalized axisymmetric wave equation.

H. G. Garnir (ed.), Singularities in Boundary Value Problems, 55–67.

The crucial difference between these two equations is that (1.2) is singular on the hypersurface $t = 0$ which is space-like with respect to its formal principal symbol, whereas (1.3) is singular on the time-like hypersurface $\partial X = \{x_1 = 0\}$. So, while it is natural to discuss a singular Cauchy problem for (1.2), a more appropriate problem for (1.3) is a singular initial-boundary value problem in which, instead of boundary data, a regularity condition is prescribed on ∂X. For (1.1), such a problem was considered, some time ago, in a paper by A.E. Heins and the author [4], and the present Note is an extension, by a different method, of the results obtained there.

The principal result which will be established is

Theorem 1. Assume that $\mathrm{Re}\mu > 0$, and let $f \in \mathcal{E}'(X)$. Then (1.3) has a unique solution $u \in \mathcal{D}'(X)$ that vanishes for $t \ll 0$ and is smooth up to the boundary ∂X. Also, if $f \in C_o^\infty(X)$, then $u \in C^\infty(\bar{X})$.

For the definition of smoothness up to ∂X, see the Remark following Proposition 4 below.

The proof of this will be effected by constructing a certain fundamental kernel $K(x,t,y,s) \in \mathcal{D}'(X \times X)$ for P that satisfies

$$\begin{aligned} &P(x,\partial_x,\partial_t)\, K = \delta(x,t,y,s) \\ &\operatorname{supp} K \subset \{(x,t,y,s) : t \geq s\} \end{aligned} \tag{1.5}$$

where the Dirac kernel $\delta \in \mathcal{D}'(X \times X)$ is defined by

$$\langle\delta(x,t,y,s),\ \phi(x,t,y,s)\rangle = \int \phi(x,t,x,t)\, dx\, dt, \qquad \phi \in C_o^\infty(X \times X) \tag{1.6}$$

The method by which K will be obtained is an adaptation of M. Riesz's method ([7]; see also [3, chapter 6]). It yields, virtually by inspection, detailed information on the reflection of singularities at the boundary, and that is probably its main advantage. A similar method for the Euler-Poisson-Darboux equation was developed by Leray and Delache [2].

The construction of K, and the proof of Theorem 1, will only be given in outline here; it is hoped to publish the details elsewhere.

The work reported on this Note had its origin in a seminar given during a semester spent at the University of Lund, and I want to thank Professor Gårding for giving me the opportunity to make this visit. I also want to thank Richard Melrose for some useful conversations concerning the uniqueness proof.

2. It is technically more convenient to work with the differential operator

$$Qu = x_1^{\mu} P(x_1^{-\mu} u) = \frac{\partial^2 u}{\partial t^2} - \sum_{i=1}^{n-1} \frac{\partial^2 u}{\partial x_i^2} + \frac{\mu(\mu-1)}{x_1^2} u \tag{2.1}$$

which is formally self-adjoint with respect to Lebesgue measure dx dt. The principal part of Q is the d'Alembertian on $\mathbb{R}^n$, so that Q has the same characteristics, and the same local dependence domains, as the wave equation. Taking into account that Q is defined on X rather than on $\mathbb{R}^n$, we accordingly introduce the forward and backward dependence domains of a point $(y,s) \in X$ by setting, respectively

$$\begin{aligned} D^+(y,s) &= \{(x,t) : t \geq s + |x-y|\}, \\ D^-(y,s) &= \{(x,t) : t \leq s - |x-y|\}. \end{aligned} \tag{2.2}$$

A connected open set $X' \subset X$ will be called a causal domain (for Q) if, for any two points (y,s) and (y',s') in X' such that $(y',s') \in D^+(y,s)$, the set $D^+(y,s) \cap D^-(y',s')$ is a compact subset of X; this will be the case if and only if x_1 is bounded away from zero on it.

Such a causal domain X' can be considered as an unbounded Lorentzian manifold equipped with the metric $dt^2 - |dx|^2$, and Q restricted to X' is then a nondegenerate linear hyperbolic operator of the second order. So it follows from standard theory that there is a unique $E \in \mathcal{D}'(X' \times X')$ such that

$$\begin{aligned} &Q(x,\partial_x,\partial_t)\, E(x,t,y,s) = \delta(x,t,y,s), \\ &\operatorname{supp} E \subset \{(x,t,y,s) \in X' \times X' : (x,t) \in D^+(y,s)\} \end{aligned} \tag{2.3}$$

As a Schwartz kernel, E induces continuous maps $C_o^\infty(X') \to C^\infty(X')$ which extend to continuous maps $\mathcal{E}'(X') \to \mathcal{D}'(X')$ that are fundamental solutions of Q : one has

$$\langle E_1\psi \otimes \phi\rangle = \int \langle E^+_{(y,s)}, \psi\rangle \phi(y,s)\,dy\,ds, \quad \phi,\psi \in C_o^\infty(X') \tag{2.4}$$

where

$$Q(x,\partial_x,\partial_t)\, E^+_{(y,s)} = \delta_{(y,s)}, \quad \operatorname{supp} E^+_{(y,s)} \subset D^+(y,s).$$

We shall presently determine E, but first make the following observation. For $x \in \mathbb{R}^{n-1}$, let $x \to \tilde{x}$ be reflection in $x_1 = 0$,

$$\tilde{x} = (-x_1, x_2, \ldots, x_{n-1}) . \tag{2.5}$$

Put

$$\Omega = \{(x,t,y,s) : |t-s| = |x-\tilde{y}| = |\tilde{x}-y|\} . \tag{2.6}$$

A simple geometrical argument, which we omit, gives :

Lemma. If $X' \subset X$ is a causal domain, then $X' \times X' \quad \Omega$.

We now introduce some notations that will be used throughout. If (x,t) and (y,s) are points of X, we write

$$\gamma = \gamma(x,t,y,s) = (t-s)^2 - |x-y|^2 \tag{2.7}$$

for the square of the Lorentzian distance, and set

$$\gamma_+ = \gamma \quad \text{if} \quad \gamma > 0, \quad \gamma_+ = 0 \quad \text{if} \quad \gamma \leq 0 \tag{2.8}$$

We also put

$$Z = Z(x,t,y,s) = \gamma/4\, x_1 y_1, \quad Z_+ = \gamma_+/4x_1 y_1 \tag{2.9}$$

We recall that, if a, b and c are complex numbers, with $c \neq 0, -1, -2, \ldots$ and ζ is a complex variable, $|\zeta| < 1$, then the hypergeometric series is, by definition,

$$F(a,b\,;c\,;\zeta) = \frac{\Gamma(c)}{\Gamma(a)\Gamma(b)} \sum_{j=1}^{\infty} \frac{\Gamma(a+j)\Gamma(b+j)}{\Gamma(c+j)\Gamma(1+j)} \zeta^j . \tag{2.10}$$

Finally, we write $H(t)$ for the Heaviside function, $H(t) = 1$ if $t > 0$, and $H(t) = 0$ if $t \leq 0$.

Proposition 1. Let $(x,t,y,s) \in \Omega$ and $\lambda \in \mathbb{C}$, and put, for $\mathrm{Re}\lambda > 0$,

$$\begin{aligned} E_\lambda(x,t,y,s) &= C_n(\lambda)\, H(t-s)\, \frac{\gamma_+^{\lambda-1}}{\Gamma(\lambda)} F(\mu,1-\mu;\lambda;Z) \\ &= C_n(\lambda)\, H(t-s)\, (4x_1y_1)^{\lambda-1} \frac{Z_+^{\lambda-1}}{\Gamma(\lambda)} F(\mu,1-\mu;\lambda;Z) \end{aligned} \tag{2.11}$$

where

$$C_n(\lambda) = 1/\pi^{\frac{1}{2}n-1}\, 2^{2\lambda+n-3}\, \Gamma(\lambda + \tfrac{1}{2}n-1) \tag{2.12}$$

Then there is an $E_\lambda \in \mathcal{D}'(\Omega \times \Omega)$ (the Reisz kernel) which is an analytic function of λ on $\mathbb{C}$, and which is equal to the second member of (2.11) when $\mathrm{Re}\lambda > 0$. This has the following properties :

$$Q(x,\partial_x,\partial_t)\,E_\lambda = Q(y,\partial_y,\partial_s)\,E_\lambda = E_{\lambda-1}\ , \quad \lambda \in \mathbb{C}\ , \tag{2.13}$$

$$E_{\lambda|\lambda=1-\frac{1}{2}n} = \delta(x,t,y,s)\ . \tag{2.14}$$

Corollary. Let $X' \subset X$ be a causal domain. Then the restriction of

$$E = E_{\lambda|\lambda=2-\frac{1}{2}n} \tag{2.15}$$

to $X' \times X'$ is the fundamental kernel of Q defined by (2.3).

Proof. This is omitted, as it is a straightforward exercise. We only point out that the first step in the determination of the Riesz kernel involves the computation of a formal power series in γ whose coefficients are got by solving a set of transport equations, and this readily yields the second member of (2.11). By (2.6) and (2.9) one has $0 \le Z < 1$ in the support of E_λ, so that the convergence, uniform on compact subsets of Ω, of the series, is assured. Everything else then follows from the general theory; see [3, chapter 6]. The Corollary is an immediate consequence of (2.13), (2.14) and the Lemma.

The singular support of $E^+_{(y,s)}$, the forward fundamental solution of Q that is defined in terms of E by (2.4), is

$$\partial D^+(y,s) = \{(x,t) : t = s + |x - y|\}\ , \tag{2.16}$$

the restriction to X of the forward null cone of the d'Alembertian in $\mathbb{R}^n$ with vertex (y,s). It is evident that, for (y,s) fixed, $E^+_{(y,s)}$ is determined uniquely in

$$\Omega(y,s) = \{(x,t) : s - |x - \tilde{y}| < t < s + |x - \tilde{y}|\}\ , \tag{2.17}$$

and the 'forward' boundary of this set is the characteristic

$$\partial^+\Omega(y,s) = \{(x,t) : t = s + |x - \tilde{y}|\}\ . \tag{2.18}$$

This is the reflection of $\partial D^+(y,s)$ in ∂X. We shall see later that the singular support of any forward fundamental solution of Q (and, similarly, of P) that is defined on all of X is the union of $\partial D^+(y,s)$ and $\partial^+\Omega(y,s)$.

3. We shall now determine a $G \in \mathcal{D}'(X \times X)$ such that $G = 0$ for $t < s$ and

$$Q(x,\partial_x,\partial_t)\, G(x,t,y,s) = \delta(x,t,y,s),\ G|\Omega = E\ . \tag{3.1}$$

To do this, we first look for a distribution G_λ that is an analytic function of λ and satisfies

$$Q(x,\partial_x,\partial_y)\, G_\lambda(x,t,y,s) = G_{\lambda-1}(x,t,y,s)\ ,$$
$$\operatorname{supp} G_\lambda \subset \{(x,t,y,s) : (x,t) \in D^+(y,s)\}\ ,\ G_\lambda|\Omega = E_\lambda\ . \tag{3.2}$$

The second form of E_λ in (2.11) suggests the ansatz

$$G_\lambda = C_n(\lambda)\ H(t-s)\ (4x_1y_1)^{\lambda-1}\ \Psi_\lambda(Z) \tag{3.3}$$

A straightforward computation gives, provided that

$$\Psi_\lambda(\zeta) \in C^2(\mathbb{R})\ ,\quad \Psi_\lambda(0) = \Psi'_\lambda(0) = 0\ , \tag{3.4}$$

the following identity :

$$Q(x,\partial_x,\partial_t)\, G_\lambda = C_n(\lambda-1)\ H(t-s)(4x_1y_1)^{\lambda-2}\ \Psi'_\lambda(Z)$$
$$+ C_n(\lambda)\ x_1^{-2}\ H(t-s)(4x_1y_1)^{\lambda-1}\ L\Psi_\lambda(Z), \tag{3.5}$$

where

$$L\Psi_\lambda(\zeta) = \zeta(1-\zeta)\ \Psi''_\lambda + (2-\lambda)(1-2\zeta)\Psi'_\lambda - (\mu+1-\lambda)(2-\mu-\lambda)\Psi_\lambda\ .$$

Now $L\Psi_\lambda = 0$ is a hypergeometric equation which admits

$$\zeta^{\lambda-1}\ F(\mu,1-\mu;\lambda;\zeta)/\Gamma(\lambda)$$

as solution if $|\zeta| < 1$; this of course agrees with, and would allow one to recover, (2.11). If one takes $\operatorname{Re}\mu > 0$, then a solution of $L\Psi_\lambda = 0$ valid for $\zeta > 1$ is

$$A\zeta^{\lambda-\mu-1}\ F(\mu+1-\lambda,\mu;2\mu;\zeta^{-1}),$$

where A is independent of ζ. When $\text{Re}\lambda > 1$, both the hypergeometric series converge at $\zeta = 1$; one can then choose A so as to make Ψ_λ continuous at $\zeta = 1$. From the well known formula

$$F(a,b;c;1) = \Gamma(c)\Gamma(c-a-b)/\Gamma(c-a)\Gamma(c-b) \ ,$$

which holds if $\text{Re}(c-a-b) > 0$, one thus obtains

$$\begin{aligned}\Psi_\lambda(\zeta) &= \frac{\zeta_+^{\lambda-1}}{\Gamma(\lambda)} F(\mu,1-\mu;\lambda;\zeta) \quad \text{if} \quad \zeta \le 1 \ , \\ &= \frac{\Gamma(\mu)}{\Gamma(\lambda-\mu)\Gamma(2\mu)} \zeta^{\lambda-\mu-1} F(\mu+1-\lambda,\mu;2\mu;\zeta^{-1}) \text{ if } \zeta \ge 1 \end{aligned} \tag{3.6}$$

It is easy to check that (3.4) holds, and also that

$$\Psi_\lambda'(\zeta) = \Psi_{\lambda-1}(\zeta) \tag{3.7}$$

provided that the $\text{Re}\lambda$ is sufficiently large. So (3.5) gives

<u>Lemma A</u>. If $\text{Re}\mu > 0$ and $\text{Re}\lambda$ is sufficiently large, then (3.3), with Ψ_λ given by (3.6), satisfies (3.2).

One can now define a distribution $\Psi_\lambda \in \mathcal{D}'(\mathbb{R})$ which coincides with (3.6) when $\text{Re}\lambda$ is sufficiently large, by analytic continuation in λ. Let $\rho \in \mathbb{C}$, and let $H_\rho(t)$ be the (well known) distribution that is equal to the locally integrable function $t \to t_+^{\rho-1}/\Gamma(\rho)$ when $\text{Re}\rho > 0$, and is defined for all $\rho \in \mathbb{C}$ by analytic continuation in ρ. Furthermore, let $H_\rho(t(1-t))$ be the pullback of this distribution under the map $t \to t(1-t)$; this is a well defined member of $\mathcal{D}'(\mathbb{R})$ because the only critical value of $t(1-t)$ is $\frac{1}{4}$ (corresponding to $t = \frac{1}{2}$), and the singular support of $H_\rho(t)$ is the origin.

<u>Lemma B</u>. Assume that $\text{Re}\mu > 0$. Then the distribution

$$\Psi_\lambda(\zeta) = H_\mu(\zeta(1-\zeta)) * H_{\lambda-\mu}(\zeta) \tag{3.8}$$

is an analytic function of λ on $\mathbb{C}$, which is equal to the second member of (3.6) when $\text{Re}\lambda$ is sufficiently large.

<u>Proof</u>. As $H_\mu(\zeta(1-\zeta))$ has compact support, the convolution exists, and it is evidently an analytic function of λ on $\mathbb{C}$. When $\text{Re}(\lambda-\mu) > 0$, it is the integral

$$\frac{1}{\Gamma(\mu)\Gamma(\lambda-\mu)} \int z_+^{\mu-1} (1-z)_+^{\mu-1} (\zeta-z)_+^{\lambda-\mu-1} dz \ . \tag{3.9}$$

That this equals the second member of (3.6) is then an easy consequence of the identity

$$F(a,b;c;\zeta) = \frac{\Gamma(c)}{\Gamma(b)\Gamma(c-b)} \int_0^1 z^{b-1} (1-z)^{c-b-1} (1-z\zeta)^{-a} dz ,$$

which holds if Rec > 0, Re(c-b) > 0, and $|\zeta| < 1$. So the Lemma is proved.

To transfer this result to $\mathbb{R}^n$ we note that, by (2.7) and (2.9), one has

$$Z = (\gamma' - (x_1-y_1)^2)/4x_1y_1, \quad \gamma' = (t-s)^2 - \sum_{i=2}^{n-1} (x_i-y_i)^2 ,$$

whence

$$4x_1y_1 \, dz = d\gamma' - \gamma'(\frac{dx_1}{x_1} + \frac{dy_1}{y_1}) - \frac{x_1^2 - y_1^2}{x_1y_1} (dx_1 - dy_1) .$$

So the critical set of $(x,t,y,s) \to Z$ is the diagonal Δ of $X \times X$, and hence $\Psi_\lambda(z(x,t,y,s)) \in \mathcal{D}'(X \times X \quad \Delta)$ is well defined as a pullback. On the other hand, $\Delta \subset \Omega$, and $G_\lambda|\Omega = E_\lambda$ when Reλ is large. So one can define a distribution $G_\lambda \in \mathcal{D}'(X \times X)$ by putting

$$G_\lambda(x,t,y,s) = \sigma E_\lambda + C_n(\lambda)(1-\sigma)H(t-s)(4x_1y_1)^{\lambda-1}\Psi_\lambda(Z), \quad (3.10)$$

where $\sigma \in C_0^\infty(\Omega)$ is such that $\sigma = 1$ on a neighbourhood of Δ. It is clear that this is an analytic function of λ on $\mathbb{C}$, and that $G_\lambda|\Omega = E_\lambda$ extends to all λ by the uniqueness of analytic continuation.

Taking into account that $C_n(\lambda)$, which is defined by (2.12), goes to zero as $\lambda \to 1-\frac{1}{2}n$, it is now not difficult to conclude from Lemmas A and B and Proposition 1 that one has

<u>Proposition 2</u>. Let Reμ > 0, and let G_λ be defined by (3.10). Then

$$G(x,t,y,s) = G(x,t,y,s)|_{\lambda = 2-\frac{1}{2}n} \quad (3.11)$$

is a fundamental kernel of Q, defined on $X \times X$, whose support is contained in

$$\{(x,t,y,s) : (x,t) \in D^+(y,s)\} \quad (3.12)$$

Remark. The singular spectrum of $\Psi_\lambda(\zeta)$ is $\{0\} \times \mathbb{R}\setminus 0 \cup \{1\} \times \mathbb{R}\setminus 0$; so it follows from properties of singular spectra of pullbacks of distributions [6] that the singular spectrum of G, restricted to $X \times X \setminus \Delta$, is contained in the union of the normal bundles of $\{t - s - |x - y| = 0, x \neq y\}$ and $\{t - s - |x - \tilde{y}| = 0\}$. As is always the case for fundamental kernels, sing spec G also contains all of the fibre of $T^*(X \times X)$ above Δ.

By a minor modification of the argument, the construction above can also be carried out for all $\mu \in \mathbb{C} \setminus \{0, -1, \ldots\}$; one can also deal with the exceptional values of μ. As Q is unchanged when μ is replaced by $1 - \mu$, one thus obtains two fundamental kernels. They are distinguished by the property that they are point-pair invariants of the group of isometries of the Lorentzian metric

$$(dt^2 - |dx|^2)/x_1^2$$

whose Laplace-Beltrami operator $\Box$ is related to Q by

$$x_1^2 Qu = x_1^{1-\frac{1}{2}n}(\Box + (\mu-\tfrac{1}{2}n)(\mu+\tfrac{1}{2}n-1)(x_1^{\frac{1}{2}n-1} u).$$

4. The differential operators P and Q are related by (2.1). So

$$K(x,t,y,s) = (y_1/x_1)^{\mu}\, G(x,t,y,s) \tag{4.1}$$

is a distribution on $X \times X$ that satisfies

$$P(x,\partial_x,\partial_t)\, K = \delta(x,t,y,s)$$

$$\operatorname{supp} K \subset \{(x,t,y,s) : (x,t) \in D^+(y,s)\} \tag{4.2}$$

The singular spectrum of K is the same as that of G. By the Remark following Proposition 2, and a property of singular spectra, K is a regular kernel; in particular, if ϕ and f are in $C_o^\infty(X)$ then

$$\langle K, \phi \otimes f\rangle = \int K.f(x,t)\phi(x,t)\, dx\, dt \tag{4.3}$$

where $f \to K.f$ is a continuous map $C_o^\infty(X) \to C^\infty(X)$ which extends by continuity to a continuous map $\mathcal{E}'(X) \to \mathcal{D}'(X)$. By a routine argument, one therefore has

Proposition 3. Let $f \in \mathcal{E}'(X)$, and put $u = K.f$. Then

$$Pu = f, \quad \operatorname{supp} u \subset D^+(\operatorname{supp} f)\,. \tag{4.4}$$

Also, if $f \in C_o^\infty(X)$, then $u \in C^\infty(X)$.

Remark. Here, $D^+(\mathrm{supp}\, f)$ is the union of the $D^+(y,s)$ with $(y,s) \in \mathrm{supp}\, f$; it is a closed subset of X. One can evidently extend the scope of Proposition 3; it holds for all $f \in \mathcal{D}'(X)$ with past-compact support.

The choice implied by the adoption of (3.6) in the construction of G, and hence of K, now gives our main result.

Proposition 4. Let $\mathrm{Re}\mu > 0$, and let u be as in Proposition 3. Then u is smooth up to ∂X.

Remark. When $f \in C_o^\infty(X)$, this just means that $u \in C^\infty(\bar{X})$. When $f \in \mathcal{E}'(X)$, there is a $\delta > 0$ such that $Pu = 0$ on $\{(x,t,y,s) : 0 < x_1 < \delta\} \cong (0,\delta) \times \mathbb{R}^{n-1}$. By the partial hypoellipticity of P, this implies that there is a C^∞ function $U : (0,\delta) \to \mathcal{D}'(\mathbb{R}^{n-1})$ such that

$$\langle u,\phi\rangle = \int \langle U(x_1),\phi(x_1,.)\rangle\, dx_1 \ , \quad \phi \in C_o((0,\delta) \times \mathbb{R}^{n-1}) \ , \tag{4.5}$$

and the Proposition asserts that U can be extended by continuity to a member of $C^\infty([0,\delta) \ ; \mathcal{D}'(\mathbb{R}^{n-1}))$; compare [5].

Proof. With $\lambda \in \mathbb{C}$, put

$$K_\lambda(x,t,y,s) = (y_1/x_1)^\mu\, G_\lambda(x,t,y,s) \ .$$

For $\mathrm{Re}\lambda$ sufficiently large and $\phi, f \in C_o^\infty(X)$, it follows from Lemma A above that, by virtue of Fubini's theorem,

$$\langle K_\lambda, \phi \otimes f\rangle = C_n(\lambda)\int_{t>s} H_\mu(\zeta(1-\zeta))\, H_{\lambda-\mu}(\gamma - 4x_1y_1\zeta)\, (2y_1)^{2\mu}\, \phi(x,t)f(y,s)\, dm\, d\zeta \tag{4.6}$$

where dm is Lebesgue measure in $X \times X$. With a change of variables, this can be put into the form

$$\langle K_\lambda, \phi \otimes f\rangle = \tfrac{1}{2}C_n(\lambda)\ \langle H_\mu(\zeta(1-\zeta)) \otimes H_{\lambda-\mu}(\eta),\ V(x,t,\eta,\zeta)\rangle$$

where

$$V = \int f(y,t-r)\, \frac{(2y_1)^\mu}{r}\, dy, \quad r = (|x-y|^2 + 4x_1y_1\zeta + \eta)^{\frac{1}{2}} \ .$$

As f has compact support, one can show that there is an $\varepsilon > 0$

such that

$$V \in C^{\infty}((-\varepsilon,\varepsilon) \times \mathbb{R}^{n-1} \times (-\varepsilon,\infty) \times (-\varepsilon,1+\varepsilon)).$$

Routine arguments then give the Proposition in the C^{∞} case. The proof in the distribution case also starts with (4.6), and is similar.

Proposition 5. Assume that $\mathrm{Re}\mu > 0$. Let $u \in \mathcal{D}'(X)$ and suppose that (i) $Pu = 0$; (ii) there is a $T \in \mathbb{R}$ such that $u = 0$ for $t < T$; (iii) u is smooth up to ∂X. Then $u = 0$.

Proof. This is by duality and essentially routine, using the fact that by (i) u is now given by (4.5) for all $\phi \in C_o^{\infty}(X)$, and that the 'time-reversed' version of Proposition 4 gives a solution of the adjoint equation ${}^tP\phi = \psi \in C_o^{\infty}(X)$ with future-compact support. Defining u^{ε}, where $\varepsilon > 0$, by

$$\langle u^{\varepsilon},\phi\rangle = \int_{\varepsilon}^{\infty} \langle U(x_1),\ \phi(x_1,.)\rangle\, dx_1, \quad \phi \in C_o^{\infty}(X) ,$$

one then shows first that $\langle u^{\varepsilon},\psi\rangle = O(\varepsilon^{2\mu})$, for any $\psi \in C_o^{\infty}(X)$, and the Proposition follows, on letting $\varepsilon \to 0$.

Proof of Theorem 1. This now follows by combining Propositions 4 and 5.

Remark. The proof of Proposition 4 sketched here breaks down if the support of f (as a subset of $\underline{X}$) meets ∂X. However, if one assumes that $f(x_1,.) \in C^{\infty}(\overline{\mathbb{R}}^+; \mathcal{D}'(\mathbb{R}^{n-1}))$, and imposes a certain compactness condition on supp f, then one can use a Fourier-Laplace transform in $(x_2,..,x_{n-1},t)$ to prove that $Pu = f$ has a solution that vanishes for $t \ll 0$ and is smooth up to ∂X.

When $f = \delta_{(y,s)}$, then $K.f = K^+_{(y,s)}$, the unique forward fundamental solution of P that is smooth up to ∂X. (In fact, this is the distribution derived from the function $(x,t) \to K_{\lambda}(x,t,y,s)$, which is locally integrable when $\mathrm{Re}\lambda \gg 0$, by analytic continuation in λ). Using the properties of hypergeometric functions, one can obtain detailed formulae which show how the singularity carried on $\partial D^+(y,s)$ is reflected at the boundary. These will now be stated for odd n; the case of even n is not dissimilar, but more complicated.

Put

$$\Omega = \{(x,t) : t - s < |x - \tilde{y}|\} ,$$

$$\Omega_1 = \{(x,t) : |x - y| < t - s < (|x - \tilde{y}|^2 + 4x_1y_1)^{\frac{1}{2}}\}$$

$$\Omega_2 = \{(x,t) : t - s > |x - \tilde{y}|\} ,$$

and

$$\tilde{\gamma} = (t-s)^2 - |x - \tilde{y}|^2 , \quad c_m = 1/2\pi^{m+\frac{1}{2}} \Gamma(\tfrac{1}{2} - m) ,$$

where n = 2m + 3, m = 0, 1,... . Then

$$K^+_{(y,s)}|\Omega = c_m (\frac{y_1}{x_1})^\mu \gamma_+^{-\frac{1}{2}-m} F(\mu,1-\mu;\tfrac{1}{2}-m; \frac{\gamma}{4x_1y_1}) \tag{4.7}$$

$$K^+_{(y,s)}|\Omega_1 = c_m (\frac{y_1}{x_1})^\mu (\tilde{\gamma}_-^{-\frac{1}{2}-m} \sin\pi\mu - \tilde{\gamma}_+^{-\frac{1}{2}-m} \cos\pi\mu) \times$$

$$F(\mu,1-\mu;\tfrac{1}{2}-m; - \frac{\tilde{\gamma}}{4x_1y_1}) \ (\text{mod } C^\infty(\Omega_1)) \tag{4.8}$$

$$K^+_{(y,s)}|\Omega_2 = c_m \frac{\Gamma(\mu)\ \Gamma(\frac{1}{2}-m)}{\Gamma(2\mu)\Gamma(\frac{1}{2}-m-\mu)} (2y_1)^{2\mu}\ \gamma^{-\frac{1}{2}-m-\mu} \times$$

$$F(\mu+m+\tfrac{1}{2},\mu;2\mu; \frac{4x_1y_1}{\gamma}) \tag{4.9}$$

For $m > 1$, $<K^+_{(y,s)},\phi>$, where $\phi \in C_o^\infty(X)$, is given by the finite part of a divergent integral.

It will be seen that there is a remarkable correspondence between (4.7) and (4.8). It is also worth observing that, if $\mu = k + \frac{1}{2}$, k = 0, 1, ... then (4.9) shows that $K^+_{(y,s)} = 0$ in X_2 : the forward fundamental solution has a lacuna after the reflection of the forward null cone in ∂X. In particular, this holds good for the axisymmetric wave equation in $\mathbb{R}^{n+1}$, if n is odd.

REFERENCES

[1] Carroll, R.W., and Showalter, R.E., "Singular and degenerate Cauchy problems", Acad. Press, New York-San Francisco-London, 1976.

[2] Delache, S., and Leray, J., "Calcul de la solution élémentaire de l'opérateur d'Euler-Poisson-Darboux et de l'opérateur de Tricomi-Clairaut, hyperbolique , d'ordre 2", Bull. Soc. Math. France 99 (1971) 313-336.

[3] Friedlander, F.G., "The wave equation on a curved space-time", Cambridge University Press, 1975.

[4] Friedlander, F.G., and Heins, A.E., "On a singular boundary value problem for the Euler-Darboux equation", J. Diff. Equ. 4 (1968) 460-491.

[5] Friedlander, F.G., and Melrose, R.B., "The wave front set of the solution of a simple initial-boundary value problem with glancing rays II", Math. Proc. Camb. Phil. Soc. 81 (1977) 97-120.

[6] Hörmander, L., "Fourier integral operators I", Acta Math. 127 (1971) 79-183.

[7] Riesz, M., "L'intégrale de Riemann-Liouville et le problème de Cauchy", Acta Math. 81 (1949) 1-203.

PSEUDO-DIFFERENTIAL OPERATORS OF PRINCIPAL TYPE

Lars Hörmander

Department of Mathematics
University of Lund
Box 725, S 22007 Lund, Sweden

We survey the existence and regularity theory for operators of principal type, with emphasis on Moyer's proof of the necessity of condition (Ψ). A brief discussion of the open problems concerning the sufficiency of condition (Ψ) is also given.

1. INTRODUCTION.

The topic of this meeting is the propagation of singularities of solutions of boundary problems. At first sight the study of operators of principal type in a manifold without boundary might therefore appear to fall outside the intended area. However, the standard procedure for reducing boundary problems for elliptic operators to pseudo-differential operators in the boundary leads to arbitrary pseudo-differential operators if one starts from the Laplacean with one additional dimension. Indeed, Egorov [11] has claimed that he can prove very strong results on operators of principal type by studying a corresponding boundary problem for the Laplacean. However, his proofs have not appeared and some of his statements are manifestly false.

Nirenberg and Treves [23] conjectured that for a pseudo-differential operator with principal symbol p solvability (see section 2) implies that Im p cannot change sign from - to + along a bicharacteristic of Re p. They called this the condition (Ψ). (For a more precise, invariant and global formulation see section 3.) The necessity of condition (Ψ) was proved by Nirenberg and Treves [22] for a special case extending the non-degenerate case studied by Hörmander [14]. The same special case was discussed by Egorov [7], and Egorov and Popivanov [12] established the necessity of (Ψ)

H. G. Garnir (ed.), Singularities in Boundary Value Problems, 69–96.

under weaker but very complicated supplementary conditions. A general proof of the necessity of (Ψ) was given by Moyer [21] in the local two dimensional case. His construction can be used to prove the necessity in full generality, globally and in higher dimensions. We shall give a complete exposition of Moyer's method in sections 5 and 6 since it does not yet seem available in the literature.

For differential operators one has a symmetry which strengthens the condition (Ψ) to the condition (P) which rules out any sign change of Im p along bicharacteristics of Re p. For operators of principal type satisfying condition (P) theorems on propagation of singularities leading to semi-global existence were proved in Hörmander [16, 17]. In section 7 we review these results and a recent theorem on propagation of singularities by Dencker [4] which completes them.

No general proof of the sufficiency of condition (Ψ) for operators of principal type is known. (Egorov [11] has claimed that he has such a proof, but as already pointed out some of his key statements are false.) In section 8 we list the cases where the sufficiency of condition (Ψ) is well established and state some conjectures. We hope that this will clarify the remaining problems and stimulate work to complete the theory of operators of principal type.

2. DEFINITION OF SOLVABILITY.

Let P be a properly supported pseudo-differential operator in a C^∞ manifold X of dimension n, and let K be a compact subset of X. The following is a very weak notion of solvability:

Definition 2.1. P is said to be solvable at K if for every f in a subspace of $C^\infty(X)$ with finite codimension one can find $u \in \mathcal{D}'(X)$ so that

$$Pu = f \tag{2.1}$$

in a neighborhood of K.

In the definition we permitted the neighborhood of K to depend on f. However, by standard arguments using Baire's theorem we shall now show that a fixed neighborhood can be used and that the definition can be made more precise in other respects as well. As usual we denote by $H_{(s)}^{loc}$ the space of distributions in X which are mapped to L^2_{loc} by any properly supported pseudo-differential operator of order s. Let $\| \ \|_{(s)}$ be a norm defining the topology of

$$H^0_{(s)}(M) = H^{loc}_{(s)} \cap \mathcal{E}'(M)$$

for any compact set M in X. Choose a fundamental decreasing system of open neighborhoods of K,

$$K \subset \ldots \subset\subset Y_2 \subset\subset Y_1 \subset\subset X$$

and then choose $Z \subset\subset X$ so that $Pu = 0$ in Y_1 if $u = 0$ in Z. ($Z \subset\subset X$ means that $\bar{Z}$ is a compact subset of X.) Fix $\varphi \in C_0^\infty(X)$ with $\varphi = 1$ in Z. Then we have $Pu = P(\varphi u)$ in Y_1 if $u \in \mathcal{D}'(X)$, and $\operatorname{supp} \varphi u \subset M = \operatorname{supp} \varphi$.

Definition 2.1 means that we can choose $f_1, \ldots, f_r \in C^\infty(X)$ so that for any $f \in C^\infty(X)$ we have

$$Pu = f + \sum_1^r a_j f_j \text{ in } Y_N \tag{2.2}$$

for some positive integer N, $a_j \in \mathbb{C}$ and $u \in \mathcal{D}'(X)$. Since u can be replaced by φu we can always choose $u \in \mathcal{E}'(M)$, hence $u \in H^0_{(-N)}$ for some N. Thus the union of the sets $F_N = \{f \in C^\infty(X);$ (2.2) is valid with $u \in H^0_{(-N)}(M)$, $\|u\|_{(-N)} + \Sigma|a_j| \leqq N\}$ is equal to $C^\infty(X)$. The weak compactness and convexity of the set of all $(u, a_1, \ldots, a_r)$ in the definition of F_N shows that F_N is convex and closed. The symmetry is obvious, so from Baire's theorem it follows that F_N has 0 as an interior point when N is large. Thus we can find $\Psi \in C_0^\infty(X)$ and N' so that

$$f \in C^\infty(X), \ \|\Psi f\|_{(N')} \leqq 1 \implies f \in F_N.$$

Using the compactness of the set of all $(u, a_1, \ldots, a_r)$ again we conclude that (2.2) has a solution $u \in H^0_{(-N)}$ for every $f \in H^{loc}_{(N')}$. Thus we have proved

<u>Proposition 2.2.</u> If P is solvable at K then one can find an integer N and a neighborhood Y of K such that for every $f \in H^{loc}_{(N)}(X)$ there is a distribution $u \in H^{loc}_{(-N)}(X)$ such that

$$Pu - f \in C^\infty(Y),$$

that is, we have solvability in Y mod C^∞.

Proposition 2.2 has a strong converse:

<u>Proposition 2.3.</u> Assume that for every $f \in H^{loc}_{(N)}(X)$ one can find $u \in \mathcal{D}'(X)$ such that

$$Pu - f \in H^{loc}_{(N+1)}(Y)$$

for some neighborhood Y of K (depending on f). Then P is locally solvable at K.

Proof. Let F_ν now denote the set of all $f \in H^0_{(N)}(\bar{Y}_1) = H$ such that

$$Pu = f + g \text{ in } Y_\nu$$

for some $g \in H^0_{(N+1)}(\bar{Y}_1)$ and $u \in H^0_{(-\nu)}(M)$ with

$$\|u\|_{(-\nu)}^2 + \|g\|_{(N+1)}^2 \leqq \nu^2. \tag{2.3}$$

Baire's theorem gives as above that F_ν contains the unit ball for large ν. The minimum of the left hand side of (2.3) is attained precisely when (u, g) is orthogonal to all (u', g') with $Pu' = g'$ in Y_ν, so g is then a linear function Tf of f. The map $T : H \to H^0_{N+1}(\bar{Y}_1)$ has norm $\leqq \nu$. Thus T defines a compact operator in H, which implies that the range of I + T has finite codimension. The equation Pu = h in Y_N has a solution for every $h \in H$ in the range of I + T. This proves the solvability at K.

In view of Propositions 2.2 and 2.3 we can also define solvability at a set in the cotangent bundle. In doing so we shall use the term sing spec u as synonymous with WF(u) to denote either a closed set in the cosphere bundle $S^*(X)$ or a cone in $T^*(X) \setminus 0$.

Definition 2.4. If K is a compact subset of the cosphere bundle $S^*(X)$ then P is said to be locally solvable at K if there is an integer N such that for every $f \in H^{loc}_{(N)}(X)$ we have $K \cap$ sing spec $(Pu - f) = \emptyset$ for some $u \in \mathscr{D}'(X)$.

From Propositions 2.2 and 2.3 it follows that solvability at a compact set $M \subset X$ in the sense of Definition 2.1 is equivalent to solvability at $S^*(X)|_M$ in the sense of Definition 2.4. Thus the two definitions are compatible. Note that solvability at a set $K \subset S^*(X)$ implies solvability at subsets of K. The invariance of Definition 2.4 under conjugation by Fourier integral operators is fairly obvious and will be discussed in section 3. However, this is a convenient place to show how solvability at $K \subset S^*(X)$ is expressed by an estimate. (For the simpler original version of the argument see also Hörmander [13, Lemma 6.1.2].)

Proposition 2.5. Let K be a compact subset of $S^*(X)$ such that P is solvable at K, and choose $Y \subset\subset X$ so that $K \subset S^*(Y)$. If N is the integer in Definition 2.4 we can find an integer ν and a pseudo-differential operator A with sing spec $A \cap K = \emptyset$ such that

$$\|v\|_{(-N)} \leqq C(\|P^*v\|_{(\nu)} + \|v\|_{(-N-n)} + \|Av\|_{(0)}), \quad v \in C_0^\infty(Y). \tag{2.4}$$

Proof. Let $Y \subset\subset Z \subset\subset X$ and let $f \in H^0_{(N)}(\bar{Z})$, which is a Hilbert space. By hypothesis we can find $u \in \mathscr{D}'(X)$ and g so that $f = Pu + g$ and $K \cap$ sing spec $g = \emptyset$. Thus

$$(f, v) = (u, P^*v) + (g, v), \; v \in C_0^\infty(Y)$$

Now we can write $g = Ag_1 + g_2$ where $g_2 \in C^\infty$, $g_1 \in L^2$ and A is a pseudodifferential operator with $K \cap \text{sing spec } A = \emptyset$. For the proof we just have to put $g_1 = Bg$ with B elliptic of sufficiently high negative order and take for A a parametrix of B conveniently cut off outside a neighborhood of sing spec g. Now we obtain if f is fixed that for some ν and C

$$|(f, v)| \leq C(\|P^*v\|_{(\nu)} + \|A^*v\|_{(0)} + \|v\|_{(-N-n)}), \; v \in C_0^\infty(Y). \quad (2.5)$$

Let V be the space $C_0^\infty(Y)$ equipped with the topology defined by the semi-norms $\|v\|_{(-N-n)}$, $\|P^*v\|_{(\nu)}$, $\nu = 1, 2, \ldots$, and $\|Av\|_{(0)}$ where A is a pseudo-differential operator with $K \cap \text{sing spec } A = \emptyset$. It suffices to use a countable sequence $A_1, A_2, \ldots$ where A_ν is non-characteristic of order ν in a set which increases to $S^*(X) \smallsetminus K$ as $\nu \to \infty$. Thus V is a metrizable space. The bilinear form $\langle f, v\rangle$ in the product of a Hilbert space and a metrizable space is therefore separately continuous, hence continuous, which means that

$$|(f, v)| \leq C \; \|f\|_{(N)} (\|P^*v\|_{(\nu)} + \|A_\nu v\|_{(0)} + \|v\|_{(-N-n)}),$$

$$f \in H^0_{(N)}(\bar{Z}), \; v \in C_0^\infty(Y),$$

for some C and ν. Taking the supremum with respect to f after dividing by $\|f\|_{(N)}$ we obtain (2.4).

3. THE CONDITION (Ψ).

Again we let P be a properly supported pseudo-differential operator in a C^∞ manifold X. We assume that the symbol is classical of order m, that is, the asymptotic sum of a homogeneous principal symbol p of order m and homogeneous symbols of order m-1, m-2, ... If r is a real valued C^∞ function in $T^*(X) \smallsetminus 0$, then a bicharacteristic of r is an integral curve of the Hamilton vector field

$$H_r = \partial r/\partial\xi \; \partial/\partial x - \partial r/\partial x \; \partial/\partial\xi.$$

on which r = 0. Note that multiplication of r by a positive (negative) function preserves the curve and (reverses) its orientation.

Definition 3.1. p is said to satisfy condition (Ψ) in the open set $Y \subset X$ if there is no positively homogeneous complex valued function q in $C^\infty(T^*(Y) \smallsetminus 0)$ such that Im qp changes sign from − to + when one moves in the positive direction on a bicharacteristic of Re qp over Y on which $q \neq 0$.

We shall say that such a bicharacteristic of Re qp is a semi-bicharacteristic of p. The necessity of condition (Ψ) is a consequence of the following theorem.

Theorem 3.2. Suppose that there is a C^∞ positively homogeneous function q in $T^*(X) \setminus 0$ and a bicharacteristic interval $t \to \gamma(t)$, $a \leqq t \leqq b$, for Re qp such that $q(\gamma(t)) \neq 0$, $a \leqq t \leqq b$, and $\operatorname{Im} qp(\gamma(a)) < 0$, $\operatorname{Im} qp(\gamma(b)) > 0$. Then P is not solvable at the projection of $\{\gamma(t), a \leqq t \leqq b\}$ in $S^*(X)$.

Corollary 3.3. If P is solvable at the compact set $K \subset X$ then K has an open neighborhood Y in X where p satisfies condition (Ψ).

Proof. By Proposition 2.2 we can find a neighborhood Y of K such that P is solvable at any compact set $M \subset T^*(Y)$. Hence the statement follows from Theorem 3.2.

The proof of Theorem 3.2 will occupy the remaining part of this section as well as sections 4, 5 and 6. The first step is to reduce p to a special form.

Lemma 3.4. If $K \subset S^*(X)$ is compact and Q is a pseudo-differential operator which is non-characteristic at K, then QP is solvable at K if and only if P is solvable at K.

Proof. If QP is solvable at K we can for every $f \in H^{loc}_{(N)}$ find u with $K \cap \text{sing spec } (QPu - f) = \emptyset$. If Q is of order μ we can take $f = Qg$ for any $g \in H^{loc}_{(N+\mu)}$, and conclude that $K \cap \text{sing spec } (Pu - g) = \emptyset$. Hence P is solvable at K. Now there is a pseudo-differential operator R such that sing spec $(RQ - I)$ does not meet K. If P is solvable at K it follows that RQP is solvable at K, so QP is solvable at K by the first part of the proof. This completes the proof of the lemma.

If Q is a pseudo-differential operator with principal symbol q and $P_1 = QP$, the principal symbol of P_1 is $p_1 = qp$ and the hypothesis of Theorem 3.2 is that $\operatorname{Im} p_1$ changes sign from $-$ to $+$ along a bicharacteristic of Re p_1. We can also modify P_1 by multiplying with an operator of order $1 - \deg P_1$ and positive principal symbol. This does not change the bicharacteristics of Re p_1, just the parametrization. Thus it is sufficient to prove Theorem 3.2 in the case where $q = 1$ and p is of degree 1. The bicharacteristics of Re p can then be considered as curves on the cosphere bundle. If the curve where Im p changes sign is closed on $S^*(X)$ we can always pick an arc which is not closed where the change of sign still occurs and this we assume done in what follows.

The proof of Lemma 3.4 also shows that solvability of P at K is equivalent to solvability of F_1PF_2 at $\chi^{-1}(K)$ if F_1 and F_2 are non-characteristic Fourier integral operators corresponding to the canonical transformations χ^{-1} and χ respectively. This means that we can replace the principal symbol p of P by $p \circ \chi$. By [5, Lemma 6.6.3] we may therefore always assume that $X = \mathbb{R}^n$ and that $\operatorname{Re} p = \xi_1$. We may also assume that the coordinates are chosen so that the bicharacteristic in Theorem 3.2 is defined by

$$a \leqq x_1 \leqq b,\ x' = (x_2, \ldots, x_n) = 0,\ \xi_1 = 0,\ \xi' = (0,\ldots,0,1) = \xi^0.$$

Global difficulties might occur in our constructions if b-a is large so we shall examine how small the intervals can be where the crucial change of sign occurs. To do so we set

$$L(x', \xi') = \inf \{t - s;\ a < s < t < b;\ \operatorname{Im} p(s,x',0,\xi') < 0 < \\ < \operatorname{Im} p(t, x', 0, \xi')\}$$

when (x',ξ') is close to $(0,\xi^0)$, and we denote by L_0 the lower limit of $L(x', \xi')$ as $(x', \xi') \to (0, \xi^0)$. For small $\varepsilon > 0$ we can choose an open neighborhood V_ε of $(0, \xi^0)$ in $\mathbb{R}^{2n-2}$ with diameter $< \varepsilon$ such that $L(x', \xi') > L_0 - \varepsilon/2$ in V_ε. For some $(x'_\varepsilon, \xi'_\varepsilon) \in V_\varepsilon$ and s_ε, t_ε with $a < s_\varepsilon < t_\varepsilon < b$ we have

$$t_\varepsilon - s_\varepsilon < L_0 + \varepsilon/2,\ \operatorname{Im} p(s_\varepsilon, x'_\varepsilon, 0, \xi'_\varepsilon) < 0 < \operatorname{Im} p(t_\varepsilon, x'_\varepsilon, 0, \xi'_\varepsilon).$$

It follows that $\operatorname{Im} p(t, x', 0, \xi')$ and all derivatives with respect to x', ξ' must vanish at $(t, x'_\varepsilon, 0, \xi'_\varepsilon)$ if $s_\varepsilon + \varepsilon < t < t_\varepsilon - \varepsilon$, for otherwise we could choose $(x', \xi') \in V_\varepsilon$ so close to $(x'_\varepsilon, \xi'_\varepsilon)$ that

$$\operatorname{Im} p(t,x',0,\xi') \neq 0,\ \operatorname{Im} p(s_\varepsilon,x',0,\xi') < 0 < \operatorname{Im} p(t_\varepsilon,x',0,\xi').$$

The required change of sign must then occur in one of the intervals (s_ε, t) and (t, t_ε) which is impossible since they are shorter than $L_0 - \varepsilon/2$.

Choose a sequence $\varepsilon_\nu \to 0$ such that $\lim s_{\varepsilon_\nu} = a_0$ and $\lim t_{\varepsilon_\nu} = b_0$ exist. Then $b_0 - a_0 = L_0$ and $\operatorname{Im} p^{(\alpha)}_{(\beta)}(t, 0, 0, \xi^0)$ vanishes for all α and β with $\alpha_1 = 0$ if $a_0 < t < b_0$. (We use the standard notation $p^{(\alpha)}_{(\beta)} = \partial^{\alpha+\beta}p/\partial x^\beta \partial\xi^\alpha$.) If $a_0 < b_0$ it follows in particular that we have a one dimensional bicharacteristic in the sense of Definition 2.3 of [17]:

Definition 3.5. A one dimensional bicharacteristic of the complex valued function p is a C^1 map $\gamma: I \to T^*(X)$, where I is an interval on $\mathbb{R}$, such that

(i) $p \circ \gamma = 0$

(ii) $0 \neq \gamma'(t) = c(t)\, H_p(\gamma(t))$ if $t \in I$.

At a one dimensional bicharacteristic the choice of the function q in Definition 3.1 is not very important by the following

Lemma 3.6. Let $\gamma: I \to T^*(X) \setminus 0$ be the inclusion of a characteristic point for p or a compact one dimensional bicharacteristic interval such that γ defines an injective map from I to $S^*(X)$. Assume that for some $q \in C^\infty$ we have

$$q \neq 0 \text{ and } \operatorname{Re} H_{qp} \neq 0 \text{ on } \gamma(I) \tag{3.1}$$

there is a neighborhood U of $\gamma(I)$ where Im qp never changes sign from - to + along a bicharacteristic of Re qp. (3.2)

Then (3.2) is valid för every q satisfying (3.1).

When $\gamma(I)$ is a point the lemma coincides with the invariance of the condition (Ψ) as stated and proved by Nirenberg and Treves [23, appendix]. We shall review their proof in section 4 (Lemma 4.5) and show that it gives a complete proof of Lemma 3.6.

Using Proposition 2.5 in Hörmander [17] we can reduce p near a compact one dimensional bicharacteristic interval to the form $p = \xi_1 + if(x, \xi')$ where f is real; the reduction involves multiplication be a non-vanishing homogeneous function and composition with a homogeneous canonical transformation. Lemma 3.6 shows that this does not affect the hypothesis on existence of sign changes, so Theorem 3.2 follows if we prove

Theorem 3.2'. Suppose that in a neighborhood of $\Gamma = \{(x_1, 0, 0, \xi^0), a_0 \leq x_1 \leq b_0\}$ the principal symbol of P has the form

$$p(x, \xi) = \xi_1 + if(x, \xi')$$

where f is real valued and vanishes of infinite order on Γ if $b_0 > a_0$. Assume also that in any neighborhood of Γ one can find an interval in the x_1 direction where f changes sign from - to + for increasing x_1. Then P is not solvable at Γ.

In the proof of Theorem 3.2' we may also assume that the lower order terms $p_0, p_{-1}, \ldots$ in the symbol are independent of ξ_1 near Γ. In fact, Malgrange's preparation theorem implies that

$$p_0(x, \xi) = q(x, \xi)(\xi_1 + if(x, \xi')) + r(x, \xi')$$

where q is homogeneous of degree -1. The term of degree 0 in the symbol of $(I - q(x, D))P$ is therefore $r(x, \xi')$. Repetition of the argument allows us to make the lower order terms successively independent of ξ_1.

We shall finally prove that in Definition 3.1 one can drop the assumption that q is homogeneous. To do so we first observe that homogeneity was neither assumed in Lemma 3.6 nor in the definition of a one dimensional bicharacteristic above. Now let q be any function in $C^\infty(T^*(Y) \setminus 0)$ such that Im qp changes sign from - to + on a bicharacteristic γ of Re qp where $q \neq 0$. As above we can find a compact one dimensional bicharacteristic interval $\Gamma \subset \gamma$ or a point Γ in γ such that the sign change occurs on bicharacteristics of Re qp arbitrarily close to it. If Γ consists of a single point then Lemma 3.6 shows that Im $q(\Gamma)p$ also changes sign from - to + on bicharacteristics of Re $q(\Gamma)p$ arbitrarily close to Γ, so (Ψ) is not even fulfilled for all constants q. If Γ is not a point we note that H_p cannot have the radial direction at any point on Γ, for the whole orbit of $H_{\mathrm{Re}\, qp}$ starting at Γ would then just be a ray where p = 0 identically. Thus the projection Γ^* on $S^*(X)$ is a smooth curve. It cannot be closed for it would then contain the whole projection of γ. Nor is it possible for Γ^* to return to the same position with a change of orientation, for a one dimensional bicharacteristic is uniquely determined by its starting point and the choice of orientation there. If $\Gamma^*(t_1) = \Gamma^*(t_2)$, $t_1 < t_2$ and the orientations are opposed, then we can for any $t_1' > t_1$ close to t_1 find t_2' with $t_1' < t_2' < t_2$ and $\Gamma^*(t_1') = \Gamma^*(t_2')$. The supremum t of such t_1' must be equal to the infimum of the corresponding t_2' which contradicts the injectivity of Γ^* near t. From the injectivity of Γ^* now proved it follows that the function c(t) in Definition 3.5 is the restriction of a homogeneous C^∞ function q_1 in $T^*(X) \setminus 0$, so Γ is a bicharacteristic of Re q_1p. By Lemma 3.6 Im q_1p changes sign from - to + on bicharacteristics for Re q_1p arbitrarily close to Γ. Thus (Ψ) is not fulfilled. Summing up, we have proved

Theorem 3.7. Each of the following conditions is necessary and sufficient for p to satisfy condition (Ψ) in Y:
(a) There is no C^∞ complex valued function q in $T^*(Y) \setminus 0$ such that Im qp changes sign from - to + when one moves in the positive direction on a bicharacteristic of Re qp over Y where $q \neq 0$.
(b) If Γ is a characteristic point or a one dimensional bicharacteristic with injective regular projection in $S^*(Y)$, then there exists a C^∞ function q in a neighborhood Ω of Γ such that Re $H_{qp} \neq 0$ in Ω and Im qp does not change sign from - to + when one moves in the positive direction on a bicharacteristic of Re qp in Ω.

Proof. Condition (a) is apparently stronger than (Ψ) but we have proved that its negation implies negation of (Ψ). On the other hand, (b) is apparently weaker but we have proved that the negation of (Ψ) implies the negation of (b). (Note that $H_{qp} \neq 0$ on Γ implies $q \neq 0$ on Γ since $p = 0$ on Γ. This is the reason why the hypothesis $q \neq 0$ was not made in condition (b).)

The interest of condition (b) is of course that it removes the need to consider arbitrary functions q. For example, if $H_{\mathrm{Re}\, p} \neq 0$ when $p = 0$ it suffices to take $q = 1$ in condition (Ψ), and this is what Nirenberg and Treves [23] meant by the invariance of condition (Ψ). The condition (a) is perhaps less important but it has been used before in Hörmander [17].

4. FLOW INVARIANT SETS AND THE INVARIANCE OF (Ψ).

A surprising feature of condition (Ψ) is that it involves the bicharacteristics of Re qp although they depend very much on q except where H_p is proportional to a real vector. In spite of this it was shown by Nirenberg and Treves [23, appendix] that the choice of q is not very important in condition (Ψ). The main point in their proof is the application of results on flow invariant sets due to Bony [2] and Brézis [3]. These play a fundamental role in the method of Moyer (see sections 5 and 6) so we shall discuss them in detail here.

Let X be a C^2 manifold and $F \subset X$ a closed subset, v a Lipschitz continuous vector field in X. We want to describe the conditions on v required for integral curves starting in F to remain in F for all later times. If F has a smooth boundary the condition is of course that v does not point out at any boundary point. The condition stated below is an adaptation of this property.

First note that if $x_0 \in F$ and $f \in C^1$, $f(x_0) = 0$ and $f \leqq 0$ in F, then we must have $vf(x_0) \leqq 0$. In fact, otherwise $f > f(x_0)$ on the integral curve to the right of x_0 so it lies outside F.

Definition 4.1. We shall write N(F) for the set of all $(x,\xi) \in T^*(X) \setminus 0$ such that one can find $f \in C^1$ with $f(x) = 0$, $df(x) = \xi$ and $f \leqq 0$ in a neighborhood of x in F.

By adding a C^1 function vanishing of second order at x we can of course always make f negative in $F \setminus \{x\}$. It is natural to think of N(F) as a generalized exterior normal "bundle".

Theorem 4.2 (Bony [2]). Let v be a Lipschitz continuous vector field in X. Then the following conditions are equivalent:

a) Every integral curve of $dx/dt = v(x(t))$, $0 \leq t \leq T$, with $x(0) \in F$ is contained in F.
b) $\langle v(x), n\rangle \leq 0$ for all $(x, n) \in N(F)$.

We have already proved that a) ⟹ b). In proving the converse we may assume that $X = \mathbb{R}^n$, for the statement is local. We need the following elementary lemma.

Lemma 4.3. Let F be a closed set in $\mathbb{R}^n$ and set

$$f(x) = \min_{z \in F} |x - z|^2$$

where $|\ |$ is the Euclidean norm. Then we have

$$f(x+y) = f(x) + f'(x, y) + o(|y|),$$

$$f'(x, y) = \min \{\langle 2y, x-z\rangle;\ z \in F,\ |x-z|^2 = f(x)\}.$$

Proof. We may assume in the proof that $x = 0$. Set

$$q_\varepsilon(y) = \min \{-2\langle y, z\rangle;\ z \in F,\ |z| \leq (f(0))^{1/2} + \varepsilon\}.$$

q_ε is a homogeneous function of degree 1, and $q \uparrow q_0$ as $\varepsilon \downarrow 0$. The limit is therefore uniform on the unit sphere, so

$$q_0(y) \geq q_\varepsilon(y) \geq q_0(y) - c_\varepsilon|y|, \quad c_\varepsilon \to 0 \text{ as } \varepsilon \to 0.$$

Now $|y - z|^2 = |z|^2 - 2\langle y, z\rangle + |y|^2$ so we have

$$f(y) \leq f(0) + q_0(y) + |y|^2.$$

On the other hand, when $|y| \leq \varepsilon$ the minimum in the definition of $f(y)$ is assumed for some z with $|z| \leq f(0)^{1/2} + \varepsilon$, hence

$$f(y) \geq f(0) + q_\varepsilon(y) + |y|^2, \quad |y| \leq \varepsilon,$$

which proves the lemma.

Proof of Theorem 4.2. With the notation in a) and Lemma 4.3 we have if $t < T$

$$\lim_{s \to t+0} (f(x(s)) - f(x(t))/(s-t) = f'(x(t), v(x(t)).$$

Since the result to be proved is local we may assume that for all x, y

$$|v(x) - v(y)| \leq C|x-y|.$$

When $z \in F$ and $|x(t) - z|^2 = f(x(t))$ we have

$2\langle v(x(t)), x(t) - z\rangle = 2\langle v(z), x(t) - z\rangle - 2\langle v(z) - v(x(t)), x(t) - z\rangle.$

The last term is $\leq 2C\, f(x(t))$ in absolute value. Since $f(x(t)) - |x(t) - \tilde{z}|^2 \leq 0$ for all $\tilde{z} \in F$, we have $(z, x(t) - z) \in N(F)$ if $x(t) \neq z$, so the first term on the right is ≤ 0 by condition b). Hence the right hand derivative of $f(x(t))$ is $\leq 2C\, f(x(t))$ so that of $f(x(t))e^{-2Ct}$ is ≤ 0. Hence $f(x(t))e^{-2Ct}$ is decreasing in every interval where it is positive, and if $f(x(0)) = 0$ it follows then that $f(x(t)) = 0$ for $0 \leq t \leq T$.

Corollary 4.4 (Brézis [3]). Let $q \in C^1(X)$ where X is a C^2 manifold and let v be a Lipschitz continuous vector field in X such that for any integral curve $t \to x(t)$ of v we have

$$q(x(0)) < 0 \Rightarrow q(x(t)) \leq 0 \text{ for } t > 0. \tag{4.1}$$

Let w be another C^1 vector field such that

$$\langle w, \operatorname{grad} q\rangle \leq 0 \text{ when } q = 0 \tag{4.2}$$

$$w = v \text{ when } q = dq = 0. \tag{4.3}$$

Then (4.1) remains valid if $x(t)$ is replaced by an integral curve of w.

Note that (4.2) is empty when $q = dq = 0$ so it is natural that another condition must be imposed then.

Proof. Let F be the closure of the union of all forward orbits for v starting at a point with $q(x) < 0$. By (4.1) we have $q \leq 0$ in F, and F contains the closure of the set where $q < 0$. Orbits of v which start in F must remain in F. If now $(x, \xi) \in N(F)$ then x is in the boundary of F so $q(x) = 0$. If $dq(x) \neq 0$ then F is bounded by the surface $q = 0$ in a neighborhood of x, and ξ is a positive multiple of $dq(x)$, thus $\langle w(x), \xi\rangle \leq 0$ by (4.2). If $dq(x) = 0$ we have $\langle w(x), \xi\rangle = \langle v(x), \xi\rangle \leq 0$ by (4.3) since v satisfies condition b) in Theorem 4.2. Hence w satisfies condition b) in Theorem 4.2 and therefore condition a) also, which proves the corollary.

We shall now prove Lemma 3.6. The proof is simplified by giving it a more general form:

Lemma 4.5. Let I be a point or a compact interval on $\mathbb{R}$, and let $\gamma: I \to M$ be an embedding of I in a sympletic manifold M as a one dimensional bicharacteristic of $p_1 + ip_2$, if I is not reduced to a point, and any characteristic point otherwise. Let

$$f_j = \sum_1^2 a_{jk}p_k, \quad j = 1, 2,$$

where $\det(a_{jk}) > 0$ on $\gamma(I)$. Assume that $H_{p_1} \neq 0$ and that $H_{f_1} \neq 0$ on $\gamma(I)$. If $\gamma(I)$ has a neighborhood U such that p_2 does not change sign from - to + along any bicharacteristic for p_1 in U, then U can be chosen so that f_2 has no such sign change along the bicharacteristics of f_1 in U.

Proof. First note that if $p_1 = p_2 = 0$ at a point in U then

$$\{p_1, p_2\} = H_{p_1} p_2 \leqq 0,$$

where { , } denotes the Poisson bracket. Hence, at the samt point,

$$\{f_1, f_2\} = \{a_{11}p_1 + a_{12}p_2, a_{21}p_1 + a_{22}p_2\} = (a_{11}a_{22} - a_{12}a_{21})\{p_1, p_2\} \leqq 0.$$

The proof is now divided into two steps, the first of which is quite trivial.

(i) Assume first that $a_{12} = 0$. Since $a_{11}a_{22} > 0$ either a_{11} and a_{22} are both positive or both are negative. Thus the bicharacteristics of $f_1 = a_{11}p_1$ are equal to those of p_1 with preserved and reserved orientation respectively, and $f_2 = a_{22}p_2$ when $p_1 = 0$ so f_2 has the same and opposite sign as p_2 respectively. This proves the lemma in this case.

(ii) By a canonical change of variables we can make $M = \mathbb{R}^{2n}$, $p_1 = \xi_1$ and $\Gamma = \gamma(I)$ equal to an interval on the x_1 axis. Let T be a vector with

$$\langle T, dp_1 \rangle = 1 \text{ and } \langle T, df_1 \rangle \neq 0 \text{ on } \Gamma.$$

Since dp_1 and df_1 do not vanish on Γ, the existence of T is obvious if Γ consists of a single point. Otherwise dp_2 is proportional to dp_1 on Γ so df_1 is proportional to dp_1. Hence we just have to take T with ξ_1 coordinate one then.

Set

$$q_2(x, \xi) = p_2((x, \xi) - \xi_1 T),$$

which means that $p_2 = q_2$ when $\xi_1 = 0$ and that q_2 is constant in the direction T. Then there is a function φ such that

$$q_2 = \varphi\, p_1 + p_2$$

so it follows from step (i) that the hypotheses in the lemma are fulfilled for $p_1 + iq_2$. We have

$$f_1 = (a_{11} - a_{12}\varphi)p_1 + a_{12}\, q_2,$$

hence

$$0 \neq \langle T, df_1\rangle = (a_{11} - a_{12}\varphi) \text{ on } \Gamma.$$

In a neighborhood of Γ we can therefore divide f_1 by $a_{11} - a_{12}\varphi$ and set

$$q_1 = f_1/(a_{11} - a_{12}\varphi) = p_1 + a_{12}(a_{11}-a_{12}\varphi)^{-1}q_2$$

which implies

$$f_j = \sum_1^2 b_{jk}q_k, \; j = 1, 2$$

where $b_{11} = a_{11} - a_{12}\varphi$, $b_{12} = 0$ and $\det b = \det a > 0$. Thus it follows from step (i) that it suffices to prove that (q_1, q_2) satisfies the hypothesis made on (p_1, p_2) in the lemma. The difficulty here is that the surfaces $p_1 = 0$ and $q_1 = 0$ are not the same; we shall identify them by projecting in the direction T.

Let U be a neighborhood of Γ where q_2 does not change sign from − to + on the bicharacteristics of p_1. Since T is transversal to the surface $f_1 = q_1 = 0$ we can choose U so small that $Y = \{(x,\xi) \in U;\ q_1(x,\xi) = 0\}$ is mapped diffeomorphically by the projection π in the direction T on $X = \{(x, \xi) \in U;\ \xi_1 = 0\}$. When $q_1 = q_2 = 0$, thus $p_1 = p_2 = 0$, we have $H_{q_1}q_2 = H_{p_1}p_2 \leqq 0$, so (4.2) is fulfilled in Y by the restriction q of q_2 to Y and $w = H_{q_1}$. At a point in Y where $q = 0$ and dq vanishes on the tangent space of Y, we have $dq_2 = 0$ since $\langle T, dq_2\rangle = 0$. Hence $w = H_{q_1} = H_{p_1}$ there so $\pi_* w = H_{p_1}$. If we apply Corollary 4.4 to $q = \pi^* q_2$ and the vector fields $v = (\pi^{-1})_* H_{p_1}$ and w we conclude that q_2 cannot change sign from − to + along a bicharacteristic of q_1 in Y. This completes the proof.

5. MOYER'S SOLUTION OF THE EICONAL EQUATION.

Let P be a pseudo-differential operator in $\mathbb{R}^n$ satisfying the hypotheses in Theorem 3.2'. To prove that P is not solvable at Γ we must construct approximate solutions of the equation $P^*v = 0$ concentrated so near Γ that (2.4) cannot hold. The principal symbol of P^* is $\xi_1 - if(x, \xi')$. By hypothesis f changes sign from − to + for some (x', ξ') arbitrarily close to $(0, \xi^0)$, $\xi^0 = (0, \ldots, 0, 1)$ when x_1 increases from a value close to a_0 to one close to b_0. We shall take v of the form

$$v_\tau(x) = e^{i\tau w(x)} \sum_0^M \varphi_j(x)\tau^{-j} \tag{5.2}$$

where Im $w \geq 0$ with equality at some point and strict inequality outside a compact set, which makes v_τ very small and φ_j irrelevant there as $\tau \to \infty$. The first step is the construction of a phase function w satisfying the eiconal equation

$$\partial w/\partial x_1 - if(x, \partial w/\partial x') = 0 \tag{5.3}$$

approximately, which is the subject of this section. To simplify the notations we shall in what follows write t instead of x_1 and x instead of x', so that (5.3) takes the form

$$\partial w/\partial t - if(t, x, \partial w/\partial x) = 0. \tag{5.4}$$

The simple case where $f = 0$, $\partial f/\partial t > 0$ at $(0, \xi^0)$ was treated by Hörmander [13, 14]. We then prescribe the boundary condition

$$w(t,x) = x_n + \langle Qx, x\rangle/2, \quad t = 0,$$

where $Q = Q_1 + iQ_2$ is a symmetric matrix. For a formal solution we have

$$w(t, x) = x_n + \langle Qx,x\rangle/2 + i\langle f'_x + Qf'_\xi, x\rangle t + (if'_t - \langle f'_\xi, f'_x\rangle - \langle Qf'_\xi, f'_\xi\rangle) t^2/2 + O(|x|^3 + |t|^3),$$

with the derivatives evaluated at $(0, \xi^0)$. This gives Im $w \geq c(|x|^2 + |t|^2)$ near 0, when $f'_\xi = 0$ if Q_2 is a large multiple of the identity, and when $f'_\xi \neq 0$ if Q_2 is a small multiple of the identity and Q_1 is chosen with $Q_1 f'_\xi + f'_x = 0$. Theorem 3.2' is well known in this case (see e.g. Hörmander [16, Th. 3.3.7]). In what follows we shall exclude this elementary case and assume that there is a neighborhood of Γ such that $\partial f/\partial t \leq 0$ when $f = 0$.

Nirenberg and Treves [22] showed that the preceding Cauchy problem with $\langle Qx, x\rangle = i|x|^2$ gives a formal solution of (5.4) with the desired properties also when $f(t, 0, \xi^0)$ has a zero of finite order when $t = 0$ where the sign changes from − to +. Also in the method of Moyer the form of w as a function of the x variables remains the same. Thus Im w is chosen strictly convex in x for fixed t, so the minimum is attained on a smooth curve $x = y(t)$. Then we have

$$\partial \text{Im } w(t, x)/\partial x = 0 \text{ when } x = y(t)$$

so we are led to looking for a solution of (5.4) which has the form

$$w(t,x) = A(t) + \langle x - y(t), \eta(t)\rangle + \sum_{2 \leq |\alpha| \leq M} w_\alpha(t)(x - y(t))^\alpha/|\alpha|!. \tag{5.5}$$

Here M is a large integer and it is convenient to use temporarily the notation $\alpha = (\alpha_1, \ldots, \alpha_s)$ for a sequence of $s = |\alpha|$ indices between 1 and the dimension n; w_α will be symmetric in these indices. If we make sure that the matrix $(\operatorname{Im} w_{jk})$ is positive definite then Im w will have a strict minimum when $x = y(t)$ as a function of the x variables, for $\eta(t)$ will be real valued.

On the curve $x = y(t)$ the equation (5.4) reduces to

$$A'(t) = \langle y'(t), \eta(t)\rangle + if(t, y(t), \eta(t)). \tag{0}$$

This is the only equation where A occurs so it can be used to determine A after y and η have been chosen. In particular

$$d \operatorname{Im} A(t)/dt = f(t, y(t), \eta(t)). \tag{0)'$$

If $f(t, y(t), \eta(t))$ has a sign change from - to + then Im A(t) will start decreasing and end increasing, so the minimum is assumed at an interior point. We can normalize the minimum value to zero and have then for a suitable interval of t that Im A > 0 at the end points and Im A = 0 at some interior point. Thus Im $w \geq 0$ with equality attained but strict inequality valid outside a compact subinterval of the curve.

Our purpose is to make (5.4) valid apart from an error of order M + 1 in x-y(t). Actually $f(t,x,\xi)$ is not defined for complex ξ, but since

$$\partial w(t,x)/\partial x_j - \eta_j(t) = \Sigma\, w_{\alpha,j}(t)(x-y(t))^\alpha/|\alpha|!$$

this is given a meaning if $f(t, x, \partial w/\partial x)$ is replaced by the finite Taylor expansion

$$\sum_{|\beta|\leq M} f^{(\beta)}(t, x, \eta(t))(\partial w(t,x)/\partial x - \eta(t))^\beta/|\beta|!.$$

(We use the notation $f^{(\beta)}_{(\alpha)}(t,x,\xi) = \partial_x^\alpha \partial_\xi^\beta f(t, x, \xi)$.) Note that to compute the coefficient of $(x-y(t))^\alpha$ we need only consider the terms with $|\beta| \leq |\alpha|$. We have

$$\partial w/\partial t = A' - \langle y', \eta\rangle + \langle x-y, \eta'\rangle + \Sigma\, w'_\alpha(x-y)^\alpha/|\alpha|! -$$
$$- \sum_k \sum_{1\leq|\alpha|\leq M-1} (x-y)^\alpha w_{\alpha,k} dy_k/dt/|\alpha|!.$$

Hence the first order terms in the equation (5.4) give

$$d\eta_j/dt - \sum_k w_{jk}(t)dy_k/dt = i(f_{(j)}(t,y,\eta) + \sum_k f^{(k)}(t,y,\eta)w_{jk}(t)) \tag{1}$$

Since y and η are real, this is a system of 2n equations

$$d\eta_j/dt - \sum_k \operatorname{Re} w_{jk}(t)dy_k/dt = - \sum_k \operatorname{Im} w_{jk}(t)\, f^{(k)}(t, y, \eta) \tag{1)'$$

$$\sum_k \operatorname{Im} w_{jk}(t)\, dy_k/dt = -f_{(j)}(t, y, \eta) - \sum_k \operatorname{Re} w_{jk}(t) f^{(k)}(t, y, \eta) \quad (1)''$$

When Im w_{jk} is positive definite we can solve these equations for dy/dt and $d\eta/dt$; at a point where $f = df = 0$ they just mean that $dy/dt = 0$, $d\eta/dt = 0$.

When $2 \leqq |\alpha| \leqq M$ we obtain from (5.4) a differential equation

$$dw_\alpha/dt - \sum_k w_{\alpha,k} dy_k/dt = F_\alpha(t, y, \eta, \{w_\beta\}) \qquad (\alpha)$$

where F_α is a linear combination of the derivatives of f of order $\leq |\alpha|$ multiplied by polynomials in w_β with $|\beta| \leqq |\alpha| + 1$. (When $|\alpha| = M$ the sum on the left hand side of (α) should of course be dropped.) Altogether (1)', (1)" and (α) is a quasilinear system of differential equations with as many equations as unknowns, so it is clear that we have local solutions with prescribed data. Now $F_\alpha(t, 0, \xi^0) = 0$ if $a_0 < t < b_0$, so when $a_0 < t < b_0$ we have the solution $y = 0$, $\eta = \xi^0$, w_α = constant. Hence we can find $\varepsilon > 0$ such that the equations (1) and (α) with initial data

$$w_{jk} = i\,\delta_{jk},\ w_\alpha = 0 \text{ when } 2 < |\alpha| \leq M,\ t = (a_0+b_0)/2 \qquad (5.6)$$

$$y = x,\ \eta = \xi \text{ when } t = (a_0+b_0)/2 \qquad (5.7)$$

have a unique solution in $(a_0-\varepsilon, b_0+\varepsilon)$ for all x, ξ such that $|x| + |\xi-\xi^0| < \varepsilon$. Moreover,

(i) $(\operatorname{Im} w_{jk} - \delta_{jk}/2)$ is positive definite

(ii) the map

$$(x, \xi, t) \to (y, \eta, t);\ |x| + |\xi-\xi^0| < \varepsilon,\ a_0-\varepsilon < t < b_0 + \varepsilon$$

is a diffeomorphism.

In the range X_ε of the map (ii) we denote by v the image of the vector field $\partial/\partial t$ under the map. Thus v is the tangent vector field of the integral curves. Note that $v = \partial/\partial t$ when $df = 0$. Since we have assumed above that $f = 0$ implies $\partial f/\partial t \leqq 0$ in X_ε, if ε is small enough, we can now apply Corollary 4.4 with $q = f$, the vector field v just defined and $w = \partial/\partial t$. This is the crucial point in Moyer's method. The conclusion is that f must have a change of sign from - to + along an integral curve of v in X_ε. In fact, otherwise $f(t, x, \xi)$ would not have a change of sign from - to + in X_ε for increasing t and fixed (x, ξ), and that contradicts the hypothesis. Recalling the discussion of the equation (0) above we have therefore proved

Proposition 5.1. Assume that the hypotheses of Theorem 3.2' are fulfilled and that in a neighborhood of Γ we have $\partial f/\partial t \leq 0$ when $f = 0$. Then one can find

(i) a curve $t \to (t, y(t), 0, \eta(t)) \in \mathbb{R}^{2n}$, $a' \leq t \leq b'$, as close to Γ as desired

(ii) C^∞ functions $w_\alpha(t)$, $2 \leq |\alpha| \leq M$, with ($\mathrm{Im}\, w_{jk} - \delta_{jk}/2$) positive definite when $a' \leq t \leq b'$

(iii) a function A with $\mathrm{Im}\, A(t) \geq 0$, $a' \leq t \leq b'$, $\mathrm{Im}\, A(a') > 0$, $\mathrm{Im}\, A(b') > 0$ and $\mathrm{Im}\, A(c') = 0$ for some $c' \in (a', b')$

such that (5.5) is a formal solution of (5.4) with error $O(|x-y(t)|^{M+1})$.

6. THE TRANSPORT EQUATIONS.

To complete the proof of Theorem 3.2' we must discuss the transport equations for the functions φ_j in (5.2) which together with the eiconal equation (5.4) make P^*v_τ very small if v_τ is defined by (5.2). However, we shall first make som preliminary general observations on functions of the form (5.2). We assume that $w \in C^\infty(X)$, $\varphi_j \in C_0^\infty(X)$, where X is an open set in $\mathbb{R}^n$, and also that $\mathrm{Im}\, w \geq 0$, $d\,\mathrm{Re}\, w \neq 0$ in X.

Lemma 6.1. For any positive integer N we have for $\tau > 1$

$$\|v_\tau\|_{(-N)} \leq C\, \tau^{-N}\,. \tag{6.1}$$

If $\mathrm{Im}\, w(x_0) = 0$ and $\varphi_0(x_0) \neq 0$ for some $x_0 \in X$ then

$$\|v_\tau\|_{(-N)} \geq c\, \tau^{-n/2\, -\, N}. \tag{6.2}$$

For every neighborhood U of $\{w'(x);\ x \in X,\ \mathrm{Im}\, w(x) = 0\}$ and every positive integer ν we have

$$|\hat{v}_\tau(\xi)| \leq C\, (1+|\xi|+\tau)^{-\nu} \text{ if } \tau > 1,\ \xi/\tau \notin U. \tag{6.3}$$

Note that $\mathrm{Im}\, w(x) = 0$ implies $\mathrm{Im}\, w'(x) = 0$ since $\mathrm{Im}\, w \geq 0$.

Proof. Since

$$\hat{v}_\tau(\xi) = \Sigma\int e^{i(\tau w(x)\, -\, \langle x,\, \xi\rangle)}\, \varphi_j(x)\, dx/\tau^j$$

and $(\tau w - \langle x, \xi\rangle)/(\tau+|\xi|)$ is in a compact set of functions with non-negative imaginary part and differential $\neq 0$, the estimate (6.3) is immediately proved by partial integrations. Combining (6.3) with the estimate

$$\int_{\tau U} |\hat{v}_\tau(\xi)|^2(1+|\xi|^2)^{-N}\, d\xi = O(\tau^{-2N})$$

which follows from the boundedness of v_τ in L^2 if $0 \notin \bar{U}$, we obtain the estimate (6.1). To prove (6.2) we assume that $x_0 = 0$ and observe that when $\psi \in C_0^\infty$ we have if $w(0) = 0$

$$\tau^n \langle v_\tau, \psi(\tau\cdot)\rangle = \int e^{i\tau w(x/\tau)}\psi(x)\Sigma\varphi_j(x/\tau)\tau^{-j}\, dx$$

$$\to \int e^{i\langle x,w'(0)\rangle}\psi(x)\varphi_0(0)\, dx$$

which is not equal to 0 for a suitable choice of ψ. Since $\|\psi(\tau\cdot)\|_{(N)} = O(\tau^{N-n/2})$ it follows that $c \leqq \tau^{N+n/2}\|v_\tau\|_{(-N)}$ which proves (6.2).

If Γ is the cone generated by

$$\{(x, w'(x)),\ x \in \cup \operatorname{supp} \varphi_j,\ \operatorname{Im} w(x) = 0\}$$

it follows from (6.3) that $\tau^k v_\tau \to 0$ in $\mathcal{D}'_\Gamma$ for any k as $\tau \to \infty$. Hence $\tau^k A v_\tau \to 0$ in $C^\infty(\mathbb{R}^n)$ if A is a pseudo-differential operator with sing spec $A \cap \Gamma = \emptyset$. (Cf. Hörmander [15, section 2.5].)

To prove Theorem 3.2' we assume that P is solvable at Γ and conclude that there is an estimate (2.4) with sing spec A disjoint from Γ. We can then use Proposition 5.1 to construct a function w of the form (5.5) with Im w > 0 except on a compact subinterval Γ_0 of a curve which is so close to Γ that it does not meet sing spec A. If we apply (2.4) to v_τ we then have a lower bound $c\tau^{-n/2-N}$ for the left hand side if $\varphi_0 \neq 0$ in Γ_0, a much smaller bound for $\|Av_\tau\|_{(0)}$ and also for $\|v_\tau\|_{(-N-n)}$. What remains is then to choose the functions φ_j so that $\varphi_0 \neq 0$ on Γ_0 and

$$\|P^* v_\tau\|_{(\nu)} = O(\tau^{-n/2-N-1}) \text{ as } \tau \to \infty.$$

It is convenient now to switch back to the notations used in section 5 where the first coordinate is called t and the others are called x. If B is a pseudo-differential operator with symbol 1 in a neighborhood of the curve $\Gamma_0 = \{(t, y(t);\ 0, \eta(t))\}$ then $(I - B)P^*\tau^k v_\tau \to 0$ in C^∞ for any k. Hence

$$\|P^* v_\tau - B\, P^* v_\tau\|_{(\nu)} = O(\tau^{-k})$$

for any k. We can choose B with so small support that the symbol of $P^* - D_t$ is of the form $F(t, x, \xi)$ where the leading term of F is $-if(t, x, \xi)$. According to Sjöstrand [24, appendix] the product

$$B(P^* - D_t - F(t, x, D))$$

is a pseudo-differential operator in the variables (t, x) with singular spectrum disjoint with Γ_0. Hence it suffices to make sure that

$$\|(D_t+F(t, x, D))v_\tau\|_{(\nu)} = O(\tau^{-N-(n+1)/2}). \tag{6.4}$$

Differentiation gives

$$D_t v_\tau = e^{i\tau w} \Sigma(\tau\varphi_j \, \partial w/\partial t + D_t\varphi_j)\tau^{-j}$$

and it was proved in Hörmander [14, Lemma 1.3.1] that

$$F(t, x, D)v_\tau \sim e^{i\tau<x, \eta(t)>} \sum_{-\infty}^{1} \sum_{\alpha} F_k^{(\alpha)}(x,\tau\eta(t)) \times$$
$$\times D^\alpha(\varphi_j\tau^{-j}e^{i\tau(w-<x,\eta(t>)}/\alpha!, \quad \tau \to \infty.$$

The general term has the bound $\tau^{k-|\alpha|/2-j}$. Carrying out the differentiation we must - as in the case of differential operators - obtain the same terms as in

$$e^{i\tau w} \Sigma \, F_k^{(\alpha)}(x, \tau w') \, D^\alpha(\varphi_j\tau^{-j})/\alpha! \tag{6.5}$$

if $F_k^{(\alpha)}(x, \tau w')$ is defined by Taylor expansion around $\tau\eta(t)$. This links up with our discussion of formal solutions of the eiconal equation in section 5 and makes the computations much more transparent. Thus we shall use notations such as those in (6.5) with the understanding that one takes a sufficiently high partial sum of the Taylor expansion of $F_k^{(\alpha)}(\tau(\eta + (w' - \eta)))$ to give an error for the product by $e^{i\tau w}$ decreasing as a high power of $1/\tau$. (See also Melin and Sjöstrand [19, pp. 156-157] where a precise interpretation in terms of almost analytic continuation is given for such formulas.)

The highest order term in P^*v_τ with respect to τ is now

$$e^{i\tau w} \, \tau(\partial w/\partial t + F_1(x, w'))\varphi_0.$$

We have chosen w in Proposition 5.1 to make the parenthesis $O(|x-y(t)|^{M+1})$, and since $e^{i\tau w} = O(e^{-c\tau|x-y(t)|^2})$ the whole term is $O(\tau^{-(M+1)/2})$. Next we consider the term of order 0. For the reasons just explained the terms involving φ_1 are $O(\tau^{-(M+1)/2})$. The terms involving φ_0 are

$$e^{i\tau w}(D_t\varphi_0 + \Sigma \, F_1^{(k)}(t, x, \partial w/\partial x)D_k\varphi_0 + c(t, x)\varphi_0)$$

where c is a function which we do not have to compute explicitly. We can make the parenthesis $O(|x-y(t)|^M)$ by taking

$$\varphi_0(x, t) = \sum_{|\alpha|<M} \varphi_{0,\alpha}(t) \, (x-y(t))^\alpha$$

and solving a system of linear differential equations for $\varphi_{0,\alpha}$. We solve them with initial conditions making $\varphi_{0,0} \neq 0$ on Γ_0.

Any inhomogeneous equation can be solved in the same way. The transport equations which occur successively for the determination of $\varphi_1, \varphi_2, \ldots$ can therefore be solved in the same manner until we have

$$P^* v_\tau = O(\tau^{(1-M)/2}).$$

Without affecting this estimate the functions φ_j can be multiplied by a cutoff function in $C_0^\infty(X)$ which is 1 in a neighborhood of Γ_0.

A moment's reflection shows that differentiation of $P^* v_\nu$ will only lead to a loss of a factor τ in the estimates. Hence we have

$$\| P^* v_\tau \|_{(\nu)} = O(\tau^{\nu + (1-M)/2})$$

which implies (6.4) if $\nu + (1-M)/2 < -n/2 - N$, that is,

$$M > 2\nu + 1 + n + 2N.$$

Note that this choice of M depends only on the constants in the estimate (2.4) so we could have made it at the beginning of the construction although the choice would not have been easy to motivate then. This contradiction completes the proof of Theorem 3.2'. (We have deliberately made the arguments in this section somewhat sketchy since they follow a well known standard pattern unlike the constructions in section 5.)

7. PROPAGATION OF SINGULARITIES WHEN CONDITION (P) HOLDS.

For a differential operator P in X, of order m, the principal symbol $p(x, \xi)$ has the symmetry property

$$p(x, -\xi) = (-1)^m p(x, \xi).$$

If $t \to (x(t), \xi(t))$ is a bicharacteristic curve of $\operatorname{Re} p(x, \xi)$ it follows that $t \to (x(t), -\xi(t))$ is also a bicharacteristic curve with the correct (reversed) orientation if m is odd (even). If the condition (Ψ) is fulfilled in X by p it follows that $\operatorname{Im} p(x, \xi)$ cannot take both positive and negative values on the bicharacteristic curve. If we apply Theorem 3.7 to $p_1 \pm i\, p_2$ it follows that p must satisfy the condition (P) defined as follows:

Definition 7.1. p is said to satisfy condition (P) if there is no C^∞ complex valued function q in $T^*(X) \setminus 0$ such that Im qp takes both positive and negative values on a bicharacteristic of Re qp where $q \neq 0$.

If p satisfies condition (P) then P is solvable at any point x such that H_p does not have the direction $\langle\xi,\partial/\partial\xi\rangle$ for any ξ with $p(x, \xi) = 0$. This was proved by Nirenberg and Treves [23] under the hypothesis that p is real analytic which was later removed by Beals and Fefferman [1] by a localization argument. Moreover, they proved that the equation $Pu = f \in H^{loc}_{(s)}$ has a local solution $u \in H^{loc}_{(s+m-1)}$ where m is the order of P, but the size of the neighborhood where the solution was obtained depended on s. This flaw was removed and solvability at large compact sets was proved in Hörmander [17] by means of a study of propagation of singularities for P^*. (It is clear that p satisfies (P) if and only if $\bar{p}$ satisfies (P), but since this is not true for condition (Ψ) we shall emphasize the distinction between P and P^* in this section also.) We shall now describe these results briefly, assuming throughtout that p satisfies condition (P).

In the characteristic set $C = p^{-1}(0) \subset T^*(X) \smallsetminus 0$ the subset C_2 where d Re p and d Im p are linearly independent is an involutive manifold of codimension 2. The flowout C_2^e of C_2 along the Hamilton fields $H_{\mathrm{Re}\, p}$ and $H_{\mathrm{Im}\, p}$ turns out to be a subset of C and is also an involutive, locally closed manifold of codimension 2. (Cf. Hörmander [17, Proposition 2.1].) Hence there is a natural foliation of C_2^e in two dimensional leaves. The quotient space $\tilde{B}$ of a leaf B obtained by identifying points on an embedded one dimensional bicharacteristic carries a unique conformal structure for which the analytic functions are the solutions of the equation $H_p\varphi = 0$. (Cf. [17, section 4].) If $u \in \mathcal{D}'(X)$ we define $\tilde{s}_u$ on $\tilde{B}$ as the supremum of all s such that $u \in H_{(s)}$ at the corresponding points in the leaf. Then Theorem 6.1 in [17] states that $\tilde{s}_u$ is a superharmonic function on $\tilde{B}$ if $P^*u \in C^\infty$ on the corresponding leaf. Hence $\tilde{s}_u = \infty$ in $\tilde{B}$ if this is true in an open subset ω. Similar results are valid if $Pu \in H^{loc}_{(t)}$ for some finite t.

If γ is a bicharacteristic interval for Re p such that Im $p < 0$ at the starting point, it follows from condition (P) that Im $p \leq 0$ in a neighborhood of γ when Re $p = 0$. This allows one to apply a theorem on propagation of singularities proved by a variant of the energy integral method (Hörmander [16, Prop. 3.5.1]). It shows that if $P^*u \in C^\infty$ at γ then $u \in C^\infty$ at γ.

Finally we shall discuss the propagation of singularities along one dimensional bicharacteristics γ. If there is some point on γ and some constant a for which dp/a is real and the Hessian of Im p/a is not semi-definite in the plane $dp = 0$, then Theorem 6.6 in [17] states that when $P^*u \in C^\infty$ then $s_u^*(x,\xi) = \sup\{s;\ u \in H_{(s)}$ at $(x, \xi)\}$ is a concave function on γ with respect to a natural affine structure. If the Hessian is different from 0 but semi-definite, it was proved that s_u^* is quasi-concave in γ, that is, for every compact interval $I \subset \gamma$

$$\min_I s_u^* = \min_{\partial I} s_u^* .$$

The same result has been proved recently by Dencker [4] for the remaining one dimensional bicharacteristics.

The preceding results on singularities, extended to the case where $P^*u \in H_{(t)}^{loc}$ for some finite t, can be used to show that condition (P) implies solvability at every compact set $K \subset X$ such that no bicharacteristic (of dimension one or two) is completely trapped in K. For the proof we refer to Hörmander [17, section 7]. (Instead of the theorem of Dencker a much weaker result ([17, Theorem 5.3]) was used there.)

8. ON THE SUFFICIENCY OF CONDITION (Ψ).

As we have seen in section 7 operators satisfying condition (P) are now quite well understood. However, as yet there are only fairly fragmentary results on operators satisfying condition (Ψ). We shall sum them up here and also add a few conjectures.

If $p = p_1 + ip_2$ satisfies condition (Ψ) then

$$\{p_1, p_2\} \leqq 0 \text{ when } p_1 = p_2 = 0. \tag{8.1}$$

In general this condition is of course much weaker than condition (Ψ), but if dp_1 and dp_2 are linearly independent we can extend $\{p_1, p_2\}$ from the manifold $p_1 = p_2 = 0$ to a non-positive function and conclude that

$$\{p_1, p_2\} \leqq a_1p_1 + a_2p_2 , \tag{8.2}$$

where $a_j \in C^\infty$. On the other hand, (8.2) always implies condition (Ψ). In fact, in the surface $p_1 = 0$ it follows from (8.2) that $H_{p_1} p_2 \leqq a_2p_2$, thus

$$H_{p_1} e^{-\psi} p_2 \leqq 0 \quad \text{if } H_{p_1}\psi = a_2.$$

If $p_2 < 0$ at a point on a bicharacteristic for p_1 it follows that $p_2 < 0$ at all later points on the bicharacteristic. It is easy to prove (Hörmander [14, Th. 1.3.4]) that (8.2) implies local solvability if H_p does not have the radial direction at any characteristic point.

Egorov [8] has also shown that there is local solvability if dp_1 does not have the radial direction and there is a smooth surface S transversal to H_{p_1} such that $p_2 \geqq 0$ (resp. $p_2 \leqq 0$) on the bicharacteristics of p_1 before (after) they intersect S. The principal symbol of p can then be reduced to the form

$$\xi_1 + if(x, \xi')$$

where $f(x, \xi') \geqq 0$ (resp. $f(x, \xi') \leqq 0$) when $x_1 \leqq 0$ (resp. $x_1 \geqq 0$). Egorov's proof is closely related to the energy integral method so it is rather similar to the proof of the propagation theorem of Hörmander [16, Prop. 3.5.1] referred to in section 7. A stronger result was proved by Menikoff [20] who only assumed that $f'_{x',\xi'} = 0$ when $f = 0$, thus $H_p = \partial/\partial x_1 - i\, \partial f(x, \xi')/\partial x_1\, \partial/\partial \xi_1$ at the characteristic points. However, since (Ψ) implies that (Ψ) is fulfilled by $p_1 + iqp_2$ for any $q \geqq 0$, it is clear that all the results mentioned only prove a very special case of the conjectured sufficiency of condition (Ψ).

Egorov [9] has made a detailed study of the repeated Poisson brackets of p_1 and p_2 and of the commutators of the Hamilton fields H_{p_1} and H_{p_2} when $p_1 + ip_2$ satisfies condition (Ψ). His analysis was simplified and somewhat extended in Hörmander [18, section 2]. Briefly the results are as follows. Let $\gamma \in T^*(X) \setminus 0$ be a zero of a function $p_1 + ip_2$ satisfying (Ψ). Denote by $k = k(\gamma)$ the largest integer such that all Poisson brackets of p_1 and p_2 with at most k factors vanish at γ, and denote by $s = s(\gamma)$ the largest integer $\geqq 0$ such that all commutators of H_{p_1} and H_{p_2} with at most s factors are linearly dependent at γ. (If no such integers exist we set $k(\gamma) = +\infty$ resp. $s(\gamma) = +\infty$.). Then

(i) if k is finite and $0 < k \leqq 2s$ we have for all C^∞ matrices (a_{jk})

$$(H_{a_{11}p_1+a_{12}p_2})^k(a_{21}p_1+a_{22}p_2)(\gamma) = c \det a(a_{11}c_1+a_{12}c_2)^{k-1}(\gamma)$$

where $c \in \mathbb{R}\setminus 0$, $(c_1, c_2) \in \mathbb{R}^2 \setminus 0$ and $c_2H_{p_1} - c_1H_{p_2} = 0$ at γ.

(ii) if $2s < k$ the commutators of H_{p_1} and H_{p_2} with at most $(1+k)/2$ factors span a two dimensional space of tangent vectors at γ, either $k = \infty$ or else k is odd and we have for all C^∞ matrices (a_{jk})

$$(H_{a_{11}p_1+a_{12}p_2})^k(a_{21}p_1+a_{22}p_2)(\gamma) = \det a\ Q(a_{11}, a_{12})(\gamma)$$

where Q is a homogeneous polynomial which is negative except on at most $(k+1)/2$ lines through 0.

These results are of course closely related to the invariance of condition (Ψ). They show that to determine if $k(\gamma) = k$ it suffices to check $(H_{\mathrm{Re}\,zp})^j \operatorname{Im} zp$ for a finite number of values of z and for $j \leqq k$. If k is finite then P^* is subelliptic at γ in the sense that $P^*u \in H^{loc}_{(s)}$ at γ implies $u \in H^{loc}_{(s+m-k/(k+1))}$ at γ (see Egorov [10] and also Hörmander [18] where some gaps in Egorov's proofs are filled). When proving solvability for P one has to prove estimates for the inverse of P^* acting on functions of compact support, and this is no problem at a subelliptic point. All remaining difficulties therefore occur where $k(\gamma) = \infty$.

When $k(\gamma) = \infty$ it follows from (ii) above that the commutators of the Hamilton fields H_{p_1} and H_{p_2} span a one or two dimensional plane at γ. If p is real analytic one can then conclude that there is a one or two dimensional bicharacteristic through γ, that is, an (isotropic) analytic manifold Γ with $\gamma \in \Gamma$ such that

a) the tangent plane is everywhere spanned by H_{p_1}, H_{p_2} and their commutators.

b) all Poisson brackets of p_1 and p_2 are equal to 0 in Γ.

The Hamilton field H_p defines a natural complex structure in Γ if Γ is two dimensional. (Cf. Hörmander [18, pp. 140-141].) Thus the situation is very close to that for operators satisfying condition (P) which we discussed in section 7. This makes it natural to conjecture that if $P^*u \in C^\infty$ then s^*_u is superharmonic (quasi-concave) in Γ if the dimension of Γ is two (one). For operators with real analytical principal part a proof of these conjectures (in an extended form parallel to [17, Theorems 6.1 and 6.7]) would again imply the sufficiency of condition (Ψ) when no bicharacteristics are trapped. The methods of [17] are applicable to prove the conjectures as soon as local solvability is established near any one dimensional bicharacteristic. Thus they are true for example in any open subset of $T^*(X) \setminus 0$ where (8.2) is valid. The proofs are very simple since the analyticity assumption allows one to find exact solutions of the equation $H_p w = 0$ (cf. [17, p. 600]).

In the C^∞ case the geometrical picture is more complicated, but one or two dimensional bicharacteristics can still be defined (cf. Definition 3.5):

Proposition 8.1. Assume that $p = p_1 + ip_2$ satisfies condition (Ψ). Let $\gamma_0 \in T^*(X) \setminus 0$ be a characteristic point and assume that all semi-bicharacteristics through γ_0 remain in the characteristic set. Then their union Γ is a one or two dimensional manifold, H_{p_1} and H_{p_2} are tangents to Γ and all Poisson brackets of p_1 and p_2 vanish in Γ. The first order differential operator H_p is locally solvable in Γ.

Recall that the semi-bicharacteristics of p are the bicharacteristics of Re qp such that $q \neq 0$. Any continuous curve composed of finitely many bicharacteristic arcs of p_1 and p_2 is a limit of semi-bicharacteristics.

Proof. a) Assume that H_{p_1} and H_{p_2} are linearly dependent at every point in Γ. Then the semi-bicharacteristics are one-dimensional bicharacteristics in the sense of Definition 3.5, so Γ is the unique maximal one dimensional bicharacteristic passing through γ_0. All commutators of H_{p_1} and H_{p_2} are tangents to Γ, so every Poisson bracket $H_{p_{i_1}} \dots H_{p_{i_{j-1}}} p_{i_j}$ must vanish in Γ.

b) Assume now that Γ contains a point γ_1 where H_{p_1} and H_{p_2} are linearly independent. We can then choose canonical coordinates in a neighborhood of γ_1 such the product of p by some non-vanishing function is of the form

$$\xi_1 + if(x, \xi')$$

and γ_1 is the point (0, 0). Thus $f(0, 0) = 0$, $df(0, 0) \neq 0$, and $f(x_1,0,0) = 0$ since the x_1 axis is a bicharacteristic of $p_1 = \xi_1$. The condition (Ψ) implies that if v is a vector with

$$\langle v, df(0)\rangle < 0 \quad \text{resp.} \quad \langle v, df(0)\rangle > 0$$

then

$$\langle v,df(x_1,0,0)\rangle \leqq 0 \text{ if } x_1 > 0 \text{ resp. } \langle v, df(x_1,0,0)\rangle \geqq 0 \text{ when } x_1 < 0.$$

Hence $df(x_1, 0, 0)$ is proportional to $df(0)$, that is, $H_{p_2}(x_1,0,0)$ is proportional to $H_{p_2}(0)$.

Now the bicharacteristic of p_2

$$dx/dt = \partial f/\partial\xi, \quad d\xi/dt = -\partial f/\partial x,$$

which starts at 0 must by assumption stay in the surface $\xi_1 = 0$. Thus the equations

$$dx'/dt = \partial f(0, x', \xi')/\partial\xi', \quad d\xi'/dt = -\partial f(0, x', \xi')/\partial x'$$

must define a curve where $\partial f(0, x', \xi')/\partial x_1 = 0$. By assumption we still have $f = 0$ on the product Γ_0 of this curve and some interval on the x_1 axis, and H_f is only multiplied by a factor when x_1 is changed. Hence H_{p_1} and H_{p_2} are tangents of Γ_0. As in case a) it follows that all Poisson brackets of p_1 and p_2 vanish in Γ_0. Thus

$$\sigma(H_{p_1}, H_{p_2}) = \langle H_{p_1}, dp_2\rangle = \{p_1, p_2\} = 0$$

so Γ_0 is isotropic. Finally, $H_p = \partial/\partial x_1 + ia(x_1,t)\partial/\partial t$ in Γ, $a \geqq 0$, so H_p is locally solvable.

We have now shown that the local flowout Γ_0 along H_{p_1} and H_{p_2} of a point in Γ where H_{p_1} and H_{p_2} are linearly independent is a two dimensional manifold with the properties stated in the proposition. This remains true for the global flowout of Γ_0 along say H_{p_1}. To prove this we just have to observe that when leaving the non-degenerate situation already discussed we arrive at an embedded one dimensional bicharacteristic of p. In a neighborhood of it we can introduce canonical coordinates which reduce p to the form $\xi_1 + if(x, \xi')$ (see [17, Prop. 2.5]). The argument above completes the proof.

Just as in the real analytic case it is natural to conjecture that Theorems 6.1 and 6.7 in [17] can be extended to the two dimensional and one dimensional bicharacteristics in Proposition 8.1. To prove the sufficiency of condition (Ψ) semi-globally one needs in addition to have some result on propagation of singularities for P^* at points where a semi-bicharacteristic leaves the bicharacteristic set. When condition (P) is fulfilled one can apply Hörmander [16, Prop. 3.5.1] to cope with this situation.

REFERENCES.

1. R. Beals and C. Fefferman, On local solvability of linear partial differential equations. Ann. of Math. 97(1973), 482-498.
2. J.M. Bony, Principe du maximum, inégalité de Harnack et unicité du problème de Cauchy pour les opérateurs elliptiques dégenerés. Ann. Inst. Fourier Grenoble 19(1969), 277-304.
3. H. Brézis, On a characterization of flow-invariant sets. Comm. Pure Appl. Math. 23(1970), 261-263.
4. N. Dencker, On the propagation of singularities for pseudo-differential operators of principal type. In preparation.
5. J.J. Duistermaat and L. Hörmander, Fourier integral operators II. Acta Math. 128(1972), 183-269.
6. Yu.V. Egorov, On the solubility of differential equations with simple characteristics. Usp. Mat. Nauk 26:2(1971), 29-41 and Russian Mathematical Surveys 26:2(1971), 113-130.
7. - , On necessary conditions for solvability of pseudo-differential equations of principal type. Trudy Mosk. Mat. Obsč. 24(1971), 29-41 and Trans. Moscow Math. Soc. 24(1971), 29-42.
8. - , On sufficient conditions for the local solvability of pseudo-differential equations of principal type. Trudy Mosk. Mat. Obsč. 31(1974), 59-83, and Trans. Moscow Math. Soc. 31(1974), 55-79.

9. Yu.V. Egorov, Subelliptic operators. Usp. Mat. Nauk 30:2(1975), 57-114, and Russian Math. Surveys 30:2(1975), 59-118.
10. - , Subelliptic operators, Usp. Mat. Nauk 30:3(1975), 57-104 and Russian Math. Surveys 30:3, 55-105.
11. - , On solvability conditions for differential equations with simple characteristics. Dokl. Akad. Nauk SSSR 229(1976), 1310-1312, and Soviet Math. Doklady 17(1976), 1194-1197.
12. - and P.R. Popivanov, Equations of principal type without solution. Usp. Mat. Nauk 29:2(1974), 176-194, and Russian Math. Surveys 29:2(1974), 176-194.
13. L. Hörmander, Linear partial differential operators. Springer Verlag, Berlin-Göttingen-Heidelberg 1963.
14. - , Pseudo-differential operators and non-elliptic boundary problems. Ann. of Math. 83(1966), 129-209.
15. - , Fourier integral operators I. Acta Math. 127(1971), 79-183.
16. - , On the existence and the regularity of solutions of linear pseudo-differential equations. L'Ens. Math. 17(1971), 99-163.
17. - , Propagation of singularities and semi-global existence theorems for (pseudo-)differential operators of principal type. Ann. of Math. 108(1978), 569-609.
18. - , Subelliptic operators. In Seminar on singularities of solutions of linear partial differential equations. Princeton University Press 1979, 127-208.
19. A. Melin and J. Sjöstrand, Fourier integral operators with complex valued phase functions. Springer Lecture Notes 459(1974), 120-223.
20. A. Menikoff, On local solvability of pseudo-differential equations. Proc. Amer. Math. Soc. 43(1974), 149-154.
21. R.D. Moyer, Local solvability in two dimensions: Necessary conditions for the principle-type case. Mimeographed manuscript. University of Kansas 1978 (9 pages).
22. L. Nirenberg and F. Trèves, On local solvability of linear partial differential equations; Part I: Necessary conditions. Comm. Pure Appl. Math. 23(1970), 1-38.
23. - , - Part II, Comm. Pure Appl. Math. 23(1970), 459-509.
24. J. Sjöstrand, Operators of principal type with interior boundary conditions. Acta Math. 130(1973), 1-51.

MIXED PROBLEMS FOR THE WAVE EQUATION

Mitsuru Ikawa

Department of Mathematics
Osaka University

Abstract. We consider the mixed problems for the wave equation with general boundary condition. First we discuss on the well posedness of the problems for a boundary operator with real valued coefficients and we show the necessary and sufficient condition for the well posedness in the sense of C^∞ when the domain is the exterior of a strictly convex object. As a consequence of the considerations on the well posedness we like to show the decay of the solutions on some additional conditions on boundary operators.

Second, we consider the decay of solutions in the exterior of convex obstacles of finite number.

1. ON THE WELL POSEDNESS OF THE MIXED PROBLEMS

1.1. Statement of the theorem

Let $\mathcal{O}$ be a bounded object in $\mathbb{R}^n$, $n=2,3$, with sufficiently smooth boundary Γ. Let us set

$$\Omega = \mathbb{R}^n - \mathcal{O} - \Gamma$$

and

$$B = \Sigma_{j=1}^n b_j(x,t)\frac{\partial}{\partial x_j} + c(x,t)\frac{\partial}{\partial t} + d(x,t)$$

where b_j, c and d are functions belonging to $C^\infty(\Gamma\times[0,\infty))$.

We consider a mixed problem

$$\begin{cases} \dfrac{\partial^2 u}{\partial t^2} - \Sigma_{j=1}^n \dfrac{\partial^2 u}{\partial x_j^2} = 0 \qquad \text{in} \quad \Omega\times(0,\infty) \end{cases}$$

H. G. Garnir (ed.), Singularities in Boundary Value Problems, 97–119.

$$(\mathrm{P})\qquad \begin{cases} Bu = 0 & \text{on} \quad \Gamma\times(0,\infty) \\ u(x,0) = u_0(x) & \\ \dfrac{\partial u}{\partial t}(x,0) = u_1(x) & \end{cases}$$

on the following assumptions:

(A-I) $\mathcal{O}$ is convex and the Gaussian curvature of Γ is bounded away from zero.

(A-II) b_j, $j=1,2,\cdots,n$ and c are real valued.

(A-III) $\Sigma_{j=1}^{n} b_j(x,t)n_j(x) = 1$ on $\Gamma\times[0,\infty)$ where $n(x)=(n_1(x),n_2(x),\cdots,n_n(x))$ denotes the unit outer normal of Γ at $x\in\Gamma$.

On the well posedness of the problem (P) we have

Theorem 1. Suppose that $n=3$, $\mathcal{O}$ and B satisfy (A-I),(A-II) and (A-III). In order that (P) is well posed in the sense of C^∞ and has a finite propagation speed it must and it suffices to hold

$$(1.1)\qquad c(x,t)<1 \qquad \text{on} \quad \Gamma\times[0,\infty).$$

1.2. Decay of solutions

By using the estimates for the proof of Theorem 1 we can show the decay of solutions.

For $u(x,t)\in C^\infty(\Omega\times[0,\infty))$ we set

$$E_m(u,R,t) = \Sigma_{|\gamma|\le m}\int_{\Omega_R}|D^\gamma_{x,t}u(x,t)|^2dx$$

$$\Omega_R = \{x;\ x\in\Omega,\ |x|<R\}.$$

We denote (12e×diameter of $\mathcal{O}$)$^{-1}$ by δ_0.

Theorem 2. Suppose that all the coefficients of B belong to $\mathcal{B}^\infty(\Gamma\times[0,\infty))$ and satisfy

$$\sup_{(x,t)\in\Gamma\times(0,\infty)} c(x,t) < 1$$

$$\liminf_{t\to\infty} \mathrm{Re}\Big(-d(x)-\frac{1}{2}\Sigma_{j=1}^{3}\frac{\partial(b_j(x,t)-n_j(x))}{\partial x_j}\Big) \geqslant d_0$$

$$\lim_{t\to\infty}\ \Sigma_{j=1}^{3}\Big(\Big|\frac{\partial b_j(x,t)}{\partial t}\Big| + \Big|\frac{\partial c(x,t)}{\partial t}\Big|\Big) = 0$$

where d_0 is a constant determined by Ω alone. Then the solution $u(x,t)$ of (P) for initial data u_0, $u_1\in C_0^\infty(\bar{\Omega})$ decays exponentially

$$E_m(u,R,t) < C_m\exp\{3\delta_0(R+2\kappa)\}\cdot\exp(-\delta_0 t/2)\cdot E_{m+2}(u,\infty,0),$$

where κ denotes the diameter of $\bigcup_{j=0,1} \text{supp } u_j$ and C_m is a constant independent of u_0, u_1.

Theorem 3. Consider the case of the boundary condition of the third kind with time independent coefficients, i.e.

$$b_j(x,t) = n_j(x), \quad j=1,2,3,$$
$$c(x,t) = 0$$
$$d(x,t) = d(x).$$

If $d(x)$ is real valued and satisfies

$$d(x) < |x-Q|^{-2} \Sigma_{j=1}^{3}(x_j - q_j)n_j(x) \qquad \text{for all } x \in \Gamma$$

for some point $Q=(q_1,q_2,q_3) \in \mathcal{O}$, the solution $u(x,t)$ of (P) for initial data u_0, $u_1 \in C_0^\infty(\bar{\Omega})$ satisfies the following inequality

$$E_m(u,R,t) < C_m \exp\{3\delta_0(R+2\kappa)\} \cdot \exp(-\gamma t) \cdot E_m(u,\infty,0)$$

where γ is a positive constant independent of u_0, u_1, and κ denotes the diameter of $\bigcup_{j=0,1} \text{supp } u_j$.

Theorem 4. Suppose that the coefficients of B are independent of t. Denote by $H(x)$ the mean curvature of Γ at x with respect to $-n(x)$. When $d(x)$ satisfies

$$d(x) < \min(|x-Q|^{-2} \Sigma_{j=1}^{3}(x_j - q_j)n_j(x), H(x)), \quad \forall x \in \Gamma$$

there exists a constant $a>0$, which depends on Ω and $d(x)$, such that if

$$\Sigma_{j=1}^{3} |b_j(x) - n_j(x)| + |c(x)| + \frac{1}{2}\Sigma_{j=1}^{3}\left|\frac{\partial(b_j-n_j)}{\partial x_j}\right| < a$$

holds the solution of (P) for initial data in $C_0^\infty(\bar{\Omega})$ decays exponentially in the form

$$(1.2) \qquad E_m(u,R,t) \leqslant C_m \exp\{3\delta_0(R+2\kappa)\} \cdot \exp(-\gamma t) \cdot E_{m+1}(u,\infty,0).$$

Theorem 5. Suppose that the coefficients of B are independent of t. For each $\eta>0$ there exists a constant C_η such that

$$\text{Re}(d(x) - \frac{1}{2}\Sigma_{j=1}^{3}\frac{\partial(b_j(x)-n_j(x))}{\partial x_j}) \le H(x) - 2\eta$$

$$|\text{Im } d(x)| \ge C_\eta$$

implies the exponential decay of the solution of (P) in the form (1.2) for initial data in $C_0^\infty(\bar{\Omega})$.

1.3. Remarks on Theorem 1

About (A-1). When Ω is not an exterior of a strictly convex object, it seems to us that the characterization of the well posed problems in the sense of C^∞ is very difficult. We like to note the results on the necessary conditions for interior domains.

Suppose that n=2 and Γ is a simple closed curve in $\mathbb{R}^2$ whose curvature never vanish. Let Ω be the interior of Γ and let B satisfy (A-II) and (A-III). Set

$$\tau(s) = \Big[b_1(x)n_2(x) - b_2(x)n_1(x)\Big]_{x=x(s)}$$

where x(s) is a representation of Γ by the arclength s.

Theorem 6 (Theorem of Ikawa[11]). In order that

$$\begin{cases} \Box u = 0 & \text{in} \quad \Omega\times(0,\infty) \\ Bu = 0 & \text{on} \quad \Gamma\times(0,\infty) \\ u(x,0) = u_0(x), & \frac{\partial u}{\partial t}(x,0) = u_1(x) \end{cases}$$

is well posed in the sense of C^∞ it is necessary to hold

$$\left|\tau(s)\right| + \left|\frac{d\tau}{ds}(s)\right| \neq 0 \qquad \text{for all s.}$$

Concerning interior domain in $\mathbb{R}^3$ see Ikawa[6].

About a class of C^∞ well posed problems for which the shape of domains does not take part, see Ikawa [3] and [5].

About (A-II). Concerning the case b_j and c are complex valued functions Miyatake characterizes L^2-well posed problems in [17] and [18].

About (A-III). Soga[22][23] treat the problem (P) without the assumptions (A-I) (A-III). Especially in [23] he shows almost necessary and sufficient condition for the well posedness in the sense of C^∞.

Eskin[3] considers more general cases than Theorem 1, and he communicated privately to me that he obtained more general results than [11].

1.4. On the proof of the sufficiency of the condition (1.1)

For the simplicity we consider the case when the coefficients of B are independent of t.

Let $\phi(x)$ be a real valued function verifying $\nabla\phi \in \mathcal{B}(\mathbb{R}^3)$ and $\sup_x \nabla\phi(x) < 1$. Set

$$A_\phi(\frac{\partial}{\partial t},\frac{\partial}{\partial x}) = (1-(\nabla\phi)^2)\frac{\partial^2}{\partial t^2} - 2\nabla\phi\cdot\nabla\frac{\partial}{\partial t} - \Delta - \Delta\phi\frac{\partial}{\partial t}$$

$$B_\phi(\frac{\partial}{\partial t},\frac{\partial}{\partial x}) = \Sigma_{j=1}^3 b_j(x)(\frac{\partial}{\partial x_j} + \frac{\partial\phi}{\partial x_j}\frac{\partial}{\partial t}) + c(x)\frac{\partial}{\partial t} + d(x).$$

First consider the following boundary value problem with a parameter $p=ik+\mu$, $\mu>0$

$$(1.3)\quad \begin{cases} A_\phi(p,\frac{\partial}{\partial x})w(x) = 0 & \text{in } \Omega \\ B_\phi(p,\frac{\partial}{\partial x})w(x) = h(x) & \text{on } \Gamma. \end{cases}$$

Note that the boundary balue problem

$$(1.4)\quad \begin{cases} A_\phi(p,\frac{\partial}{\partial x})w(x) = 0 & \text{in } \Omega \\ w(x) = 0 & \text{on } \Gamma \end{cases}$$

has a unique solution in $\bigcap_{m\geq 0} H^m(\Omega)$ for $g\in C^\infty(\Gamma)$ and an estimate

$$(1.5)\quad \Sigma_{|\ell+\ell'|\leq m} |p|^\ell |||w(x)|||_{\ell'} \leq C_{m,\phi} \|g\|_{m+1}$$

when $\operatorname{Re} p \geq \mu_\phi$, where μ_ϕ is a constant depend on ϕ. In (1.5) $|||w|||_\ell$ and $\|g\|_\ell$ denote the Sobolev norm $\|w\|_{H^\ell(\Omega)}$ and $\|g\|_{H^\ell(\Gamma)}$ respectively. Denote the solution of (1.4) as

$$w(x) = U_\phi(p,g,;x).$$

Define an operator $\mathcal{B}_\phi(p)$ from $C^\infty(\Gamma)$ into $C^\infty(\Gamma)$ by

$$\mathcal{B}_\phi(p)g = B_\phi(p,\frac{\partial}{\partial x})U_\phi(p,g;x)\Big|_\Gamma.$$

The following estimate is essential for the proof of the well posedness of (P).

<u>Theorem 1.1.</u> Suppose that (A-I), (A-II) and (A-III) hold. Then for $\operatorname{Re} p \geq \mu_\phi$ we have

$$(1.6)\quad -\operatorname{Re}(\mathcal{B}_\phi(p)g,g)_m \geq (\mu c_0(\phi) - C)\|g\|_m^2 - C_{\phi,m}\|g\|_0^2$$

for all $g\in C^\infty(\Omega)$, where

$$c_0(\phi) = \inf_{x\in\Gamma} (\sqrt{1-|\phi_s|^2} - v|\phi_s| - c(x))$$

$$\phi_s = \nabla\phi - (\nabla\phi\cdot n)n$$

$$v = (\Sigma_{j=1}^3 (b_j(x) - n_j(x))^2)^{1/2}$$

and $C_{\phi,m}$ is a constant independent of p and g.

Set

$$b_j'(x) = 2n_j(x) - b_j(x),$$

$$d'(x) = \overline{d(x)} + \Sigma_{j=1}^{3} \frac{\partial}{\partial x_j}(n_j(x)-b_j(x))$$

$$B' = \Sigma_{j=1}^{3} b'(x)\frac{\partial}{\partial x_j} + c(x)\frac{\partial}{\partial t} + d'(x).$$

Then we have

$$(1.7) \qquad (\mathcal{B}_\phi(p)g,h)_0 = (g,\mathcal{B}'_{-\phi}(\overline{p})h)_0, \quad \forall g,h\in C^\infty(\Gamma).$$

Since

$$\Sigma_{j=1}^{3} b_j'(x)n_j(x) = 1$$

$$\Sigma_{j=1}^{3}(b_j'(x)-n_j(x))^2 = \Sigma_{j=1}^{3}(b_j'(x)-n_j(x))^2$$

we have also

$$(1.8) \qquad -\mathrm{Re}(\mathcal{B}'_{-\phi}(\overline{p})h,h)_m \geqslant (\mu c_0(\phi)-C)\,\|h\|_m^2 - C_{\phi,m}\|h\|_0^2.$$

Therefore taking account that

$$(1.9) \qquad \sup_{x\in\Gamma}|\phi_s| < \inf_{x\in\Gamma}\{-c(x)v(x)+\sqrt{1+|v(x)^2-c(x)^2|}\}(1+v(x)^2)^{-1}$$

implies $c_0(\phi)>0$ we have the existence of the solution g in $C^\infty(\Gamma)$ of the equation

$$\mathcal{B}_\phi(p)g = h$$

for all $h\in C^\infty(\Gamma)$ and an estimate

$$\Sigma_{\ell+\ell'\le m}|p|^\ell\|g\|_{\ell'} \le C_{\phi,m}\|h\|_{m+1}$$

holds for $\mu>\mu_\phi'$.

For $f(x,t)\in C^\infty(\Gamma\times R)$ verifying

$$(1.10) \qquad \mathrm{supp}\, f\subset\Gamma\times(t_0,\infty), \quad t_0>0$$

$$e^{-\mu t}f(x,t)\in H^m(\Gamma\times R), \qquad \mu\ge\tilde{\mu}_f$$

where $\tilde{\mu}_f$ is a constant, define $w(x,t)$ by

$$w(x,t) = \int_{-\infty}^{\infty} e^{(\mu+ik)t}U_\phi(\mu+ik,\mathcal{B}_\phi(\mu+ik)^{-1}\hat{f}(\cdot,\mu+ik);x)dk$$

where

$$\hat{f}(x,p) = \int_{-\infty}^{\infty}e^{-pt}f(x,t)dt$$

and $\mu\ge\max(\mu_\phi,\tilde{\mu}_f)$.

It is easy to verify

$$(1.11)\qquad \begin{cases} A_\phi(\frac{\partial}{\partial t},\frac{\partial}{\partial x})w(x,t) = 0 & \text{in } \Omega\times(-\infty,\infty) \\ B_\phi(\frac{\partial}{\partial t},\frac{\partial}{\partial x})w(x,t) = f(x,t) & \text{on } \Gamma\times(-\infty,\infty) \\ \text{supp } w \subset \Omega\times[t_0,\infty). \end{cases}$$

This implies that the mixed problem

$$(P_\phi)\qquad \begin{cases} A_\phi w = g(x,t) & \text{in } \Omega\times(0,\infty) \\ B_\phi w = f(x,t) & \text{on } \Gamma\times(0,\infty) \\ w(x,0) = w_0(x), \quad \frac{\partial w}{\partial t}(x,0) = w_1(x) \end{cases}$$

has a unique solution in $H^m_\mu(\Omega\times(0,\infty))$ if $g\in H^m_\mu(\Omega\times(0,\infty))$, $f\in H^m_\mu(\Gamma\times(0,\infty))$ and $w_0\in H^{m+2}(\Omega)$, $w_1\in H^{m+1}(\Omega)$, where $w\in H^m_\mu(\Omega\times(0,\infty))$ means $e^{-\mu t}w\in H^m(\Omega\times(0,\infty))$. Moreover if $w_0=w_1=0$ and $g=0, f=o$ for $t\le t_0$, the solution w satisfies $w=0$ for $t\le t_0$. By using the Holmgren's transformation we can show the local uniqueness of the problem (P_ϕ) for ϕ verifying (1.9). Since the local uniqueness holds we can use the method of sweeping out. Finally we have

Theorem 1.2. Suppose that the coefficients of B are independent of t and (1.1) holds. Then the phenomenon governed by (P) has a finite propagation speed and its maximum speed is

$$\max(1, \sup_{x\in\Gamma}(1+v(x)^2)(-c(x)v(x)+\sqrt{1+|v(x)^2-c(x)^2|})^{-1}.$$

The problem (P) is nothing but (P_0). Then (P) is well posed in the space $H^m_\mu(\Omega\times(0,\infty))$. The well posedness in the sense of C^∞ follows immediately from the well posedness in the Sobolev spaces and the finiteness of the propagation speed.

The necessity of (1.1) can be shown by following the reasoning of Kajitani [13], who applyed the idea of Ivrii to the mixed problems(see, for example, §9 of Ikawa[8]).

1.5. Sketch of the proof of Theorem 1.1

To prove Theorem 1.1, first we construct a parametrix of the problem (1.4) and calculate the symbole of $\mathcal{B}_\phi(p)$ as a pseudo-differential operator on Γ. For the construction of a parametrix of (1.4) it is essential to construct an asymptotic solution of

$$(1.12)\qquad A_\phi(ik+\mu,\frac{\partial}{\partial x})u(x) = 0 \qquad \text{in } \Omega$$

for an oscillatory boundary data

(1.13) $$u(x) = \exp\{ik(\theta(x,\eta,\beta)-\phi(x))\}\, v(x)$$

where θ a function belonging to $C^\infty(\Omega\times S^1\times[-\beta_0,\beta_0])$, $\beta_0 > 0$ determined by

$$\begin{cases} (\nabla\theta)^2 + \rho(\nabla\rho)^2 \equiv 1 & (\text{mod } \beta^\infty) \quad \text{in } \overline{\Omega}\cap\mathcal{U} \\ \nabla\theta\cdot\nabla\rho \equiv 0 & (\text{mod } \beta^\infty) \quad \text{in } \overline{\Omega}\cap\mathcal{U} \\ \rho = -\beta & \qquad\qquad\quad \text{on } \Gamma\cap\mathcal{U} \\ (\nabla\theta)(0,\eta,0) = \eta_1\dfrac{\partial}{\partial\sigma_1} + \eta_2\dfrac{\partial}{\partial\sigma_2} \end{cases}$$

$x(\sigma)=x(\sigma_1,\sigma_2)$ is a representation of Γ near $s_0\in\Gamma$, and $\mathcal{U}$ is a neighborhood in $\mathbb{R}^3$ of s_0. We construct an asymptotic solution of (1.12) in the form

(1.14) $$u(x;p,\eta,\beta) = \exp\{ik(\nu(x,\eta,\beta,p)-\phi(x))\}\,\frac{1}{H(k^{2/3}\hat{\zeta}(x,\eta,\beta,p))}$$

$$\cdot\{H(k^{2/3}\zeta(x,\eta,\beta,p)g_0 + \frac{1}{ik^{1/3}}\, H'(k^{2/3}\zeta(x,\eta,\beta,p)g_1\}$$

where

$$H(z) = Ai(-z) + i\, Bi(-z).$$

We require that ν and ζ satisfy

$$\begin{cases} (\nabla\nu + \dfrac{\mu}{ik}\nabla\phi - \dfrac{1}{ik^{1/3}}R\nabla\hat{\zeta}\,)^2 + \zeta(\nabla\zeta)^2 \equiv (1+\dfrac{\mu}{ik})^2 \\ \qquad\qquad\qquad\qquad\qquad\qquad \text{mod}(|p|^{-\infty} + \beta^\infty) \\ (\nabla\nu + \dfrac{\mu}{ik}\nabla\phi - \dfrac{1}{ik^{1/3}}R\,\nabla\hat{\zeta})\cdot\nabla\zeta \equiv 0 \qquad \text{mod}(|p|^{-\infty} + \beta^\infty) \\ \nu(x(\sigma),\eta,\beta,p) = \theta(x(\sigma),\eta,\beta), \end{cases}$$

and g_0, g_1 satisfy

$$2(\nabla\nu + \frac{\mu}{ik}\nabla\phi - \frac{1}{ik^{1/3}}R\nabla\hat{\zeta})\cdot\nabla g_0 + \nabla(\nabla\nu + \frac{\mu}{ik}\nabla\phi - \frac{1}{ik^{1/3}}R\nabla\hat{\zeta})g_0$$

$$+ 2\zeta\nabla\zeta\cdot\nabla g_1 + \zeta\Delta\zeta g_1 + (\nabla\zeta)^2 g_1 + \frac{1}{ik}g_0 \equiv 0$$

$$\text{mod}(|p|^{-\infty} + \beta^\infty)$$

$$2(\nabla\nu + \frac{\mu}{ik}\nabla\phi - \frac{1}{ik^{1/3}}R\nabla\hat{\zeta})\cdot\nabla g_1 + \nabla(\nabla\nu + \frac{\mu}{ik}\nabla\phi - \frac{1}{ik^{1/3}}R\nabla\hat{\zeta})g_1$$

$$+ 2\nabla\zeta\cdot\nabla g_0 + \Delta\zeta g_0 + \frac{1}{ik}\Delta g_1 \equiv 0 \qquad \text{mod}(|p|^{-\infty} + \beta^\infty)$$

$$g_0 + \frac{1}{ik^{1/3}} R(k^{2/3}\zeta) g_1 \equiv v(x) \quad \text{on} \quad \Gamma$$

where

$$R(z) = \frac{H'(z)}{H(z)} .$$

For detailed proof, see Ikawa[7],[8],[9],[10].

2. DECAY OF SOLUTIONS OF THE WAVE EQUATION IN THE EXTERIOR OF CONVEX OBSTACLES OF FINITE NUMBER.

2.1. Statement of the theorems

Let $\mathcal{O}_j, j=1,2,\cdots\cdots,J$, be compact and convex sets in $\mathbb{R}^3$ with smooth boundary Γ_j. Suppose that

$$\mathcal{O}_j \cap \mathcal{O}_{j'} = \phi \qquad \text{for } j \neq j'.$$

Set

$$\mathcal{O} = \bigcup_{j=1}^{J} \mathcal{O}_j, \quad \Gamma = \bigcup_{j=1}^{J} \Gamma_j$$

and

$$\Omega = \mathbb{R}^3 - \mathcal{O} - \Gamma.$$

We like to consider the decay of solutions of the mixed problem

$$(\mathrm{P}) \quad \begin{cases} \square u = 0 & \text{in} \quad \Omega\times(0,\infty) \\ Bu = 0 & \text{on} \quad \Gamma\times(0,\infty) \\ u(x,0) = u_0(x), \quad \dfrac{\partial u}{\partial t}(x,0) = u_1(x). & \end{cases}$$

Here we treat as boundary operator B the following two operators

$$B_1 u = u$$

and

$$B_2 u = \frac{\partial u}{\partial n} + \sigma(x) u$$

where $\sigma(x)$ is a real valued C^∞ function defined on Γ. Set

$$\Sigma_j^+ = \{(x,\xi);\ x \in \Gamma_j,\ \xi \in \mathbb{R}^3 \text{ such that } |\xi|=1,\ n(x)\cdot\xi \geq 0\}$$

$$\Sigma^+ = \bigcup_{j=1}^{J} \Sigma_j^+ .$$

For $(x,\xi) \in \Sigma^+$, denote by $\mathcal{X}(x,\xi)$ the broken ray according to the geometric optics starting from x in the direction ξ, and by

$X_1(x,\xi)$, $X_2(x,\xi)$, $\cdots\cdots$, $X_j(x,\xi)$, $\cdots\cdots$, the reflecting points of the broken ray, and by $\Xi_j(x,\xi)$ the direction of reflected ray at $X_j(x,\xi)$. We denote by $\#\mathfrak{X}(x,\xi)$ the number of reflecting points of the broken ray $\mathfrak{X}(x,\xi)$.

Definition. We say that a broken ray $\mathfrak{X}(x,\xi)$ is closed if for some j

$$(X_j(x,\xi),\Xi_j(x,\xi)) = (x,\xi).$$

holds.

We divide $\mathcal{O}$ into groups of obstacles

$$\mathcal{G}_\ell = \{\mathcal{O}_j;\ j=j_{\ell 1}, j_{\ell 2}, \cdots\cdots, j_{\ell m_\ell}\}$$

with the following properties:

(i) $\bigcup_\ell \mathcal{G}_\ell = \mathcal{O}$.

(ii) $\mathcal{G}_\ell \cap \mathcal{G}_{\ell'} = \phi$, $\quad \ell \neq \ell'$

(iii) for each ℓ the number of closed broken rays which start from $\partial\mathcal{G}_\ell$ and fit $\mathcal{O}-\mathcal{G}_\ell$ is finite.

(iv) for each ℓ, if we divide $\mathcal{G}_\ell$ in such a way $\mathcal{G}_\ell = \mathcal{G}'_\ell \cup \mathcal{G}''_\ell$, $\mathcal{G}'_\ell \cap \mathcal{G}''_\ell = \phi$, $\mathcal{G}'_\ell, \mathcal{G}''_\ell \neq \phi$, the number of closed broken rays which start from $\mathcal{G}'_\ell$ and fit $\mathcal{O}-\mathcal{G}'_\ell$ are infinite.

Definition. We say that $\mathcal{O} = \{\mathcal{O}_j; j=1,2,\cdots,J\}$ satisfies Condition SC when it holds that
(i) The Gaussian curvature of Γ never vanishes.
(ii) For all $x \in \Gamma$

$$\operatorname{dis}(x, \mathcal{G}_\ell - \mathcal{O}_j) > (m_\ell - 2)/2 \cdot \min(\kappa_1(x), \kappa_2(x))$$

where $\kappa_j(x)$, $j=1,2$, are the principal curvatures of Γ at x with respect to $-n(x)$, and ℓ and j are determined by x according to a relation $x \in \mathcal{O}_j \subset \mathcal{G}_\ell$.

Theorem 7. Suppose that $\mathcal{O}$ satisfies Condition SC. When $B=B_1$ there exists a constant $\alpha_1 > 0$ determined by $\mathcal{O}$ with the following properties:

Let R and κ be positive numbers and let m be a positive integer. For any initial data u_0, $u_1 \in C_0^\infty(\overline{\Omega})$ such that

$$\bigcup_{j=0}^{1} \operatorname{supp} u_j \subset \overline{\Omega}_\kappa$$

we have an estimate of the solution

$$(2.1) \qquad \sum_{|\gamma|\le m} \sup_{x\in\Omega_R} |D^{\gamma}_{x,t}u(x,t)|$$

$$\le C_{R,\kappa,m}\, e^{-\alpha_1 t} \{\|u_0\|_{m+5,L^2(\Omega)} + \|u_1\|_{m+4,L^2(\Omega)}\},$$

where $C_{R,\kappa,m}$ is a positive constant depending on R, κ and m but independent of u_0, u_1.

Definition. We say that $\sigma(x)$ satisfies Condition A when the relations

$$\Delta v = 0 \qquad \text{in} \quad \Omega$$

$$\frac{\partial v}{\partial n} + \sigma v = 0 \qquad \text{on} \quad \Gamma$$

$$\sup_{x\in\Omega} \{|x||v(x)| + |x|^2|\nabla v(x)|\} < \infty$$

imply that $v\equiv 0$.

Remark. For fixed $\mathcal{O}$ there exists a positive constant $\delta>0$ such that

$$\sigma(x) < \delta, \qquad x\in\Gamma$$

implies Condition A. For the case of J=1, see Asakura[1] and for the case of J=2, see §2 of Ikawa[12].

Theorem 8. Suppose that $\mathcal{O}$ satisfies Condition SC. For B_2 with $\sigma(x)$ verifying Condition A, there exists $\alpha_2>0$ determined by $\mathcal{O}$ and σ with the following properties:

For any initial data u_0, $u_1 \in C_0^{\infty}(\overline{\Omega})$ such that

$$\bigcup_{j=0}^{1} \operatorname{supp} u_j \subset \overline{\Omega}_{\kappa}$$

we have an estimate of the solution

$$(2.2) \qquad \sum_{|\gamma|\le m} \sup_{x\in\Omega_R} |D^{\gamma}_{x,t}u(x,t)|$$

$$\le C_{R,\kappa,m}\, e^{-\alpha_2 t} \{\|u_0\|_{m+5,L^2(\Omega)} + \|u_1\|_{m+4,L^2(\Omega)}\}.$$

2.2. Remarks on the theorems

It seems to us that until now studies on the uniform decay of solutions of the wave equation in the exterior domain are made mainly concerning the existence of a function p(t) such that

$$(2.3)\qquad \begin{cases} E(u,R,t) \le p(t)E(u,\infty,0) \\ p(t) \longrightarrow 0 \qquad \text{as} \quad t \longrightarrow \infty \end{cases}$$

for all u_0, $u_1 \in C_0(\Omega_K)$, where

$$E(u,R,t) = \frac{1}{2} \int_{\Omega_R} (|\nabla u(x,t)|^2 + |\frac{\partial u}{\partial t}(x,t)|^2)dx.$$

About the necessary condition on the obstacles for the existence of such $p(t)$ we know the work of Ralston[21]. Roughly speaking [21] shows that if the obstacles admit a trapped ray there is no $p(t)$ verifying (2.3).

If $\mathcal{O}$ consists of more than one obstacle there is always a trapped ray. Therefore, when $\mathcal{O}$ consists of several obstacles, in order to estimate the uniform rare of the decay of solutions it is necessary to consider the decay in a weaker form than (2.3). Indeed, Walker[24] shows that

$$\lim_{t\to\infty} p_{\alpha,R}(t) = 0$$

holds for any $\alpha > 0$, where

$$p_{\alpha,R}(t) = \sup_{u_0, u_1 \in C_0^\infty(\Omega_R)} E(u,R,t)^{1/2} (\|u_0\|_{1+\alpha} + \|u_1\|_\alpha)^{-1}.$$

But we cannot obtain from its proof any more informations about qualitative nature of $p_{\alpha,R}(t)$ for general domains.

We should like to remark on the assumption of the strict convexity of obstacles. If only the convexity of obstacles is assumed, we cannot have in general the estimate of the form (2.1) or (2.2). Namely we have convex obstacles $\mathcal{O}_1$ and $\mathcal{O}_2$ such that

$$\limsup_{t \to \infty} p_{m,R}(t)t^{m+\varepsilon} > 0$$

holds for all positive integer m and positive constant ε.

2.3. Reduction of the problem

It is well known that the boundary value problem with a parameter $p=\mu+ik$ for boundary data $g \in C^\infty(\Gamma)$

$$(2.4)\qquad \begin{cases} (\Delta - p^2)u = 0 & \text{in } \Omega \\ Bu = g & \text{on } \Gamma \end{cases}$$

has a solution uniquely in $H^2(\Omega)$ if $\mu \ge \mu_0$, where μ_0 is a positive constant, and an estimate

$$(2.5)\qquad \|u\|_{m+2,L^2(\Omega)} \le C_m \Sigma_{j=1}^m |p|^{m-j} \|g\|_{j+3/2,L^2(\Omega)}$$

holds.

Denote by $V(p)$ a mapping which corresponds for $g \in C^\infty(\Gamma)$ the solution of (2.4). Then we have that

$$(2.6) \quad \begin{cases} V(p) \text{ is analytic in } \operatorname{Re} p > \mu_0 \text{ as } \mathcal{L}(C^\infty(\Gamma), C^\infty(\bar{\Omega}))\text{-valued} \\ \text{function.} \end{cases}$$

To prove the exponential decay of the solution of (P) we have to show that $V(p)$ can be prolonged analytically into a region $\{p;\ \operatorname{Re} p \geq -\alpha\}$, $\alpha>0$. Admit now

<u>Proposition 2.1.</u> Suppose that $\mathcal{O}$ satisfies Condition SC. Then for any integer N we can construct an operator $\mathcal{W}^{(N)}$ from $C^\infty(\Gamma\times\mathbb{R})$ into $C^\infty(\bar{\Omega}\times\mathbb{R})$ with the following properties:

For $h(x,t) \in C_0^\infty(\Gamma\times(0,1))$

$$(2.7) \qquad \mathcal{W}^{(N)} h(x,t) = 0 \qquad \text{for } t < 0$$

$$(2.8) \qquad \square \mathcal{W}^{(N)} h = 0 \qquad \text{in } \Omega\times\mathbb{R}$$

$$(2.9) \qquad |\mathcal{W}^{(N)} h|_m(\Omega_R, t) \leq C_{N,m,R,\varepsilon}\, e^{-(c_0-\varepsilon)t} \|h\|_{m+4, L^2(\Gamma\times\mathbb{R})}$$

$$(2.10) \qquad |D_t(B\mathcal{W}^{(N)} h - h)|_m(\Gamma,t) \leq C_{N,m,\varepsilon}\, e^{-(c_0-\varepsilon)t} \|h\|_{-N+m+2, L^2(\Gamma\times\mathbb{R})}$$

where c_0 is a positive constant determined by $\mathcal{O}$ and

$$|w|_m(\omega,t) = \sum_{|\gamma|<m} \sup_{x\in\omega} |D^\gamma_{x,t} w(x,t)| .$$

Define for $\operatorname{Re} p > -c_0$ $\mathcal{Z}^{(N)}(p)$ by

$$(\mathcal{Z}^{(N)}(p)h)(x) = \int_{-\infty}^{\infty} e^{-pt} (\mathcal{W}^{(N)} h)(x,t)\,dt.$$

We have from (2.9) that $\mathcal{Z}^{(N)}(p)h \in C^\infty(\bar{\Omega})$ in $\{p;\ \operatorname{Re} p > -c_0\}$ and

$$(2.11) \qquad |\mathcal{Z}^{(N)}(p)h|_m(\Omega_R) \leq C_{N,m,R,\varepsilon}\,(c_0-\varepsilon+\operatorname{Re} p)^{-1} \|h\|_{m+4}.$$

By using the energy estimate of (P) we have from (2.7) and (2.8) $\mathcal{Z}^{(N)}h \in \bigcap_{m>0} H^m(\Omega)$ and an estimate

$$(2.12) \qquad \|\mathcal{Z}^{(N)} h\|_{m,L^2(\Omega)} \leq C_m (\operatorname{Re} p)^{-1} \|h\|_{m+1, L^2(\Gamma\times\mathbb{R})}.$$

It follows from (2.8) that

$$(2.13) \qquad (p^2 - \Delta)\,\mathcal{Z}^{(N)}(p)h = 0 \qquad \text{in } \Omega.$$

We have from (2.10) for $m<N-2$

$$(2.14) \qquad |\mu+ik|\ |B\,\mathcal{Z}^{(N)}(p)h - \hat{h}|_m(\Gamma) < C_{N,m,\varepsilon} \|h\|_{L^2(\Gamma\times\mathbb{R})}.$$

Take a function $m(t) \in C_0^\infty(0,1)$ such that

$$m(t) \geq 0 \text{ and } \int m(t)dt = 1.$$

Set

$$m_q(t) = e^{iqt} m(t).$$

Since

$$\hat{m}_q(ik+\mu) = \hat{m}(i(k-q)+\mu)$$

there exist $a_0 > 0$ and $C_0 > 0$ such that

$$|\hat{m}_q(ik+\mu)| > C_0 \quad \text{in } D_q = \{ik+\mu; |k-q| \leq a_0, \mu_0+1 \geq \mu \geq -c_0\}.$$

Define $\tilde{V}_q^{(N)}(p)$ for $p \in D_q$ as an operator from $C^\infty(\Gamma)$ into $C^\infty(\bar{\Omega})$ by

$$\tilde{V}_q^{(N)}(p)g = \hat{m}_q(p)^{-1} \cdot \mathcal{Z}^{(N)}(p)h, \quad h = g(x)m_q(t)$$

Then from (2.14) we have

$$(2.15) \quad |B\tilde{V}_q^{(N)}(p)g - g|_m(\Gamma) \leq C_{m,\varepsilon} |p|^{-1} \|g\|_{L^2(\Gamma)}$$

for $p \in D_q \cap \{p;\ \mathrm{Re}\ p \geq -(c_0-\varepsilon)\} = D_{q,\varepsilon}$. Then for large q

$$A_q^{(N)}(p) = I + \Sigma_{j=1}^{\infty}(I - B\tilde{V}_q^{(N)}(p))^j$$

is well defined for $p \in D_{q,\varepsilon}$ and it is a bounded operator from $C^\infty(\Gamma)$ into $C^\infty(\Gamma)$. Define $V_q^{(N)}(p)$ for $p \in D_q$ by

$$V_q^{(N)}(p) = \tilde{V}_q^{(N)}(p)A_q^{(N)}(p).$$

Then it holds that for all $g \in C^\infty(\Gamma)$ and $p \in D_q$

$$\begin{cases} (p^2-\Delta)V_q^{(N)}(p)g = 0 & \text{in } \Omega \\ BV_q^{(N)}(p)g = g & \text{on } \Gamma . \end{cases}$$

And (2.12) implies that

$$V_q^{(N)}(p)g \in \bigcap_{m>0} H^m(\Omega), \quad \text{for } \mathrm{Re}\ p > 0.$$

Note that the analyticity of $V_q^{(N)}(p)$ as $\mathcal{L}(C^\infty(\Gamma), C^\infty(\bar{\Omega}))$-valued function in D_q follows from those of $\mathcal{Z}^{(N)}(p)$, $A_q^{(N)}(p)$. The uniqueness of the solutions of (2.4) assures that $V_q^{(N)}(p)$ and $V(p)$ coincide in $D_q \cap \{\mathrm{Re}\ p \geq \mu_0\}$. Taking account of this fact and the analyticity of each $V_q^{(N)}(p)$, $V_q^{(N)}(p)$ is independent of N and q. Then by using the estimates (2.11), (2.15) and

$$\|m_q\|_{m,L^2(0,1)} \leq C_m |q|^m,$$

we have

Proposition 2.2. Suppose that $\mathcal{O}$ satisfies Condition SC. Then $V(p)$ can be prolonged analytically into a region $\{\mu+ik;\ \mu\geq-(c_0-\varepsilon),\ |k|\geq C_\varepsilon\}$, where ε is arbitrary positive constant and C_ε is a positive constant depending on ε and an estimate

$$|V(p)g|_m(\Omega_R) \leq C_{m,R,\varepsilon}\ \Sigma_{j=1}^{m+3}\ |p|^{m+3-j}\ \|g\|_{j,L^2(\Omega)},\quad \forall g\in C^\infty(\Gamma)$$

holds.

Combining this proposition and the results of Chapter V of Lax-Phillips [14] we have

Theorem 2.2. Suppose that $\mathcal{O}$ satisfies Condition SC and that σ verifies Condition A if $B=B_2$. Then $V(p)$ can be prolonged analytically into a region containing

$$\{p;\ \mathrm{Re}\ p \geq -\alpha\}$$

for some $\alpha>0$. And we have for $\mathrm{Re}\ p\geq-\alpha$

$$(2.16)\qquad |V(p)g|_m(\Omega_R) \leq C_{m,R}\ \Sigma_{j=1}^{m+3}\ |p|^{m+3-j}\ \|g\|_{j,L^2(\Omega)},\quad g\in C^\infty(\Gamma)$$

Let us show Theorems 7 and 8 with the aid of Theorem 2.2. Note that the solution of the problem

$$(2.17)\qquad \begin{cases} \Box z(x,t) = 0 & \text{in } \Omega\times\mathbb{R} \\ Bz(x,t) = h(x,t) & \text{on } \Gamma\times\mathbb{R} \\ \mathrm{supp}\ z\subset\Omega\times(0,\infty) \end{cases}$$

for $h\in C_0^\infty(\Gamma\times(0,\kappa))$ is represented as

$$(2.18)\qquad z(x,t) = \int_{-\infty}^{\infty} e^{(ik+\mu)t}(V(ik+\mu)\hat{h}(\cdot,ik+\mu))(x)\,dk$$

for large $\mu>0$, where

$$\hat{h}(x,p) = \int_{-\infty}^{\infty} e^{-pt}h(x,t)\,dt.$$

Theorem 2.2 assures us that the path of integration (2.18) can changed to $\mathrm{Re}\ p = -\alpha$, namely, we have

$$z(x,t) = \int_{-\infty}^{\infty} e^{(ik-\alpha)t}(V(ik-\alpha)\hat{h}(\cdot,ik-\alpha))(x)\,dk.$$

By using the estimate (2.16) we have from this formula

$$(2.19)\qquad |z|_m(\Omega_R,t) \leq C_{m,R}e^{-\alpha t}\ e^{\alpha\kappa}\ \|h\|_{m+4,L^2(\Gamma\times\mathbb{R})}.$$

Let $\tilde{u}_0$, $\tilde{u}_1$ be functions in $C^\infty(\mathbb{R}^3)$ such that

$$\tilde{u}_j(x) = u_j(x)\qquad \text{for } x\in\Omega$$

and

$$\|\tilde{u}_j\|_{m,L^2(\mathbb{R}^3)} \leq C_m\ \|u_j\|_{m,L^2(\Omega)},$$

where C_m is a constant independent of u_j, and let $w(x,t)$ be the solution of the Cauchy problem

$$\begin{cases} \Box w = 0 & \text{in } \mathbb{R}^3\times(0,\infty) \\ w(x,0) = \tilde{u}_0, \quad \frac{\partial w}{\partial t}(x,0) = \tilde{u}_1. \end{cases}$$

Set

$$h(x,t) = -Bw(x,t)|_{\Gamma\times\mathbb{R}}.$$

By the Huygens'principle we have

$$\operatorname{supp} h \subset \Gamma\times(0,\kappa+d_0)$$

where d_0 denotes the diameter of $\mathcal{O}$. Denoting the solution of (2.17) for this h by $z(x,t)$, the solution of (P) is represented as

$$u(x,t) = w(x,t) + z(x,t).$$

Since

$$\|h\|_{m+4,L^2(\Gamma\times\mathbb{R})} \leq C_m\{\|u_0\|_{m+5,L^2(\Omega)} + \|u_1\|_{m+4,L^2(\Omega)}\}$$

and

$$w(x,t) = 0 \qquad \text{for } |x| < t - \kappa,$$

we see that $u(x,t)$ decays in the form (2.1) or (2.2). Thus the theorems are proved.

2.4. Construction of asymptotic solutions

In order to construct $\mathcal{W}^{(N)}$ of Proposition 2.1 it is essential to construct asymptotic solutions for oscillatory boundary data.

For $\phi(x)$ such that $|\nabla\phi|=1$

$$\mathcal{C}_\phi(x) = \{y;\ \phi(y) = \phi(x)\}.$$

Hereafter the principal curvatures of $\mathcal{C}_\phi(x)$ means those of with respect to $-\nabla\phi$.

Definition. Let $s_0 \in \Gamma$ and $\mathcal{U}$ be a neighborhood of s_0 in $\mathbb{R}^3$ and let $\psi(x)$ be a real valued function defined on $\Gamma\cap\mathcal{U}$. We say that ψ satisfies Condition C if there exists a function $\phi(x)$ defined in some neighborhood $\mathcal{U}'$ of s_0 such that

$$\phi(x) = \psi(x) \qquad \text{in } \Gamma\cap\mathcal{U}'$$

$$|\nabla\phi(x)| = 1 \qquad \text{in } \mathcal{U}'$$

$$\frac{\partial\phi}{\partial n}(x) \geq c > 0$$

and the principal curvatures of $\mathcal{C}_\phi(x)$ are positive.

Proposition 2.3. Let u be an oscillatory boundary data of the form

$$u(x,t;\eta,\beta,k) = e^{ik(\theta(x,\eta,\beta)-t)} f(x,t;k) \tag{2.20}$$

where θ is the function introduced in Section 1.5, or

$$u(x,t;k) = e^{ik(\psi(x)-t)} f(x,t;k) \tag{2.21}$$

where ψ is a function verifying Condition C. Suppose that

$$\text{supp } f \subset \Gamma_1 \times [T,T+1]$$

where Γ_1 is a small neighborhood of $s_0 \in \Gamma$ in Γ. Then for any N positive integer there exists a function $z^{(N)}(x,t;k) \in C^\infty(\bar{\Omega}\times\mathbb{R})$ with the following properties:

$$\text{supp } z^{(N)} \subset \bar{\Omega}\times[T,\infty) \tag{2.22}$$

$$|z^{(N)}|_m(\Omega_R,t) \leq C_{N,m,R,\varepsilon}\, e^{-(c_0-\varepsilon)(t-T)} \cdot k^{m+1} \Sigma_{j=0}^N k^{-j} |f|_{m+2j+40}(\Gamma_1\times\mathbb{R}) \tag{2.23}$$

$$\Box z^{(N)} = 0 \qquad \text{in} \quad \Omega\times\mathbb{R} \tag{2.24}$$

$$|Bz^{(N)} - u|_m(\Gamma,t) \leq C_{N,m,\varepsilon}\, e^{-(c_0-\varepsilon)(t-T)} k^{-N+m} |f|_{m+2N+40}(\Gamma_1\times\mathbb{R}) \tag{2.25}$$

where c_0 is a positive constant determined by $\mathcal{O}$ only, ε is an arbitrary positive constant, $C_{N,m,R,\varepsilon}$ depends on N,m,R,ε and $\{|\psi|_m(\Gamma_1);m=1,2,\cdots\}$.

By using the Fourier transformation of boundary data h we can easily construct $w^{(N)}$ of Proposition 2.1 from this proposition. Then in the remainder we like to sketch the reasoning of the above proposition in the case where $\mathcal{O}$ consists of two strictly convex obstacles and $B=B_1$.

(a) On asymptotic solutions in the free space

Let D be an open set in $\mathbb{R}^3$ abd let $\phi\in C^\infty(D)$ verifying $|\nabla\phi|=1$. Set

$$w(x,t;k) = e^{ik(\phi(x)-t)} v(x,t;k) \tag{2.26}$$

$$v(x,t;k) = \Sigma_{j=0}^N v_j(x,t)k^{-j}. \tag{2.27}$$

When v_j, j=0,1,2, ⋯ ,N, satisfy

$$2\frac{\partial v_j}{\partial t} + 2\nabla\phi\cdot\nabla v_j + \Delta\phi v_j = -i\Box v_{j-1} \tag{2.28}$$

where $v_{-1}=0$, we have

$$w(x,t;k) = e^{ik(\phi(x)-t)} k^{-N} \Box v_N. \tag{2.29}$$

Concerning the solution of the equation

$$2\frac{\partial h}{\partial t} + 2\nabla\phi\cdot\nabla h + \Delta\phi h = g$$

we have a representation formula

(2.30) $$h(x+\ell\nabla\phi(x),t+\ell) = \{\frac{G(x,\ell)}{G(x,0)}\}^{1/2} + \int_0^\ell \{\frac{G(x,\ell)}{G(x,s)}\}^{1/2} g(x+s\nabla\phi(x),t+s)ds,$$

where $G(x,\ell)$ denotes the Gaussian curvature of $\mathcal{C}_\phi(x+\ell\nabla\phi(x))$ at $x+\ell\nabla\phi(x)$. From this formula we have estimates of the solutions of the transport equation.

<u>Lemma 2.4.</u> Denote by $\kappa_j(x)$, $j=1,2$, the principale curvatures of $\mathcal{C}_\phi(x)$ at x. Suppose that $\kappa_1(x) \geq \kappa_2(x) > 0$. Then it holds that for $\ell\geq 0$

(2.31) $$|h|_m(x+\ell\nabla\phi(x),t+\ell) \leq (1+\kappa_2(x))^{-1}|h|_m(x,t) + \int_0^\ell |g|_m(x+s\nabla\phi(x),t+s)ds + C_m|\nabla\phi|_m(x)\{|h|_{m-1}(x,t) + |h|_{m-1}(x+\ell\nabla\phi(x),t+\ell) + |g|_{m-1}(x+\ell\nabla\phi(x),t+\ell) + \int_0^\ell |g|_{m-1}(x+s\nabla\phi(x),t+s)ds\}.$$

Especially for $g=0, m=0$ we have

(2.32) $$|h(x+\ell\nabla\phi(x),t+\ell)| \leq (1+\kappa_2(x))^{-1}|h(x,t)|.$$

(b) Reflection of asymptotic solutions.

Let an asymptotic solution of the form (2.26)

$$w_-(x,t;k) = e^{ik(\phi_-(x)-t)} v_-(x,t;k),$$

$$v_-(x,t;k) = \Sigma_{j=1}^N v_{-,j}(x,t)k^{-j}$$

be given near of $s_0 \in \Gamma$ and let

$$-\frac{\partial\phi_-}{\partial n} \geq c > 0.$$

Define w_+ by

$$w_+(x,t;k) = e^{ik(\phi_+(x)-t)} v_+(x,t;k),$$

$$v_+(x,t;k) = \Sigma_{j=1}^N v_{+,j}(x,t)k^{-j}.$$

We require that

$$|\nabla\phi_+|=1 \text{ and } \phi_+(x) = \phi_-(x), \frac{\partial\phi_-}{\partial n} = -\frac{\partial\phi_\pm}{\partial n} \text{ on } \Gamma$$

and $v_{+,j}$ satisfy (2.28) and $v_{+,j}=-v_{-,j}$ on $\Gamma\times\mathbb{R}$. Then we have

$$|\Box(w_- + w_+)|_m(\Omega_R\times R) \le C_{N,m,R}\, k^{-N+m}$$

$$B(w_- + w_+) = 0 \qquad \text{on} \quad \Gamma\times\mathbb{R}.$$

Moreover we have

Lemma 2.5. Let $K_1(x) \ge K_2(x) > 0$ be the principal curvatures of Γ at x and $\kappa_1^-(x) \ge \kappa_2^-(x) > 0$ be those of $\mathscr{C}_{\phi_-}(x)$. Then the principal curvatures of $\mathscr{C}_{\phi_+}(x)$ verify

$$\kappa_2^+(x) \ge 2K_2(x) + \kappa_2^-(x) \qquad \text{at } x\in\Gamma.$$

Lemma 2.6. For $x\in\Gamma$ we have

$$(2.33)\qquad |v_{+,j}|_m(x,t) \le |v_{-,j}|_m(x,t) + |v_{-,j-1}|_m(x,t) + C_m|\nabla\phi_-|_m\{|v_{-,j-1}|_m(x,t) + |v_{-,j}|_{m-1}(x,t) + |\Box v_{+,j-1}|_{m-1}(x,t)\}\,.$$

(c) Reflection of grazing rays.

Using the idea of Ludwig[15] or the considerations in Chapter 1, we can construct explicitly an reflected wave of an incident wave with grazing rays. The reflected wave is represented as a superposition of asymptotic solutions written with the Airy function, θ, ρ and g_0, g_1. From this representation we can estimate the absolute value of reflected waves. And the support of reflected waves is contained in a small neighborhood of the support of reflected rays of geometric optics.

(d) On the broken rays according to geometric optics.

Let $a_j\in\mathscr{O}_j$, $j=1,2$, be the points such that

$$|a_1-a_2| = \text{dis}(\mathscr{O}_1, \mathscr{O}_2) = d_1.$$

Denote by $S_j(\delta)$ the connected component containing a_j of

$$\Gamma_j\cap\{x;\text{dis}(x,L)<\delta\}\ , \qquad \delta>0$$

where L is the line passing a_1 and a_2.

Lemma 2.7. For δ_2 sufficiently small, there is a positive integer K depending on δ_2 such that $x\in\Gamma-S(\delta_2)$ and $\mathscr{X}(x,\xi)\cap S(\delta_2) = \phi$ imply that

$$\#\mathscr{X}(x,\xi)\le K$$

where $S(\delta_2)=S_1(\delta_2)\cup S_2(\delta_2)$.

Corollary. If we choose δ_3 as $\delta_3>\delta_2$ and sufficiently close to

δ_2, it holds that for any $(x,\xi)\in\Sigma^+$ verifying $x\in S(\delta_3)$ and $X_1(x,\xi)\in S(\delta_3)-S(\delta_2)$

$$\#\mathcal{X}(x,\xi) \leq K+1.$$

(e) Construction $z^{(N)}$.

Applying (a), (b) and (c) and using the relation with the support of solutions of transport equations and the broken rays, we have immediately

Lemma 2.8. For oscillatory data $u(x,t;k)$ of (2.20) or of (2.21), if

$$\#\mathcal{X}(x,\xi) \leq K+1, \qquad x\in \mathrm{Proj}_x \operatorname{supp} f, \tag{2.34}$$

we can construct $\tilde{z}^{(N)}$ satisfying (2.22), (2.23), (2.25) and

$$|\Box\tilde{z}^{(N)}|_m(\Omega_R,t) \leq C_{N,m,R}\, k^{-N}\, |f|_{2N+2+m}(\Gamma_1\times R)$$

$$\tilde{z}^{(N)} = 0 \qquad \text{for } |x|\leq t-(T+2d_0(K+1)).$$

With the aid of solutions of the Cauchy problem and the Huygens' principle, we may derive from the above lemma Proposition 2.1 on the assumption (2.34).

Let $\upsilon_{ij}(x)$, $i,j=1,2$, be functions defined on Γ_i such that

$$\upsilon_{ij}(x) = \begin{cases} 1 & x\in S_i(\delta_2) \\ 0 & x\notin S_i(\delta_3) \end{cases}$$

and $\upsilon_{i1}+\upsilon_{i2}=1$ on Γ_i.

First suppose that

$$\operatorname{supp} f \subset S_1(\delta_2)\times(0,1).$$

Let $z_1^{(N)}$ be asymptotic solution constructed for $u(x,t;k)$ according to the method (b) and set

$$z^{(N)}(x,t;k) = e^{ik(\phi(x)-t)}\, v_1(x,t;k)$$

$$v_1(x,t;k) = \Sigma_{j=0}^{N} v_{1j}(x,t)k^{-j}.$$

Set

$$u_{2j}^{(N)} = \upsilon_{2j} Bz_1^{(N)}|_{\Gamma_2\times R}.$$

$z_2^{(N)}$ be the asymptotic solution for boundary data $u_{21}^{(N)}$ on Γ_2. When $z_\ell^{(N)}$ is defined we set

$$u_{\ell+1,j}^{(N)} = \upsilon_{\varepsilon j} Bz^{(N)}|_{\Gamma_\varepsilon\times R}, \qquad \varepsilon = \begin{cases} 1 & \ell\ \text{:even} \\ 2 & \ell\ \text{:odd} \end{cases}.$$

And $z^{(N)}$ is an asymptotic solution with an oscillatory boundary data $u^{(N)}_{\ell+1,1}$ of the form

$$z^{(N)}_{\ell+1}(x,t;k) = e^{ik(\phi_{\ell+1}(x)-t)} \Sigma^{N}_{j=0} v_{\ell+1,j}(x,t)k^{-j}.$$

About the sequence $z^{(N)}_{\ell}$, $\ell=1,2,\cdots\cdots$ we have

(i) $\{\phi_\ell,\ \ell=1,2,\cdots\cdots\}$ is a bounded set in $C^\infty(\bar{\Omega})$.

(ii) the principal curvatures of $\mathcal{C}_{\phi_\ell} > 2K_2$.

(iii) $\operatorname{supp} z^{(N)}_{\ell} \cap \bar{\Omega}_R \times \mathbb{R} \subset \bar{\Omega}_R \times [\ell d_1, \ell d_0+R+1]$

(iv) $\# \mathfrak{X}(x,\nabla\phi_\ell(x)) \leq K+1,\quad x \in \mathrm{Proj}_x \operatorname{supp} u^{(N)}_{\ell 2}$.

(v) $|v_{\ell j}|_m(\Omega_R,t) \leq C_{j,m}(1+2d_1K_{min})^{-\ell}|f|_{m+2j}(\Gamma_1\times\mathbb{R})$

where $K_{min}=\inf_x K_2(x)$.

Indeed, (ii) is a direct consequence of Lemma 2.5, and (iv) follows from Corollary of Lemma 2.7 and the definition of $u^{(N)}_{\ell 2}$. And to show (v), with the aid of Lemmas 2.4 and 2.6 we use the method of induction with respect to ℓ first for j=0 and m=0, and for j=0,m=1 use the method of induction with respect to ℓ. Thus we show first (v) for j=0, and for j=1,2, $\cdots\cdot$. (iii) is easily shown by taking account of the formula (2.30) and the equation (2.28).

Then applying Lemma 2.8 construct $\tilde{z}^{(N)}_{\ell}$ for the boundary data $u^{(N)}_{\ell 2}$. We see that

$$z^{(N)} = \Sigma^{\infty}_{\ell=1} (z^{(N)}_{\ell} + \tilde{z}^{(N)}_{\ell})$$

is well defined and satisfies (2.22), (2.23) and (2.25). Instead of (2.24) it holds that

$$|\Box z^{(N)}|_m(\Omega_R,t) \leq C_{N,m,R}\, e^{-(c_0-\varepsilon)(t-T)}\, |f|_{m+2N+40}.$$

Then by using the solution of the Cauchy problem and the Huygens' principle we can obtain $z^{(N)}$ verifying (2.22) $\sim$ (2.25). Here $c_0=\log(1+2d_1K_{min})/d_0$.

References

[1] Asakura F., On the Green's function $\Delta-\lambda^2$ with the boundary condition of the third kind in the exterior domain of a bounded obstacle, J.Math.Kyoto Univ., 12(1978), 615-625.

[2] Eskin G., Seminaire Goulaouic-Schwartz, 1979-1980.

[3] Ikawa M., Mixed problem for the wave equation with an oblique derivative boundary condition, Osaka J.Math.7(1970) 495-525.
[4] ———, Remarques sur les problèmes mixtes pour l'équation des ondes, Colloque international du C.N.R.S.,Astérisque 2 et 3,217-221.
[5] ———, Problèmes mixtes pas nécessairement L^2-bien posés pour les équations strictement hyperboliques, Osaka J.Math., 12(1975),69-115.
[6] ———, Sur les problèmes mixtes pour l'équation des ondes, Publ.Res.Inst.Math.Sci.Kyoto Univ.,10(1975),669-690.
[7] ———, Problèmes mixtes pour l'équation des ondes,Publ. Res.Inst.Math.Sci.Kyoto Univ.,12(1976),55-122.
[8] ———, Problèmes mixtes pour l'équation des ondes.II, Publ.Res.Inst.Math.Sci.Kyoto Univ.,13(1977),61-106.
[9] ———, Mixed problems for the wave equation.III, Exponential decay of solutions, Publ.Res.Inst.Math.Sci. Kyoto Univ.,14(1978),71-110.
[10] ———, Mixed problems for the wave equation.IV, J.Math. Kyoto Univ.,19(1979),375-411.
[11] ———, On the mixed problems for the wave equation in an interior domain.II, Osaka J.Math.,17(1980),253-279.
[12] ———, Decay of solutions of the wave equation in the exterior of two convex obstacles, to appear in Osaka J.Math.
[13] Kajitani K., A necessary condition for the well posed hyperbolic mixed problem with variable coefficients, J.Math. Kyoto Univ.,14(1974),231-242.
[14] Lax P.D. and Phillips R.S., Scattering theory, Academic press,New York,(1967).
[15] Keller J.B.Lewis R.M. and Seckler B.D., Asymptotic solution of some diffraction problems, Comm.Pure Appl.Math.,9(1956), 207-265.
[16] Ludwig D., Uniform asymptotic expansion of the fiels scattered by a convex object at high frequencies, Comm.Pure Appl.Math.,20(1967),103-138.
[17] Miyatake S., Mixed problems for hyperbolic equations of second order with first order complex boundary operators, Japan J.Math.,1(1975),111-158.
[18] ———, A sharp form of the existence theorem for hyperbolic mixed problems of second order, J.Math.Kyoto Univ. 17(1977),199-223.
[19] Morawetz C.S., Decay for solutions of the exterior problems for the wave equation, Comm.Pure Appl.Math.,28(1975),229-264.
[20] Morawetz C.S., Ralston J. and Strauss W.A., Decay of solutions of the wave equation outside nontrapping obstacles, Comm.Pure Appl.Math.,30(1977),447-508.
[21] Ralston J., Solutions of the wave equation with localized energy, Comm.Pure Appl.Math.,22(1969),807-823.
[22] Soga H., Mixed problems in a quater space for the wave

equation with a singular oblique derivative, Publ.Res.Inst. Kyoto Univ.,15(1979),357-399.

[23] ———, Mixed problems for the wave equation with a singular oblique derivative, Osaka J.Math.,17(1980),199-232.

[24] Walker H.F., Some remarks on the local energy decay of solutions of the initial-boundary value problem for the wave equation in unbounded domains,J.Diff.Equ.,23(1977) 459-471.

[25] Kohigashi N., On the strong hyperbolicity of mixed problems with constant coefficients in a quarter space,(in Japanese) Master's thesis, Osaka Univ., 1979.

MICROLOCAL ANALYSIS OF BOUNDARY VALUE PROBLEMS WITH APPLICATIONS TO DIFFRACTION

Kiyoomi Kataoka

Department of Mathematics, University of Tokyo

The subject is to study boundary value problems microlocally in the framework of hyperfunctions. We introduce the concept of mildness for hyperfunctions, which expresses a wide class of hyperfunctions having boundary values. Mild hyperfunctions have many interesting properties, for example, singularity properties, topological properties,..etc. By employing these properties, boundary value problems and the Green formula are naturally microlocalized. As applications, we have the solvability of boundary value problems for microdifferential operators semi-hyperbolic in one side of the boundary. We also have a kind of regularity up to the boundary for operators anti-semi-hyperbolic in one side of the boundary and for diffractive operators.

0. INTRODUCTION

In this article we survey the results obtained by the author in (13) and (14). Let $P(x,D)$ be a differential operator with analytic coefficients of order m defined in $M=\{x\in \mathbb{R}^n; |x|<r\}$. Suppose that $N=\{x\in M; x_1=0\}$ is non-characteristic with respect to P. Then Komatsu-Kawai (16) and Schapira (20) independently proved that any hyperfunction solution of $P(x,D)u=0$ in $\{x\in M; x_1>0\}$ has a unique extension $\tilde{u}(x)$ with support in $M_+=\{x\in M; x_1\geq 0\}$ and "the boundary values" $(f_0(x'),..,f_{m-1}(x'))\in\Gamma(N,\mathcal{B}_N)^m$ such that $\tilde{u}=u$ in $\{x\in M; x_1>0\}$ and

$$P\tilde{u} = \sum_{j=0}^{m-1} f_j(x')\delta^{(j)}(x_1).$$

H. G. Garnir (ed.), Singularities in Boundary Value Problems, 121–131.

Here we want to stress as for $u(x)$ that every normal derivative up to infinite order has a boundary value on $x_1=+0$. More precisely for every differential operator $J(x,D)$ (of finite or infinite order) defined in a neighborhood of the boundary, $J(x,D)u(x)$ has a boundary value $J(x,D)u|\{x_1=+0\}$ as a hyperfunction on N. In order to generalize this property apart from solutions of any differential equations, the author has introduced the concept of mildness in (13). Let $f(x)$ be a hyperfunction defined in $\{x\in M; 0<x_1<\delta, |x'-x_0'|<\delta\}$. Then $f(x)$ is said to be mild at $(0,x_0')\in N$ from the positive side of N if $f(x)$ is written as a sum of boundary values of holomorphic functions defined in such domains as

$$D(x_0',\eta_0';\varepsilon)=\{z\in\mathbb{C}^n; |z_1|<\varepsilon, \langle \operatorname{Im}z',\eta_0'\rangle >\varepsilon|\operatorname{Im}z'|+ \frac{1}{\varepsilon}(|\operatorname{Im}z_1|+(\operatorname{Re}(-z_1))_+), |z'-x_0'|<\varepsilon \}. \tag{0.1}$$

Here ε is a small positive number, η_0' is a unit vector of $\mathbb{R}^{n-1}$ and $(x)_+=x$ if $x\geqq 0$, $=0$ if $x<0$. More exactly speaking, there exist a small positive number ε and some holomorphic functions $F_1(z),\ldots, F_k(z)$ defined on $D(x_0',\eta_1';\varepsilon),\ldots,D(x_0',\eta_k';\varepsilon)$ such that the equality

$$f(x)=\sum_{j=1}^{k} F_j(x_1,x'+i0\eta_j') \tag{0.2}$$

holds on $\{0<x_1<\varepsilon, |x'-x_0'|<\varepsilon\}$. As for the uniqueness of such a representation, we have the edge of the wedge theorem of Martineau's type for these holomorphic functions. This is more delicate than the ordinary edge of the wedge theorem. In the case of $\dim M=2$, hence $\eta'=\pm 1$, it is stated as follows: Let $F_\pm(z)$ be any holomorphic functions defined on $D(0,\pm 1;\varepsilon)$ satisfying

$$f(x)=F_+(x_1,x_2+i0)-F_-(x_1,x_2-i0)=0$$

in $\{0<x_1<\varepsilon, |x_2|<\varepsilon\}$ (where $0<\varepsilon<<1$) as a hyperfunction. Then F_+ and F_- are both analytically extensible to a neighborhood of

$$\{z\in\mathbb{C}^n; \operatorname{Im}z=0, 0\leqq \operatorname{Re}z_1<\varepsilon, |\operatorname{Re}z_2|<\varepsilon\},$$

and coincide there with each other. In particular it is important that F_+ (or F_-) is prolonged analytically through the real boundary N. By employing the expression in (0.2) and this theorem, we can justify many operations on mild hyperfunctions. For example, the boundary value of a mild hyperfunction is defined by

$$f(+0,x')=\sum_{j=1}^{k} F_j(0,x'+i0\eta_j'). \tag{0.3}$$

Actually, the intersection of $D(x_0',\eta_0';\varepsilon)$ with $\{z_1=0\}$ constitutes a wedge in $\mathbb{C}^{n-1}$. Furthermore, as seen in the above argument, such a definition is justified by the edge of the wedge theorem. We denote by $\mathcal{B}_{N|M+}$ the sheaf on N of mild hyperfunctions from the positive side of N (in (13) and (14), we used the notation $\hat{\mathcal{B}}_{N|M+}$,

$\hat{\mathcal{C}}_{N|M+}$ instead of $\mathring{\mathcal{B}}_{N|M+}$, $\mathring{\mathcal{C}}_{N|M+}$ respectively). There are two important examples of mild hyperfunctions. One is any hyperfunction solution of a differential equation defined only in one side of a non-characteristic boundary. The other is any hyperfunction (defined through the boundary) with real analytic parameter which is transversal to the boundary. Following the usual procedure, we microlocalize $\mathring{\mathcal{B}}_{N|M+}$ on iS*N. That is, a sheaf $\mathring{\mathcal{C}}_{N|M+}$ is naturally constructed on the cotangential spherical bundle iS*N and it satisfies the following exact sequence.

$$0 \longrightarrow \mathcal{A}_M|_N \longrightarrow \mathring{\mathcal{B}}_{N|M+} \longrightarrow \pi_{N*}\mathring{\mathcal{C}}_{N|M+} \longrightarrow 0$$

Here $\pi_N : iS^*N \to N$ is the projection and $\mathcal{A}_M$ is the sheaf of analytic functions on M. Furthermore it is very useful that the sheaves $\mathring{\mathcal{C}}_{N|M+}$ and $\mathring{\mathcal{B}}_{N|M+}$ are soft sheaves on iS*N and N respectively. The main purpose of introducing "mildness" is to microlocalize the Green formula. We will survey this formula and the related topics in 1. To obtain this formula, we must prepare several operations on mild hyperfunctions and calculation rules for them. For example, the sheaf homomorphisms :

$$\text{Trace} : \mathring{\mathcal{B}}_{N|M+} \ni f(x) \longrightarrow f(+0,x') \in \mathcal{B}_N,$$

$$(\text{or} \quad \mathring{\mathcal{C}}_{N|M+} \longrightarrow \mathcal{C}_N),$$

$$\text{ext} : \mathring{\mathcal{B}}_{N|M+} \ni f(x) \longrightarrow f(x)Y(x_1) \in \mathcal{H}^0_{M+}(\mathcal{B}_M),$$

$$(\text{or} \quad \mathring{\mathcal{C}}_{N|M+} \longrightarrow \rho_*(\mathcal{C}_M|iS^*M \underset{M}{\times} N \setminus iS^*_N M))$$

are the most important; particularly ext is injective and ext(f) is often called "the canonical extension of f". Here $Y(x_1)$ is the Heaviside function, $\mathcal{C}_M$ ($\mathcal{C}_N$) is the sheaf of microfunctions in M (or N) and $\mathcal{H}^0_{M+}(\mathcal{B}_M)$ is the sheaf of hyperfunctions in M with support in M+ ; ρ is the following projection

$$\rho : iS^*M \underset{M}{\times} N \setminus iS^*_N M \ni (0,x';i\eta) \to (x';i\eta') \in iS^*N. \qquad (0.4)$$

The product operations are also important :

$$\mathring{\mathcal{B}}_{N|M+} \underset{\mathbb{C}}{\otimes} \cdots \underset{\mathbb{C}}{\otimes} \mathring{\mathcal{B}}_{N|M+} \longrightarrow \mathring{\mathcal{B}}_{N|M+},$$

$$\mathring{\mathcal{B}}_{N|M+} \underset{\mathbb{C}}{\otimes} \mathcal{H}^0_{M+}(\mathcal{B}_M)|_N \longrightarrow \mathcal{H}^0_{M+}(\mathcal{B}_M)|_N.$$

In these definitions, however, some assumptions on singular support are required; here, we mean by the singular support the support in $\mathring{\mathcal{C}}_{N|M+}$, or for sections of $\mathcal{H}^0_{M+}(\mathcal{B}_M)|_N$, we use the "reduced singular support", that is,

$$\rho\text{-SS}(f) = \rho(SS(f) \cap \{x_1 = 0\} \setminus \{\pm i dx_1\}).$$

As the other operations we have "substitution", "integration along fibers", which are operations with respect to the variables tangential to the boundary. Furthermore the calculation rules are important , for example, the Leibniz rule for calculating the derivatives of products, commutativities among the operations etc.. However these rules are very simple because we can treat mild hyperfunctions as if they are hyperfunctions (defined through the boundary) depending analytically on the parameter transversal to the boundary.

As for the topological properties of mild hyperfunctions, it is proven that for any mild hyperfunction $f(x_1,x')$ with proper support in x', the limit of $f(x_1,x')$ as $x_1 \to +0$ exists in the space of analytic functionals of x'. Furthermore this limit coincides with Trace(f).

The boundary value problems are naturally microlocalized by the way of $\mathring{C}_{N|M+}$. Let $P(x,D)$ be a microdifferential operator of the form

$$D_1^m + P_1(x,D')D_1^{m-1} + \ldots\ldots + P_m(x,D') \tag{0.5}$$

(where $D_j = \partial/\partial x_j$) of order m defined in a neighborhood of $(0,x_0';_*,$ $i\eta_0') \in iS^*M$. Then the sheaf homomorphism :

$$\mathring{C}_{N|M+} \ni u(x) \longrightarrow (Pu(x), Bu(x')) \in \mathring{C}_{N|M+} \oplus C_N^m, \tag{0.6}$$

$$Bu(x') = (u(+0,x'),\ldots,(D_1^{m-1}u)(+0,x'))$$

is well-defined on a neighborhood of $(x_0';i\eta_0')$ and this is injective. On the other hand this homomorphism is not surjective in general. In fact, as shown by Schapira (21) and Kataoka (12), the boundary values of solutions of Pu=0 always satisfy several equations if P has any elliptic factor. For semihyperbolic operators, the homomorphism in (0.6) is surjective.

Definition : Let $P(x,D)$ be a microdifferential operator of the form in (0.5). Then P is said to be "semihyperbolic in $\{x_1>0\}$ at $(x_0';i\eta_0') \in iS^*N$" if the principal symbol

$$\sigma(P) = \zeta_1^m + p_1(x,i\eta')\zeta_1^{m-1} + \ldots + p_m(x,i\eta')$$

of P does not vanish on $\{(x;\zeta_1,i\eta') \in \mathbb{R}^n \times \mathbb{C} \times iS^{n-1}; 0 \leq x_1 < \varepsilon, |x'-x_0'| < \varepsilon,$ $|\eta'-\eta_0'| < \varepsilon, \mathrm{Re}\zeta_1 > 0\}$ for sufficiently small $\varepsilon > 0$.

Theorem A : Let $P(x,D)$ be a microdifferential operator as above. Assume that $P(x,D)$ is semihyperbolic in $\{x_1>0\}$ at $(x_0';i\eta_0')$ $\in iS^*N$. Then the sheaf homomorphism in (0.6) is bijective in a neighborhood of $(x_0';i\eta_0')$.

Theorems of this type have been obtained by many authors (2),

(9),(25),(5),(18), though they assumed that P is a differential operator or P is semihyperbolic in both sides of the boundary or Pu=f=0. Another subject of the microlocal analysis of boundary value problems is to study the propagation of the microanalyticity of solutions up to the boundary for several operators. Recall the following injection :

$$\text{ext} : \mathring{\mathcal{C}}_{N|M+} \ni u(x) \rightarrow u(x)Y(x_1) \in \rho_*(\mathcal{C}_M|iS^*M \underset{M}{\times} N \setminus iS^*_N M).$$

Hence, particularly a germ of $\mathring{\mathcal{C}}_{N|M+}$ at $(x'_0;i\eta'_0)$ defines a section of microfunctions on $\{(x_1,x';i\eta_1,i\eta')\in iS^*M; 0<x_1<\varepsilon, |x'-x'_0|<\varepsilon, |\eta'-\eta'_0|<\varepsilon, \eta_1\in\mathbb{R}\}$ for small $\varepsilon>0$. This section of microfunctions is just the inner singularity of u(x). Therefore we can formulate the propagation of the microanalyticity of solutions up to the boundary as follows : Let P(x,D) be a microdifferential operator of the form in (0.5). Then we say that the microanalyticity of any solution in $\mathring{\mathcal{C}}_{N|M+}$ of Pu=0 propagates up to the boundary at $(x'_0;i\eta'_0)\in iS^*N$ if there is no $\mathring{\mathcal{C}}_{N|M+}$-solution of Pu=0 at $(x'_0;i\eta'_0)$ without inner singularity except for 0-solution. In connection with this subject, we have the following theorems.

Theorem B : Let P(x,D) be a microdifferential operator of the form in (0.5). Assume that P(x,D) is semihyperbolic in $\{x_1>0\}$ at $(x'_0;-i\eta'_0)\in iS^*N$. Then the microanalyticity of any $\mathring{\mathcal{C}}_{N|M+}$-solution of Pu=0 propagates up to the boundary at $(x'_0;i\eta'_0)$.

This is obtained as the dual version of Theorem A by employing the microlocal Green formula. This is a generalization of the theorems of Kaneko (5) and of Schapira (21).

Theorem C : Let P(x,D) be a second-order microdifferential operator of the form in (0.5) with real principal symbol defined in a neighborhood of $p_0=(0,x'_0;i\eta_0)\in iS^*M$. Assume that P is diffractive at p_0 in the positive side of N, that is,

$$\begin{cases}\sigma(P)(p_0)=\{\sigma(P),x_1\}(p_0)=0, \ \{\{\sigma(P),x_1\},\sigma(P)\}(0,x'_0,\eta_0)<0,\\ \text{and} \quad d\sigma(P)\wedge dx_1\wedge\omega(p_0)\neq 0.\end{cases} \qquad (0.7)$$

Here $\sigma(P)$ is the principal symbol of P, { , } is the poisson bracket :

$$\{f,g\}(x,\xi)=\sum_{j=1}^{n}\left(\frac{\partial f}{\partial\xi_j}\frac{\partial g}{\partial x_j}-\frac{\partial f}{\partial x_j}\frac{\partial g}{\partial\xi_j}\right),$$

and $\omega=\xi_1dx_1+\ldots+\xi_ndx_n$ is the fundamental 1-form. Then the microanalyticity of any $\mathring{\mathcal{C}}_{N|M+}$-solution of Pu=0 propagates up to the boundary at $(x'_0;i\eta'_0)$. In other words, all the boundary values of a solution of Pu=0 are microanalytic at $(x'_0;i\eta'_0)$ if and only if the solution vanishes as a section of microfunctions on a neighborhood of $\gamma\setminus\{p_0\}$ ($\subset\{x_1>0\}$). Here γ is the bicharacteristic strip through p_0.

In a recent paper of Sjöstrand (24) a similar theorem is obtained . However he considers only solutions satisfying a boundary condition. Therefore his result is weaker than ours, though the methods are completely different from each other. To prove this theorem, we need almost all the results on mild hyperfunctions, Theorem B and Bony's and Schapira's results on non-microcharacteristic operators ((1),(22)). We will give a brief sketch of this theorem in 2.

Theorem D : Let $P(x,D)$ be a microdifferential operator of the form in (0.5) defined in a neighborhood of $(0,x_0';*,i\eta_0')\in iS^*M$. Let $N'=\{x\in M;x_1-\phi(x')=0\}$ be another analytic hypersurface such that $\phi(x_0')=0$, $\mathrm{grad}\phi(x_0')=0$ and $\phi(x')\geqq 0$ for every x'. Write $P(x,D_x)$ in the coordinate system $y_1=x_1-\phi(x'),y'=x'$. Then $P(x,D_x)$ is decomposed in a neighborhood of $T=\{(x;i\eta dx)\in iS^*M;x_1=0,x'=x_0',\eta'=\eta_0',\eta_1\in\mathbb{R}\}=\{(y;i\xi dy);y_1=0,y'=x_0',\xi'=\eta_0',\xi_1\in\mathbb{R}\}$ as follows :

$$P(x,D) = R(y,D_y)S(y,D_y).$$

Here R is invertible on T and S is a microdifferential operator of order m written in the form

$$S(y,D_y) = D_{y_1}^m + S_1(y,D_{y'})D_{y_1}^{m-1} + \dots + S_m(y,D_{y'}).$$

Assume that the microanalyticity of any $\mathring{C}_{N'|M'_+}$-solution of $Su=0$ propagates up to the boundary N' at $(y';i\xi')=(x_0';i\eta_0')$. Here $M_+'=\{y_1\geqq 0\}$. Then the microanalyticity of any $\mathring{C}_{N|M_+}$-solution of $Pu=0$ propagates up to the boundary N at $(x';i\eta')=(x_0';i\eta_0')$.

By combining Theorem D with Theorem B, we obtain a new class of the operators which admit the propagation of the microanalyticity of solutions up to the boundary. For example,

$$P = D_1^2 - (x_1-x_2^m)D_3^2 \quad (\text{with } m\geqq 2)$$

on $\{(x_2,x_3;i\eta_2,i\eta_3)\in iS^*N;x_2=0,\eta_3\neq 0\}$. As for another important class of such operators, we know the result on non-microcharacteristic operators by Schapira (22).

1. THE MICROLOCAL GREEN FORMULA

In (6), Kaneko employed the Green formula effectively to prove a theorem on the singularities of the boundary values of hyperfunction solutions for some boundary value problems. His idea is the following : Construct a suitable fundamental solution with parameter for the adjoint boundary value problem. Next, make a kind of inner products between this fundamental solution and the given solution. Then from this "inner product" we will get some information on the singularities of the boundary values of the given solution. The difficulty is in defining this "inner product". Because, in his case, the given solution and the fundamental solution are both defined only in the positive side of the boundary, but we must define canonically the product of them as a hyperfunction with support in $\{x_1 \geq 0\}$. In this situation, he saw through the mildness of the fundamental solutions which he had constructed and the calculation rules for mild hyperfunctions. The author has been strongly stimulated by this idea. Let us introduce the microlocal Green formula.

Theorem : Let $P(x,D)$ be a microdifferential operator of order m of the form in (0.5) defined in a neighborhood of $(0,x_0';*,i\eta_0') \in iS^*M$. Let $u(x)$ be a germ of $\mathring{\mathcal{C}}_{N|M_+}$ at $(x_0';i\eta_0')$ such that $Pu=0$ and let $v(x,y')$ be a germ of $\mathring{\mathcal{C}}_{N'|M'_+ n-1}$ at $(x_0',x_0';-i\eta_0',i\eta_0')$ such that $P^*(x,D_x)v=0$. $M'_+=M_+\times\mathbb{R}^{n-1}$ and $N'=N\times\mathbb{R}^{n-1} \ni (x',y')=(x_2,\dots,x_n,y_2,\dots,y_n)$ and $P^*(x,D)$ is the formal adjoint of $P(x,D)$. Then we mean by the "microlocal Green formula" the one

$$
\begin{aligned}
0 &= \int P(x,D)u(x)\cdot \mathrm{ext}(v(x,y'))dx \\
&= \int u(x)\cdot P^*(x,D)\mathrm{ext}(v(x,y'))dx \\
&= \sum_{j=0}^{m-1} (-1)^j \cdot \int (D_1^j u)(+0,x')w_j(x',y')dx',
\end{aligned}
$$

which holds at $(y';i\tau')=(x_0';i\eta_0')$ in the sense of microfunctions in y', where we put

$$
P^*(x,D_x)\mathrm{ext}(v(x,y')) = \sum_{j=0}^{m-1} w_j(x',y')\delta^{(j)}(x_1).
$$

Two conditions are required for the microlocal Green formula. One is the well-definedness of the products $P_j(x,D')D_1^{m-j}u(x)\cdot \mathrm{ext}(v)$, etc. as microfunctions in (x,y') at $(x_0',x_0';0,i\eta_0')$. The other is the integrability of the microfunctions

$$
P_j(x,D')D_1^{m-j}u(x)\cdot \mathrm{ext}(v(x,y')),
$$

etc.. For example, the following is a sufficient condition for satisfying these requirements. "$SS(w_j(x',y')) \subset \{(x',y';i\eta',i\tau') \in iS^*N'; x'=y', \eta'+\tau'=0\}$ for every j and $SS(\mathrm{ext}(u)) \subset \{(x;i\eta) \in iS^*M; x_1 = 0\}$" (the second statement is equivalent to the following : u has no inner singularity near $(x_0';i\eta_0')$).

Actually the assumptions on u assure the conditions on products and the integrability of each term. This theorem was used effectively in the proof of Theorem B. It will be also available for the microlocal analysis of mixed problems.

2. A BRIEF SKETCH OF THEOREM C

Let $P(x,D)$ be a second-order microdifferential operator of the form

$$P = D_1^2 + P_1(x,D')D_1 + P_2(x,D')$$

with real principal symbol defined in a neighborhood of $p_0=(0,x_0';i\eta_0)\in iS^*M$. Then P is said to be "diffractive at p_0 in the positive side of N" if (0.7) is satisfied. Actually in this case, the bicharacteristic strip γ passing through p_0 is strictly tangent to $\{x_1=0\}$ from $\{x_1>0\}$. Now by using a quantized contact transformation which preserve the boundary, we reduce P and p_0 to the following form

$$P = D_1^2 - (x_1-x_2)A(x,D'),\ p_0= (0;idx_n) \qquad (2.1)$$

In particular n must be more than 3. Here $A(x,D')$ is a second-order microdifferential operator defined in a neighborhood of p_0 such that $\sigma(A)(x,\xi')$ is real and positive for every real x and ξ'. The simplest example is Sato's operator :

$$P = D_1^2 - (x_1-x_2)D_3^2.$$

At first, we assume that $A(x,D')$ is a differential operator defined in $\{|x|<\varepsilon\}$ and that $\sigma(A)(x,\xi')\geqq 0$ holds in $\{|x|<\varepsilon,\xi'\in\mathbb{R}^{n-1}\}$. Furthermore we assume that $u(x)$ is a hyperfunction solution of $Pu=0$ in $U=\{0<x_1<\varepsilon,|x|<\varepsilon\}$. Note that P is hyperbolic in $\{x_1-x_2>0,|x|<\varepsilon'\}$ for some smaller $\varepsilon'>0$ to $d(x_1-x_2)$-codirection. Here we can extend u to $\{x_1-x_2>0,|x|<\delta\}$ as a solution of $Pu=0$ for sufficiently small $\delta>0$. Consequently u is a solution of $Pu=0$ defined on

$$V = \{x\in\mathbb{R}^n;|x|<\delta\}\cap\{x_1>0 \text{ or } x_1-x_2>0\}.$$

As the simplest case, we consider the case that u is a real analytic solution in U. Then, easily to see, the extension of u to V is also real analytic. Therefore by Theorem B, we know that all the boundary values of u on $\{x_1-x_2=+0\}$ are real analytic in a neighborhood of the origin. In particular, by the Cauchy-Kowalevsky theorem, u is extended through $\{x_1-x_2=+0\}$ near the origin as an analytic function. Hence all the boundary values of u on $\{x_1=+0\}$ are real analytic at the origin. This shows that the analyticity of solutions propagates up to the boundary. However this case is too simple. There are two difficulties to proceed to the general cases. The first one is in extending the solution to the domain like V. For the solution and the operator are defined only microlocally, in particular, the positivity of $\sigma(A)(x,\xi')$ holds only microlocally. The other one is in relating the boundary values on $x_1=+0$ with the boundary values on $x_1-x_2=+0$. The first one has been overcome by the following theorem.

Theorem : Let $u(x)$ be any $\mathring{\mathcal{C}}_{N|M+}$-solution of $Pu=0$ at $p_0'=(0;idx_n)$ $\in iS^*N$. Then for every neighborhood W of p_0', there is a section $\tilde{u}$ of $\mathring{\mathcal{C}}_{N|M+}$ on W with support compactly contained in W such that $P\tilde{u}=0$ on W and $\tilde{u}=u$ at p_0'.

It is important that $\tilde{u}(x)$ is a globally defined solution. The second difficulty has been overcome by the following argument: Rewrite V as follows.

$$V=\{|x|<\delta\}\cap\{x_1-tx_2>0 \text{ for some } t\in(0,1)\}.$$

Set

$$D=\{(x,t)\in\mathbb{R}^n\times\mathbb{R};\ |x|<\delta,\ x_1-tx_2>0,\ 0<t<1\}. \qquad (2.2)$$

Then we can identify the solution u with the solution in D of the following system of equations :

$$\begin{cases} (D_{x_1}^2-(x_1-x_2)A(x,D_{x'}))v(x,t)=0, \\ \\ D_t v(x,t)=0. \end{cases} \qquad (2.3)$$

Note that the boundary values of v on $x_1-tx_2=+0$ combine the boundary values of u on $x_1=+0$ with the boundary values of u on $x_1-x_2=+0$. To speak more explicitly, we consider $v_0(x',t)=v|\{x_1-tx_2=+0\}$ and $v_1(x',t)=(D_1v)|\{x_1-tx_2=+0\}$. Then the boundary values of the pair (v_0,v_1) on $t=+0$ or $t=1-0$ coincide with the boundary values of u on $x_1=+0$ or $x_1-x_2=+0$ respectively. Here it is the most important that the pair (v_0,v_1) satisfy a system of microdifferential equations of the form

$$\{D_tI_2 - B(t,x',D_{x'})\}\begin{pmatrix} v_0 \\ v_1 \end{pmatrix}=0 \qquad (2.4)$$

in $\{0<t<1\}$. Here B is the 2×2-matrix of microdifferential operators. This is easily obtained as the tangential equations for (2.3). In fact we know from the explicit form of B that this system of equations is just the systems of equations treated by Bony (1) and Schapira (22). So we can directly apply their results to our case. Simply speaking, our idea is the following : The microanalyticity of any solution of (2.4) propogates from $t=1-0$ up to $t=+0$ along the involutory submanifold

$$L=\{(t,x';i\tau dt+i\eta' dx');x_2=0,\tau=0\}.$$

Hence the microanalyticity of the boundary values of u on $x_1-x_2=+0$ assures the microanalyticity of the boundary values of u on $x_1=+0$. Thus the second difficulty has been overcome. The other parts of the proof are essentially the same as in the simplest case.

REFERENCES

(1) Bony,J.M.:"Extension du theorème de Holmgren", Sem. Goulaouic-Schwartz, 1975-76, exposé 17.
(2) Bony,J.M., Schapira,P.:"Solutions hyperfonctions du problème de Cauchy", Lecture Notes in Math., Springer 287(1973), pp. 82-98.
(3) Bony,J.M, Schapira,P.:"Propagation des singularités analytiques pour les solutions des équations aux derivées partielles", Ann. Inst. Fourier 26(1976), pp. 81-140.
(4) Friedlander,F.G., Melrose,R.B.:"The wavefront set of a simple initial-boundary value problem with glancing rays II", Math. Proc. Camb. Phil. Soc., 81(1977), pp. 97-120.
(5) Kaneko,A.:"Singular spectrum of boundary values of solutions of partial differential equations with real analytic coefficients", Sci. Pap. Coll. Gen. Educ. Univ. Tokyo 25(1975), pp. 59-68.
(6) Kaneko,A.:"Estimation of singular spectrum of boundary values for some semihyperbolic operators", J. Fac. Sci. Univ. Tokyo 27, No.2(1980), pp. 401-461.
(7) Kashiwara,M., Kawai,T.:"On the boundary value problem for elliptic system of linear differential equations, I", Proc. Japan Acad. 48(1972), pp. 712-715.
(8) Kashiwara,M., Kawai,T.:"On the boundary value problem for elliptic system of linear differential equations, II", Proc. Japan Acad. 49(1973), pp 164-168.
(9) Kashiwara,M., Kawai,T.:"Microhyperbolic pseudo-differential operators I", J. Math. Soc. Japan 27, No. 3(1975), pp. 359-404.
(10) Kashiwara,M., Oshima,T.:"Systems of differential equations with regular singularities and their boundary value problems", Ann. Math. 106(1977), pp. 145-200.
(11) Kataoka,K.:"On the theory of Radon transformations of hyperfunctions", to appear in J. Fac. Sci. Univ. Tokyo.
(12) Kataoka,K.:"A micro-local approach to general boundary value problems", Publ. Res. Inst. Math. Sci. Kyoto Univ. 12,Suppl. (1977), pp. 147-153.
(13) Kataoka,K.:"Micro-local theory of boundary value problems I", J. Fac. Sci. Univ. Tokyo 27, No. 2(1980), pp. 355-399.
(14) Kataoka,K.:"Micro-local theory of boundary value problems II, - Theorem on Regularity up to the Boundary for Reflective and Diffractive Operators-", to appear in J. Fac. Sci. Univ. Tokyo.

(15) Kataoka,K.:"Microlocal theory of boundary value problems III", in preparation.
(16) Komatsu,H., Kawai,T.:"Boundary values of hyperfunction solutions of linear differential equations", Publ. Res. Inst. Math. Sci. 7(1971), pp. 95-104.
(17) Melrose,R.B.:"Equivalence of glancing hypersurfaces", Inv. Math. 37(1976), pp. 165-191.
(18) Ôaku,T.:"Micro-local Cauchy problems and local boundary value problems", Proc. Japan Acad. 55(1979), pp. 136-140.
(19) Sato,M., Kawai,T., Kashiwara,M.:"Microfunctions and pseudo-differential equations", Lecture Notes in Math.,Springer 287 (1973), pp. 265-529.
(20) Schapira,P.:"Hyperfonctions et problèmes aux limites elliptiques", Bull. Soc. Math. France 99(1971), pp. 113-141.
(21) Schapira,P.:"Propagation au bord et réflexion des singularités analytiques des solutions des équations aux dérivées partielles", Sem. Goulaouic-Schwartz 1975-76, exposé 6.
(22) Schapira,P.:"Propagation au bord et réflexion des singularités analytiques des solutions des équations aux dérivées partielles II", Sem. Goulaouic-Schwartz 1976-77, exposé 9.
(23) Sjöstrand,J.:"Sur la propagation des singularités analytiques pour certains problèmes de Dirichlet", C. R. Acad. Sc. Paris t. 288(1979), pp. 673-674.
(24) Sjöstrand,J.:"Propagation of analytic singularities for second-order Dirichlet problems. II", Comm. Partial Differential Equations 5(2) (1980), pp. 187-207.
(25) Tahara,H.:"Fuchsian type equations and Fuchsian hyperbolic equations", Jap. J. Math. 5, No. 2(1979), pp. 245-347.

TRANSFORMATION METHODS FOR BOUNDARY VALUE PROBLEMS

Richard B. Melrose

Massachusetts Institute of Technology

The theory of pseudodifferential and Fourier integral operators provides a very powerful tool for the examination of many questions concerning linear partial differential operators on manifolds. These methods have also been successfully applied to boundary value problems. Certain aspects of the standard theory of boundary value problems from this point of view are unsatisfactory however. For example the usual notion of wavefront set is not well-adapted to the microlocalization of distributions near the boundary. In particular in the treatment of the propagations of singularities for boundary value problems it was found necessary to introduce a new notion of wavefront set:

$$WF_b(u) \subset T^*M \cup T^*\partial M$$

which was not universally applicable. In these lectures it is shown that WF_b has a natural extension to all distributions on a manifold with boundary.

The emphasis here is on material not covered in [1]. Namely the direct definition of the compressed cotangent bundle which carries the extended notion of wavefront set:

$$WF_b(u) \subset \tilde{T}^*M \setminus 0$$

and applications of the related calculus of 'totally character-

---

[1] Differential boundary value problems of principal type. in Seminar on singularities of solutions of linear partial differential equations Ed L. Hörmander P.U.P 1979.

H. G. Garnir (ed.), Singularities in Boundary Value Problems, 133–168.

istic' pseudodifferential and Fourier integral operators. Details of the calculus and properties of $\tilde{T}^*M$ can be found in [2]. The first two applications, to totally characteristic operators of real principal type and to boundary value problems with gliding rays will be contained in [3]. Treatment of the degenerate elliptic operators in the last lecture is part of work in progress on the analysis of strictly pseudoconvex domain and will appear elsewhere.

Contents:

I. Compressed cotangent bundle

Let M be a C^∞ manifold with boundary. In examining the behaviour of singularities of distributions on M, in particular of solutions to boundary value problems one needs first to define what a smooth distribution should be. It is important to realize that choice is involved here, even though the claims of

$$C^\infty(M) = \{f: M \longrightarrow \mathbb{R};\ f \text{ is continuous with all derivatives}\}, \tag{I.1}$$

seem at first overwhelming. There are certain drawbacks, at least for the philosophical, with this choice. First, if $f \in C^\infty(M)$ and P is a differential operator then in order to consider Pf to be continuous one should consider $C^\infty(M) \subset \mathcal{D}'(\mathring{M})$ as a space of distributions in the interior with extension properties to the closure. This involves treating ∂M as an 'infinity' rather than as a finite boundary. Secondly, and more candidly, it is not adequate for the purpose of these lectures.

Now, for the purist , the space of smooth distributions on M is clearly

$$\dot{C}^\infty(M) = \{f \in C^\infty(M);\ f \text{ and all derivatives vanish at the boundary}\}. \tag{I.2}$$

---

[2] Transformation of boundary value problems I. to appear.

[3] Transformation of boundary value problems II. in preparation.

Indeed, on a compact manifold without boundary, M', one can characterize the condition $u \in C^\infty(M')$ by demanding

$$Pu \in C^o(M') \quad \forall \text{ differential operators } P.$$

The space $C^o(M')$ is not even playing an important rôle here and can be replaced by any 'finite regularity space', for example some Sobolev space $H^s(M')$. If $\partial M \neq \emptyset$ and we apply P to $u \in C^\infty(M)$, including the boundary jumps then

$$\text{(I.3)} \qquad Pu \in H^s(M) \ \forall P \in \mathrm{Diff}(M) \Leftrightarrow u \in \dot{C}^\infty(M).$$

It is perhaps in this sense that $\dot{C}^\infty(M)$ is the ultimate space of smooth distributions. However, the difficulties with such a choice are manifest - it is hard to prove reasonable results.

Now, to formalize the choice made during these lectures we must first recall the definition of the space, $\dot{\mathcal{D}}'(M)$, of supported distributions on M. The line bundle Ω of densities on M is certainly C^∞ over M so we can define $C^\infty(M,\Omega)$ by complete analogy with (I.1), and topologize it with the usual C^k-seminorms, assuming for simplicity that M is compact. Then

$$\dot{\mathcal{D}}'(M) = (C^\infty(M, \Omega))' .$$

Since densities can be invariantly integrated we have

$$\text{(I.4)} \qquad C^\infty(M) \hookrightarrow \dot{\mathcal{D}}'(M),$$

in the usual way. Differential operators act on $\dot{\mathcal{D}}'(M)$, but this action does not commute with (I.4) if the action of P on $C^\infty(M)$ disregards boundary terms.

Returning to (I.3) we observe that this condition can be replaced by the demand

$$\text{(I.5)} \qquad V_1^{k_1} \cdots V_N^{k_N} u \in H^s(M) \quad \forall k_j$$

where $V_j \in C^\infty(M,TM)$ are a finite set of vector fields spanning TM everywhere, assuming M compact. When $\partial M \neq \emptyset$ we have already noted that (I.5) leads to the space $\dot{C}^\infty(M)$. However we can consider the subspace

$$\text{(I.6)} \qquad \mathcal{V}(M) = \{V \in C^\infty(M,TM);\ V \text{ is tangent to } \partial M\}.$$

(I.7) Definition. The space $\dot{A}(M) \subset \dot{\mathcal{D}}'(M)$ consists of those distributions u for which there exists a fixed $s \in \mathbb{R}$ such that (I.5) holds for every finite set of $V_j \in \mathcal{V}$.

Now, it turns out that $\dot{A}(M)$ is a well-known space of distribution. Following, at least in spirit, Hörmander [1] we can introduce the space of Lagrangian distributions associated to the conormal bundle $N^*\partial M \subset \dot{T}^*M$. This space $I^\bullet(M,N^*\partial M)$ consists of distributions $u \in \dot{\mathcal{D}}'(M)$ which are C^∞ in $\mathring{M}$ and in local coordinates (x,y), $y \in \mathbb{R}^n$ $x \geq 0$ on M $(\dim M = n+1)$ take the form:

$$u(x,y) = (2\pi)^{-1} \int e^{ix\xi} a(y,\xi)\,d\xi \qquad x < \varepsilon \tag{I.8}$$

where $a(y,\xi)$ is a symbol of type $S^m_{1,0}(\mathbb{R}^n,\mathbb{R})$.

<u>Theorem</u> $\dot{A}(M) = I^\bullet(M,N^*\partial M)$.

<u>Proof.</u> It should be noted that a standard identification has been used in the statement above. Elements of $I^\bullet$ are distributions defined on a covering of M by open sets, including open extensions across the boundary, but always having support in M. It is shown in [2] that such distributions can be identified with elements of $\dot{\mathcal{D}}'(M)$.

First we show $I^\bullet(M,N^*\partial M) \subset \dot{A}(M)$. The condition (I.5), with the $V_j \in \mathcal{V}$, is clearly local in nature so only needs to be verified in any local coordinates near ∂M, (x,y) as above. Then $V \in \mathcal{V}$ is of the form

$$V = \sum_{1\leq j\leq n} \alpha_j(x,y) D_{y_j} + \alpha_0(x,y)\, x\, D_x. \tag{I.9}$$

A simple inductive argument shows that (I.5) need only be checked for $V_j = D_{y_k}$ or $x D_x$. If u is of the form (I.8) then

$$D^\alpha_y (xD_x)^\kappa u = (2\pi)^{-1} \int e^{ix\xi} (-D_\xi \xi)^k D^\alpha_y a(y,\xi). \tag{I.10}$$

The symbol here is of the same order as m and since any distribution (I.8), with m fixed, lies in some fixed Sobolev space this proves the first inclusion.

The converse is similar, simply defining a by (I.8) when u is localized near $x = 0$. Details can be found in [3].

It should be noted that this theorem really has nothing to do with manifolds with boundary. For completeness a more general result of more or less "folk theorem" status is proved in the appendix.

As a consequence of this theorem we know that

$$WF(u) \subset N^*\partial M \quad \forall u \in \dot{A}(M) \tag{I.11}$$

and indeed, (I.5) and Definition I.7, are just quantitative strengthenings of this. The position adopted here is that elements of $\dot{A}(M)$ are largely ignorable, for the purist we note how to improve matters:

$$\text{If } u \in \dot{A}(M),\ WF(u) = \phi \iff u \in \dot{C}^\infty(M). \tag{I.12}$$

Next we turn to the geometric investigation of the space

$\mathcal{V}$ of C^∞ vector fields tangent to ∂M. Note from (I.9) that in any local coordinates, an element of $\mathcal{V}$ is specified by precisely n C^∞ functions, individually quite arbitrary. This has a more abstract formulation.

(I.13) Proposition. There is a natural vector bundle $\tilde{T}M$ over M and vector bundle mapping

$$\tilde{\iota}: \tilde{T}M \longrightarrow TM$$

such that the induced map $\tilde{\iota}_* : C^\infty(M,\tilde{T}M) \longrightarrow C^\infty(M,TM)$ is an identification of its range with $\mathcal{V}$.

Proof. We simply define $\tilde{T}M$ and leave all other details to the reader (see [3]). For $m \in M$ we define the fibre at m to be

$$\tilde{T}_m M = \mathcal{V}/\sim$$

where $\sim$ is an equivalence relation on $\mathcal{V}$. If $m \in \mathring{M}$ then

(I.14) $\quad V \sim W \iff Vf(m) = Wf(m) \quad \forall f \in C^\infty(M).$

If $m \in \partial M$ we add to (I.14) the extra condition

(I.15) $\quad V \sim W$ (I.14) and $d(Vg)(m) = d(Wg)(M)$

$$\forall g \in C^\infty(M) \text{ with } g \equiv 0 \text{ on } \partial M.$$

It should be noted that $\tilde{\iota}$ is by no means a vector bundle isomorphism. In fact its rank is $n = \dim M$ except on the fibres $\tilde{T}_m M$, $m \in \partial M$ where it has rank n-1. Clearly the range of $\tilde{\iota}$ is precisely

$$T\mathring{M} \cup T\partial M \subset TM.$$

Since an element $V \in \mathcal{V}$ can be considered as a section of $\tilde{T}M$ it is equivalent to a linear function on the dual $\tilde{T}^*M = (\tilde{T}M)^*$. This function is the symbol of V as a first order differential operator and we shall see later that a natural theory of micro-localization on the space $\tilde{T}^*M$ is readily developed.

The dual to the mapping of Proposition I.13 is a C^∞ vector bundle mapping

$$\tilde{\iota}^*: T^*M \longrightarrow \tilde{T}^*M$$

with the same rank properties as $\tilde{\iota}$. In view of (I.16) it is easy to see:

(I.16) Lemma. The range of ι^* is canonically identified with $T^*\partial M \cup T^*\mathring{M}$.

This latter topological space is the carrier of the notion, WF_b, used for example in [4]. For the moment we consider the subspace

$$C^\infty(\tilde{T}^*M) \subset C^\infty(T^*M), \tag{I.17}$$

where we delete the pull-back notation as cumbersome. The form (I.9) of $V \in \mathcal{V}$ in local coordinates (x,y) shows that such coordinates induce canonical dual coordinates (x,y,λ,η) in $\tilde{T}^*M$ in which a point $p \in \tilde{T}^*M$ is of the form

$$p = \sum_{1 \leq j \leq n} \eta_j dy^j + \frac{\lambda}{x} dx.$$

If (x,y,ξ,η) are the usual canonically dual coordinates in T^*M

$$\tilde{\imath}^*(x,y,\xi,\eta) = (x,y,x\xi,\eta)$$

i.e. $\lambda = x\xi$. This easily leads to a characterization of (I.17): $\phi \in C^\infty(T^*M)$ is in $C^\infty(\tilde{T}\,^*M)$ if, and only if, in and canonically dual coordinates (x,y,ξ,η)

$$\partial_x^k \phi\Big|_{x=0} \text{ is a polynomial in } \xi \text{ of degree at most } k. \tag{I.18}$$

To conclude this first lecture we shall consider further important spaces inside $\dot{\mathcal{D}}'(M)$. The traditional method of functional analysis is always to consider the dual to any spaces which might appear. Since $\dot{A}(M)$ consist of distributions which are smooth except in the normal direction its dual $\dot{A}'(M)$ may be expected to have elements smooth in the normal direction. First we need to discuss the topology of $\dot{A}(M)$. The locally convex topology we use has first the seminorms of $C^\infty(\mathring{M})$. Next we note that for each fixed s the space $\dot{A}^{(s)}(M)$ satisfying (I.5) for $V_j \in \mathcal{V}$ and that fixed s is topologized by the seminorms

$$\| V_1^{k_1} \cdots\cdots V_n^{k_n} u \|_s .$$

Then we take the inductive limit over s.

(I.19) <u>Lemma.</u> $\dot{C}^\infty(M) \subset \dot{A}(M)$ is dense and continuous.

Tensoring everything with the density bundle Ω over M we recall that the dual of $\dot{C}^\infty(M,\Omega)$ is the space $\mathcal{D}'(M)$ of distributions on $\mathring{M}$ extendable across the boundary, i.e., of finite order and growth at the boundary. Restriction to the interior gives an exact sequence

(I.20) $$0 \longrightarrow \dot{\mathcal{D}}'(M,\partial M) \hookrightarrow \dot{\mathcal{D}}'(M) \longrightarrow \mathcal{D}'(M) \longrightarrow 0$$

where

$$\dot{\mathcal{D}}'(M,\partial M) = \{u \in \dot{\mathcal{D}}'(M);\ \mathrm{supp}\ u \subset \partial M\}.$$

In view of Lemma I.19 we have an injection

(I.21) $$\dot{A}'(M) \hookrightarrow \mathcal{D}'(M).$$

We shall denote by $\dot{\mathcal{B}}(M) \subset \dot{\mathcal{D}}'(M)$ the space of those $u \in \dot{\mathcal{D}}'(M)$ with

$$u\big|_{\mathring{M}} \in \dot{A}'(M).$$

Alternatively to Lemma I.19 we also have

(I.22) Lemma $C^\infty(M) \hookrightarrow \dot{A}(M)$ is continuous with dense range.

Again Tensoring with Ω we have an injection

(I.23) $$\dot{A}'(M) \hookrightarrow \dot{\mathcal{D}}'(M).$$

To reduce the already heavy notation we shall use the injection (I.23) as an identification.

(I.24) Proposition. If $u \in \mathcal{B}(M)$ or $\dot{A}'(M)$ and P is a differential operator with C^∞ coefficients the restriction

$$Pu\big|_{\partial M} \in \mathcal{D}'(\partial M)$$

is well-defined.

To define these traces take $\psi \in C^\infty(\partial M, \Omega(\partial M))$ and observe that

(I.25) $$\delta \otimes \psi \in \dot{A}(M,\Omega)$$

where δ is the Dirac 'measure' on ∂M. The adjoint of P, ${}^tP$ acts on distributional densities and

$${}^tP(\delta \otimes \psi) \in \dot{A}(M,\Omega).$$

Thus if either $u \in \dot{\mathcal{B}}(M)$ or $\dot{A}'(M)$ the pairing $\langle u, {}^tP(\delta \otimes \psi)\rangle = v(\psi)$ is defined by continuity. This gives a continuous functional, $Pu|_{\partial M}$. It is useful to note that the difference between $\dot{A}'(M)$, $\mathcal{B}(M)$ is in whether singularities are permitted at the boundary:

(I.26) $$\dot{\mathcal{B}}(M) = \dot{A}(M) \oplus \dot{\mathcal{D}}'(M, \partial M).$$

These rather formal considerations have the following consequence, an extended version of a theorem of Peetre

(I.27) Proposition. Suppose P is a differential operator on M and

$$P:\dot{A}(M) \longrightarrow \dot{A}(M)$$

is surjective modulo $\dot{C}^{\infty}(M)$. Then any $u \in \dot{\mathcal{D}}'(M)$ satisfying $Pu \in \dot{\mathcal{B}}(M)$ has $u \in \dot{\mathcal{B}}(M)$, so has well-defined boundary values of every order.

Examples of operators satisfying the hypothesis of this proposition are non-characteristic operators.

II. Totally characteristic operators.

The space $\mathcal{V}(M)$ introduced in (I.6) consists of vector fields, which can be considered as first order differential operators. The space Diff(M) of differential operator on M is locally finitely generated, over $C^{\infty}(M)$, by vector fields. Let us define

$$\mathrm{Diff}_b(M) \subset \mathrm{Diff}(M)$$

to be the space of differential operators locally finitely generated by $C^{\infty}(M)$ and $\mathcal{V}$. Away from the boundary this is no condition at all, but in local coordinates (x,y) at a boundary point if $P \in \mathrm{Diff}(M)$ then

$$\text{(II.1)} \qquad P \in \mathrm{Diff}_b(M) \iff P = \sum_{|\alpha|+k\le m} a_{\alpha,k}(x,y) D_y^{\alpha}(xD_x)^k,$$

where the coefficients $a_{\alpha,k} \in C^{\infty}$.

In this lecture we shall indicate how to microlocalize $\mathrm{Diff}_b(M)$ to arrive at the ring (assuming M to be compact) $L_b^{*}(M)$ of totally characteristic pseudofifferential operators. Since P in (II.1) is actually a pseudodifferential operator in the usual sense it is reasonable to write its kernel as an oscillatory integral:

$$\text{(II.2)} \quad k(z,z') = (2\pi)^{-n-1}\int e^{i(x-x')\xi + i(y-y')\cdot\eta} a(x,y,x\xi,\eta)\,d\xi\, d\eta.$$

Here we write $z = (x,y)$, $z' = (x',y')$. The amplitude $a(x,y,\lambda,\eta)$ in (II.2) is a polynomial in λ,η so there is no difficulty in making sense of (II.2). We wish to consider, more generally, the case

$$\text{(II.3)} \qquad a(x,y,\lambda,\eta) \in S^m_{1,0}(Z \times \mathbb{R}^{n+1}_{(\lambda,\eta)})$$

where $Z = \overline{\mathbb{R}^+} \times \mathbb{R}^n$ and determine suitable conditions under which (II.2) defines an operator.

Note that, even with (II.3), the amplitude in (II.2) is not a true 'symbol'. It satisfies estimates

$$(II.4)\quad |D_y^\alpha D_\eta^\beta D_x^k D^\ell \tilde{a}| \leq C|x|^\ell (1+|\xi|)^k (1+|\eta|+|x\xi|)^{m-k-\ell-|\beta|}$$

where $\tilde{a}(x,y,\xi,\eta) = a(x,y,x\xi,\eta)$. Certainly these are symbol estimates, of the standard type, if $x > 0$. Thus:

(II.5) Lemma. Given (II.3), (II.2) defines a distribution $k \in \mathcal{D}'(\overline{\mathbb{R}_x^+ \times \mathbb{R}^n} \times \mathbb{R}_{z'}^{n+1})$.

Now, we need to consider the behaviour of k as $x \longrightarrow 0$, first for $x' \neq 0$. To do this, set

$$(II.6)\qquad s = \frac{x'}{x}\ , \quad \lambda = x\xi$$

in (II.2). This gives,

$$(II.7)\qquad k(\frac{x'}{s},y,x',y') = \frac{1}{x}\,\kappa(s,x,y,y')$$

where

$$(II.8)\quad \kappa(s,x,y,y') = (2\pi)^{-1}\int e^{i(1-s)\lambda+i(y-y')\cdot\eta}a(x,y,\lambda,\eta)\,d\lambda d\eta.$$

The explicit appearance of x here is as a parameter. The important x-dependence of k comes from s-dependence of κ and this occurs simply through the inverse Fourier transformation of a symbol.

(II.9) Proposition. Given (II.3), κ in (II.8) is singular only at $s = 1$, $y = y'$ and is rapidly decreasing as $s \to \infty$.

Remembering the transformation (II.6) we see that this gives the following simple description of k near $x = 0$ for $x' > 0$.

(II.10) For $(x',y',y) \in K \subset\subset \mathbb{R}^+ \times \mathbb{R}^n \times \mathbb{R}^n$ $k(x,y,x',y')$ given by (II.2), (II.3) is C^∞ in $0 \leq x \leq x'/2$ and vanishes with all its derivatives at $x = 0$.

Naturally we need also to consider what happens as $x' \longrightarrow 0$ for $x > 0$. This reveals a basic asymetry between x and x' already obvious from the choice of left-reduced form (II.2) for the kernel k. Clearly we need to analyse κ near $s = 0$. From (II.8) it is already clear that κ is C^∞ in all variables near $s = 0$, but if symmetry is to be restored between x,x' we must impose some conditions on a. A suitable choice is the following set of transmission conditions

$$(II.11)\qquad \int e^{i\lambda}\lambda^k\, a(x,y,\lambda,\eta)\,d\lambda \equiv 0 \quad \forall k \in \mathbb{N}\,.$$

clearly this guarantees

(II.12) For $(x,y,y') \in K \subset\subset \mathbb{R}^+ \times \mathbb{R}^n \times \mathbb{R}^n$ $k(x,y,x',y')$ given by (II.2), (II.3), (II.11) is C^∞ in $0 \leq x' \leq \frac{x}{2}$ and vanishes with all its derivatives at $\overline{x'} = \overline{0}$.

The conditions (II.11) may at first appear to be severely restrictive but are in fact 'residual' in the following sense.

(II.13) Proposition. Given $a' \in S^\infty_{1,0}(Z \times \mathbb{R}^{n+1})$ there exists $b \in S^{-\infty}_{1,0}(Z \times \mathbb{R}^{n+1})$ such that $a = a' - b$ satisfies (II.11).

Thus, with (II.3),(II.11) we can define k to be (II.2) in $x,x' > 0$ and to be given by smooth extension in $x,x' \geq 0$ except at the corner itself, $x = x' = 0$. To see what is desirable there consider the form of an operator A with useful properties:

(II.14) $$A: \dot{\mathcal{E}}'(Z) \longrightarrow \dot{\mathcal{D}}'(Z)$$

is continuous and has the properties

(II.15) $$A(C^\infty_c(Z)) \subset C^\infty(Z),$$

(II.16) $$A^*: \dot{\mathcal{E}}'(Z) \longrightarrow \dot{\mathcal{D}}'(Z) \quad \text{(by continuity)}$$

(II.17) Proposition. An operator (II.14) with the restriction property (II.15) and adjoint having the extension property (II.16) is uniquely specified by the restriction of its Schwartz kernel to the open quarter space $\mathring{Z} \times \mathring{Z}$.

The Schwartz kernel of A is here regarded as a distribution supported on the quarter space $Z \times Z$. To do this extend A using

(II.18) $$A: C^\infty_c(\mathbb{R}^{n+1}) \xrightarrow{\;|_Z\;} C^\infty_c(Z) \xrightarrow{\;A\;} \dot{\mathcal{D}}'(Z) \hookrightarrow \mathcal{D}'(\mathbb{R}^{n+1})$$

to construct $k_A \in \mathcal{D}'(\mathbb{R}^{n+1} \times \mathbb{R}^{n+1})$ and observe that k_A vanishes in $x < 0$ or $x' < 0$. Proposition II.17 shows the uniqueness part of

(II.19) Proposition. If (II.3), (II.11) hold then there is a unique operator A as in (II.14), (II.15), (II.16) with kernel given by (II.2).

Proof. To define A it suffices to construct A^*, formally, with the property (II.15). Using the expression (II.7) for k_A one easily deduces the representation:

(II.20) $$A^*\phi(x',y') = \int_0^\infty \int K(\frac{1}{t}, x't, y, y')\phi(x't, y)\,dy\,\frac{dt}{t}.$$

For $\phi \in C_c^\infty$ this can be used to check that $A^*\phi$ is C^∞ down to $x' = 0$. For the remainder of the proof the reader is referred to [3]. The important point to note here is the convolution form of (II.20).

This defines the linear space $L_b^m(Z)$ of totally characteristic operators, at least if we admit into $L_b^m(Z)$ operators with kernels

$$\text{(II.21)} \qquad k_A \in \dot{C}^\infty(Z \times Z)$$

which are the purists' smoothing operators:

$$\text{(II.21)} \Leftrightarrow A\colon \dot{\mathcal{E}}'(Z) \longrightarrow \dot{C}^\infty(Z).$$

The kernels k of operators in $L_b^m(Z)$ do not have singular support confined to the diagonal. For this reason they are not, strictly speaking, pseudodifferential operators. However, the extra singularities are innocuous enough:

$$\text{(II.22)} \qquad \text{If } A \in L_b^m(Z),\ WF(k_A) \subset N^*\Delta \cup N^*(\partial Z \times \partial Z),$$

where $\Delta \subset Z \times Z$ is the diagonal and $\partial Z \times \partial Z$ is the corner, the submanifold $x = x' = 0$. These distributions are examples of 'singular' Lagrangian distributions. It can be shown directly from (II.7), (II.8) that near the diagonal the form of k_A is unchanged by an arbitrary coordinate transformation of Z, the same in the two factors. Away from Δ but still near the corner the form is the same after arbitrary coordinate changes in the two factors, always preserving ∂Z of course.

This immediately leads to the definition of $L_b^m(M)$ for any manifold with boundary M, in terms of the kernels.

(II.23) Theorem. If M is a compact manifold with boundary $L_b^\infty(M)$ is a ring with symbol filtration:

$$\text{(II.24)} \qquad \begin{array}{ccccccc} & & \uparrow & & & & \\ 0 & \hookrightarrow & L_b^m(M) & \xrightarrow{\sigma_m} & S^m(\tilde{T}^*M)/S^{m-1}(\tilde{T}^*M) & \longrightarrow & 0 \\ & & \uparrow & & & & \\ 0 & \hookrightarrow & L_b^{m-1}(M) & \xrightarrow{\sigma_{m-1}} & S^{m-1}(\tilde{T}^*M)/S^{m-2}(\tilde{T}^*M) & \longrightarrow & 0 \\ & & \uparrow & & & & \end{array}$$

the enclosed sequence being exact. Each $A \in L_b^\infty(M)$ maps $\dot{\mathcal{D}}'(M)$, $C^\infty(M)$, $\dot{C}^\infty(M)$, $\dot{\mathcal{A}}(M)$, $\dot{\mathcal{A}}'(M)$, $\dot{\mathcal{B}}(M)$, $\dot{\mathcal{D}}'(M,\partial M)$, $\mathcal{D}'(M)$, $L^2(M)$ to itself. The ring is closed under the taking of adjoint and

$$\text{(II.25)} \qquad \sigma_m(A^*) = \overline{\sigma_m(A)},\ \sigma_{m+m'}(A\cdot B) = \sigma_m(A)\cdot\sigma_{m'}(B).$$

If $u \in \dot{\mathcal{A}}'(M)$ (or $\dot{\mathcal{B}}(M)$) then there exists $A_\partial \in L^m(\partial M)$ with

(II.26) $Au\big|_{\partial M} = A_\partial(u\big|_{\partial M})$,

(II.27) $\sigma_m(A_\partial) = \sigma_m(A)\big|_{T^*\partial M}$.

The residual space $L_b^{-\infty}(M) = \cap L_b^m(M)$ of the calculus consists of those $A \in L_b^\infty(M)$ with

(II.28) $A: \dot{\mathcal{D}}'(M) \longrightarrow \dot{A}(M)$.

This cumbersome statement resembles closely the form a similar result for pseudodifferential operators on a manifold with out boundary would take, except for (II.26), (II.27). In particular (II.28) is in keeping (more truly motivates) the treatment of $\dot{A}(M)$ as a space of ignorable distributions.

Suppose we define, for $u \in \dot{\mathcal{D}}'(M)$

$$\text{sing.supp}_b(u) = M \setminus \{m \in M; m \in B \text{ open and } \exists\ v \in \dot{A}(B) \text{ with } v = u \text{ in } B\}.$$

where of course $\dot{A}(B) = C^\infty(B)$ if $m \notin \partial M$. Similarly, following the pattern established by Hörmander [1] we can define the characteristic set:

$$\Sigma_b(A) = \{p \in \tilde{T}^*M\backslash 0;\ \sigma_m(A) \text{ is not elliptic in a cone around } p\},$$

and then for any $u \in \dot{\mathcal{D}}'(M)$

(II.29) $WF_b(u) = \underset{A}{\cap}\{\Sigma_b(A); Au \in \dot{A}^*M)\}$.

In local coordinates the law for the composition of symbols is the usual one: If A, B are given by (II.2) with symbols a,b as in (II.3), (II.11) and one at least is properly supported then $C = A\cdot B$ is, modulo $L_b^{-\infty}(Z)$ of the form (II.2) with symbol

(II.30) $$c \sim \sum_{\alpha,p\leq k} \frac{1}{\alpha!p!(k-p)!}\partial_\eta^\alpha\partial_\lambda^k a(x,y,\lambda,\eta)\lambda^p x^{k-p} D_y^\alpha D^{k-p} D_\lambda^p b(x,y,\lambda,\eta).$$

This is just the standard formula for pseudodifferential operators. Its locality means the operators in $L_b^m(M)$ can be microlocally inverted at elliptic points. In particular $WF_b(u)$ lies precisely above sing. supp_b (u):

(II.31) If $u \in \dot{\mathcal{D}}'(M)$, $Wf_b(u) = \emptyset \iff u \in \dot{A}(M)$.

The operators in $L_b^\infty(M)$ are microlocal and if $u \in \dot{A}'(M)$ then

(II.32) $WF(Au|_{\partial M}) \subset WF_b(u) \cap T^*\partial M.$

As an essentially trivial application of these ideas we note the following regularity result. On $Z = \mathbb{R}^+ \times \mathbb{R}$ consider the differential operator

$$P = xD_x + iD_y.$$

Since $\sigma_1(P) = \lambda + i\eta$ is elliptic everywhere,

(II.33) $u \in \dot{\mathcal{D}}'(Z),\ Pu \in \dot{A}(Z) \Rightarrow u \in \dot{A}(Z).$

Note that the change of coordinates

$$x = e^{-s} \quad , \quad D_s = - xD_x$$

transforms P to the Cauchy Riemann operator $-D_s + i\,Dy$. Suppose that $f(s+iy)$ is holomorphic in a half strip $s \geq 1/\varepsilon$, $|y| \leq \varepsilon$ and is of exponential type:

$$|f(s + iy)| \leq C\,e^{Rs} .$$

It follows that $u(x,y) = f(s + iy)$ is a solution of $Pu = 0$ in $0 < x < \delta$, $|y| < \varepsilon$ and is of finite order at $x = 0$, i.e., is extendible to an element of $u' \in \dot{\mathcal{D}}'((0,\delta) \times (-\varepsilon,\varepsilon))$. An easy computation shows that u' can be chosen so that $Pu' = 0$. We conclude from (II.33) that

$$f(s + iy) = (2\pi)^{-1} \int e^{ie^{-s}\xi} a(y,\xi)d\xi,$$

$a \in S^m_{1,0}(\mathbb{R} \times \mathbb{R})$ for some m.

III. Boundary canonical transformations.

In this lecture we shall consider some aspects of the 'symplectic' geometry of the space $\tilde{T}^*M$. As usual emphasis is placed on the more practical problems of constructing canonical transformations with suitable properties preparatory to the examination of Fourier integral operators and their use in the analysis of certain boundary value problems.

The natural vector bundle mapping

$$\tilde{\imath}^*: \quad T^*M \longrightarrow \tilde{T}^*M$$

is an isomorphism away from ∂M, so can be used to transfer the canonical 1-form on T^*M to $\tilde{T}^*M$, where it is singular. In canonical dual coordinates (x,y,ξ,η) on T^*M, $x = 0$ on ∂M, the 1-form is

$$\alpha = \xi\, dx + \sum_{j=1}^{n} \eta_j\, dy_j$$

and since $\tilde{\imath}^*(x,y,\xi,\eta) = (x,y,\lambda,\eta)$, $\lambda = x\xi$, the canonical 1-form on $\tilde{T}^*M$ is

$$\tilde{\alpha} = \lambda\, d \log x + \sum_{j=1}^{n} \eta_j\, dy_j \tag{III.1}$$

(III.2) <u>Definition.</u> A boundary canonical transformation from M to M' is a C^∞ map from an open cone $\Gamma \subset \tilde{T}^*M\backslash 0$:

$$\tilde{\chi}: \Gamma \longrightarrow \tilde{T}^*M'\backslash 0$$

such that $\tilde{\chi}^*\, \tilde{\alpha}' = \tilde{\alpha}$.

Since $\tilde{\alpha}$,$\tilde{\alpha}'$ are singular only over the respective boundaries, ∂M and $\partial M'$, it follows easily that $\tilde{\chi}$ maps $\Gamma \cap \partial\tilde{T}^*M$ into $\partial\tilde{T}^*M'$.

III.3) <u>Lemma.</u> If $\tilde{\chi}$ is a boundary canonical transformation from M to M' and (x,y,λ,η), (X,Y,Λ,H) are canonical coordinates in $\tilde{T}^*M$, $\tilde{T}^*M'$ then

$$\begin{cases} \tilde{\chi}^*(X) = \mu_0(x,y,\lambda,\eta)x \\ \tilde{\chi}^*(\Lambda) = \lambda + x\, \mu_1(x,y,\lambda,\eta) \end{cases} \tag{III.4}$$

with $\mu_0 > 0$. In particular the projection of $\tilde{\chi}$ onto T^*M

$$\chi: \tilde{\imath}^{*-1}(\Gamma) \longrightarrow T^*M'\backslash 0 \tag{III.5}$$

is well-defined C^∞ canonical transformation.

Note that the fact that $\tilde{\chi}$ is a boundary canonical transformation can be restated in the form (III.5) once it is known that $\tilde{\chi}$ itself is C^∞ from $\tilde{T}^*M$ to $\tilde{T}^*M'$. There are two simple consequences of this lemma to note straight away, first that a boundary canonical transformation is always locally a diffeomorphism and second that $\tilde{\chi}$ leaves the surface $x=\lambda=0$ invariant. That is, $\tilde{\chi}$ restricts to a C^∞ map

$$\partial\tilde{\chi}: T^*\partial M \cap \Gamma \longrightarrow T^*\partial M'. \tag{III.6}$$

The first example of the construction of a boundary canonical transformation to simplify a given problem concerns the analogue of an operator of real-principal type in $L_b^1(M)$. Suppose that $p \in C^\infty(\tilde{T}^*M\backslash 0)$ is real-valued and homogeneous of degree one. If $\bar{\rho} \in \partial\tilde{T}^*M\backslash 0$ is a zero of p

$$p(\bar{\rho}) = 0 \quad , \; p \text{ real-valued}, \tag{III.7}$$

we wish to consider a normal form for p under boundary canonical transformation. As has already been shown the function $\lambda \in C^\infty(\partial\tilde{T}^*M)$ is well-defined, independently of coordinates, and is even invariant under boundary canonical transformations. We therefore add the generic condition:

(III.8) dp, $d\lambda$ are independent on $\partial\tilde{T}^*M$ at $\bar{\rho}$

If $\bar{\rho} \in T^*\partial M \subset \partial\tilde{T}^*M$ we can add a further invariant condition:

(III.9) $dp(\bar{\rho})$, $\alpha_{\partial M}$ are independent on $T^*\partial M$ if $\bar{\rho} \in T^*\partial M$.

(III.10) Proposition. If $p \in C^\infty(\tilde{T}^*M \setminus 0)$ satisfies (III.7), (III.8,), (III.9) at $\bar{\rho} \in \partial\tilde{T}^*M$ then there exists a boundary canonical transformation

$$\tilde{\chi}\colon \tilde{T}^*M \supset \Gamma \longrightarrow \tilde{T}^*Z$$

defined in a conic neighbourhood of $\bar{\rho}$ such that $\tilde{\chi}(\bar{\rho}) = (0,0,\bar{\lambda},\bar{\eta})$, $\bar{\eta} = (\bar{\eta}_1, 0,\ldots,0)$ and

$$p = \tilde{\chi}^*(\eta_1).$$

To start the proof of this Proposition observe that if $\bar{\rho} \in T^*\partial M \setminus 0$ then condition (III.9) means that in any canonical dual coordinates on $\tilde{T}^*M$,

(III.11) $$p = g_0(y,\eta) + \mu_0 x + \mu_1\lambda.$$

with $dg_0(y,\eta)$ independent of $\eta\cdot dy$. Standard results on the existence of canonical transformations show that

$$\partial\chi\colon T^*\partial M \longrightarrow T^*\mathbb{R}^n,$$

a canonical transformation, can be chosen so that $g_0 = (\partial\chi)^*\eta_1$. Defining $\tilde{\chi}$ by

$$\tilde{\chi}(x,y,\lambda,\eta) = (x,\partial\chi(y),\lambda,\partial\chi(\eta))$$

clearly gives a boundary canonical transformation. Thus, we can assume that $g_0 = \eta_1$ in (III.11).

Now, to construct $\tilde{\chi}$ we demand that

$$\tilde{\chi}^*(\eta_1) = \eta_1 + \mu_0 x + \mu_1\lambda.$$

We then proceed to construct the canonical transformation $\chi\colon T^*Z \longrightarrow T^*Z$. To do this choose C^∞ functions $X = \chi^*(x)$, $\Xi = \chi^*(\xi)$ to satisfy the usual Poisson bracket conditions:

$$\text{(III.12)} \quad \begin{cases} \{H_1,\chi\} = \{H_1,\Xi\} = 0 \\ \Xi = \xi,\ X = x \quad \text{on} \quad y_1 = 0 \end{cases}$$

where $H_1 = \tilde{\chi}^*(\eta_1) = \emptyset$ by definition. We claim that $\Lambda = X\Xi = \tilde{\chi}^*(\lambda)$ is C^∞ in a neighbourhood of the base point in $\tilde{T}^*Z$. In fact the Hamilton vector field of a function p can be written

$$\text{(III.13)} \quad \begin{aligned} H_p &= \frac{\partial p}{\partial \xi}\partial_x - \frac{\partial p}{\partial x}\partial_\xi + \sum_{j=1}^{n} \frac{\partial p}{\partial \eta_j}\partial_{y_j} - \frac{\partial p}{\partial y_j}\partial_{\eta_j} \\ &= x\frac{\partial p}{\partial \lambda}\partial_x - x\frac{\partial p}{\partial x}\partial_\lambda + \sum_{j=1} \left(\frac{\partial p}{\partial_{\eta j}}\partial_{y_j} - \frac{\partial p}{\partial y_j}\partial_{\eta_j}\right). \end{aligned}$$

Thus, Λ satisfies the first-order partial differential equation

$$\text{(III.14)} \quad \sum_{j=1}^{n} \left(\frac{\partial p}{\partial_{\eta_j}}\partial_{y_j}\Lambda - \frac{\partial p}{\partial y_j}\partial_{\eta_j}\Lambda\right) + x\frac{\partial p}{\partial \lambda}\frac{\partial \Lambda}{\partial x} - x\frac{\partial p}{\partial x}\partial_\lambda \Lambda = 0$$

in which the vector field is C^∞ and transversal to the surface $y_1 = 0$. Thus Λ is uniquely specified and C^∞. The same is true for the functions Y_j, H_k solving

$$\text{(III.15)} \quad \begin{aligned} &\{H_1, Y_1\} = \delta_{1j}, \quad \{H_1, H_k\} = 0 \qquad (k > 1) \\ &Y_j = y_j, \quad H_k = \eta_k \qquad \text{on} \qquad y_1 = 0 \end{aligned}$$

This constructs the transformation

$$\tilde{\chi}(X,Y,\Lambda,H) = (x,y,\lambda,\eta)$$

since it has a C^∞ inverse which is a boundary canonical transformation in view of (III.12), (III.15).

The case $\bar{\rho} \notin T^*\partial M$, i.e., $\bar{\lambda} \neq 0$ if $\bar{\rho} = (0,y,\bar{\lambda},\eta)$ is the base point in canonically dual coordinates, is similar. Condition (III.8) means that

$$\text{(III.16)} \qquad dp \neq 0 \quad \text{on} \quad x = 0, \quad \lambda = \bar{\lambda} \quad \text{at} \quad \bar{\rho}.$$

when $\bar{\lambda} \neq 0$ this submanifold of codimension two in $\tilde{T}^*M$ is not a homogeneous symplectic submanifold, as is the case for $\bar{\lambda} = 0$, giving $T^*\partial M$. However it is symplectic with 2-form

$$\text{(III.17)} \qquad \omega = \sum_{j=1} d\eta_j \wedge dy_j.$$

Now, (III.16) means that symplectic coordinates (Y,H), keeping the form (III.17) for ω, can be introduced so that

(III.18) $$p = H_1 \quad \text{on} \quad x = 0, \lambda = \overline{\lambda} .$$

Let us extend (Y,H) to be independent of x and homogeneous of degree one. Choosing $\Lambda = \lambda$ we then have

$$\{Y_j,\Lambda\} = \{H_j,\Lambda\} = 0 \ , \quad \{H_j, Y_k\} = \delta_{jk}$$

so choose X to solve

(III.19) $$\{\Lambda,X\} = X \ , \ \{H_j,X\} = \{Y_j,X\} = 0.$$

with, of course $X = 0$ on $x = 0$. Although the first of these equations,

$$x \frac{\partial X}{\partial x} = X$$

is singular at $x = 0$, its solutions are all of the form $X = r(y,\lambda,\eta)x$ and then r need only be chosen to satisfy

$$\{H_j,r\} = \{Y_j,r\} = 0 \quad , \quad r = 1 \ \text{at} \ \overline{\rho}.$$

After this boundary canonical transformation has been carried out we have

(III.20) $$p = \eta_1 + x\mu(x,y,\lambda,\eta).$$

Setting $p = H_1$ and proceeding to solve the same system of equations as before can readily be shown to yield the desired boundary canonical transformation.

The second example of reduction to normal form has been described, briefly, in [5]. Suppose that, on T^*Z,

(III.21) $$p = \xi^2 + xr(x,y,\eta) + \eta_1\eta_n$$

is a polynomial in ξ with real, C^∞ coefficients near $\eta = (0,\ldots,0,1)$, where r is homogeneous of degree one in η.

(III.22) <u>Proposition</u>. If $r \neq 0$ near $(0,0,0,\overline{\eta}) \in \tilde{T}^*Z$, $\overline{\eta} = (0,\ldots,0,1)$ there is a boundary canonical transformation

$$\tilde{\chi}: \tilde{T}^*Z,\overline{\rho} \longrightarrow \tilde{T}^*Z,\overline{\rho}$$

such that under the associated canonical transformation

(III.23) $$p = a(x,y,x\xi,\eta) \ \chi^*(\xi^2 \pm x\eta_n^2 + \eta_1 \eta_n),$$

with the sign that of r.

The main part of the construction of $\tilde{\chi}$ is the construction of a suitable boundary transformation, (III.6). This has been carried out in [6], here we shall show how that result applies and then discuss the construction of $\tilde{\chi}$. The main point to observe is that, in the sense of [6], the hypersurfaces $\{x = 0\}$, $\{p = 0\}$ are glancing at $\bar{\rho} = (0,0,0,\bar{\eta}) \in T^*Z$ under the hypothesis $r \neq 0$. Here, we assume that some extension of p, still of the form (III.21), has been made into $x < 0$. The Poisson brackets are

$$\begin{aligned} &\{p,x\} = 2\,\xi \quad (= 0 \quad \text{at} \quad \bar{\rho}) \\ \text{(III.24)} \qquad &\{p,\{p,x\}\} = -2r \quad \text{on} \quad x = 0 \quad (\neq 0 \quad \text{at} \quad \bar{\rho}) \\ &\{x,\{p,x\}\} = -2 \end{aligned}$$

These conditions, together with $p = x = 0$ and $\bar{\rho}$ are the definition of glancing for hypersurface in a symplectic manifold.

We shall concentrate, from now on, on the case $r > 0$; the other case is not essentially different. The symplectic geometry of glancing hypersurface is shown, in [6], to have no invariants. That is, (III.23) can be arranged with $\chi = \chi'$ some canonical transformation and a C^∞ though not necessarily of the special form indicated. Although this transformation, χ', need not be a boundary canonical transformation it certainly preserves $x = 0$ and so projects to a canonical transformation

$$\text{(III.25)} \qquad \partial\chi\colon T^*\mathbb{R}^n,(0,\bar{\eta}) \longrightarrow T^*\mathbb{R}^n,(0,\bar{\eta}).$$

We shall take this as the boundary transformation of $\tilde{\chi}$. It is important to note that $\partial\chi$ is by no means arbitrary.

Explicitly, the function p defines a singular symplectic map on $T^*\mathbb{R}^n$, the boundary map. This is defined by taking

$$\delta_+(y,\eta) = (y',\eta') \qquad (\text{if} \quad \eta_1 \leq 0)$$

where $(0,y',\xi',\eta')$ is the other point in $x = 0$ on the H_p-curve through $(0,y,\xi,\eta)$ where $p(0,y,\xi,\eta) = 0$, i.e., $\xi^2 = -\eta_1\eta_n$, and $\xi \geq 0$. Then,

$$\text{(III.26)} \qquad \delta_+ \cdot \partial\chi = \partial\chi \cdot \delta'_+$$

where $\delta'_+(y,\eta) = \exp(-(-\zeta)^{\frac{1}{2}} H_\zeta)$, with $\zeta = \eta_1\eta_n^{-1/3}$ is the boundary map for the simple example occurring in (III,23).

The importance of (III.26) rests on the fact that

$$\text{(III.27)} \qquad \begin{aligned} H_p\, g &= 0 \qquad \text{on} \quad p = 0 \\ g &= g_0 \qquad \text{on} \quad p = x = 0 \end{aligned}$$

has a smooth solution g for smooth data g_0 if, and only if,

(III.28) $$\delta_+^*(g_0\big|_{\xi=(-\eta_n\eta_1)^{\frac{1}{2}}}) = g_0\big|_{\xi=-(-\eta_n\eta_1)^{\frac{1}{2}}}.$$

The intertwining relation (III.26) shows that, in the coordinates introduced by $\partial\chi$ trivially extended to preserve x,ξ, in which we work from now on, (III.28) takes a simple form:

(III.29) $$g_0 = f(y_2,\ldots,y_{n-1},y_n + \frac{\eta_1 y_1}{3\eta_n}, \eta_1,\ldots,\eta_n,\xi + y_1\eta_n)$$
$$(\text{on } p = x = 0).$$

Using this observation we see that the function $\xi + y_1\eta_n$ gives initial data for a C^∞ function on $p = 0$ solving

$$\begin{cases} H_p R = 0 & \text{on } p = 0 \\ R = \xi + y_1\eta_n & \text{on } x = p = 0. \end{cases}$$

Any C^∞ function on $p = 0$ extends to a linear function of ξ, since p is quadratic. We choose such a real extension

(III.30) $$R = a(x,y,\eta)\xi + b(x,y,\eta)$$

and note that $a > 0$. Since we want $R = \Xi + Y_1 H_n$ we extend $\partial\chi$ to χ by defining $(X,Y,H) = \chi^*(x,y,\eta)$ to satisfy

(III.31) $$\begin{cases} H_R X = 1,\ H_R Y_i = 0 \quad 1 \le j < n,\ H_R H_k = 0 \ 1 < k \le n \\ H_R H_1 = -H_n, H_R Y_n = Y_1;\ {}^*X,Y,H) = (0,y,\eta) \quad \text{on } x = 0. \end{cases}$$

Since H_R is transversal to $x = 0$ the solution to (III.31), together with $\Xi = R + Y_1 H_n$ defines a canonical transformation. From (III.31) and the definition of R it follows that on $p = 0$,

$$H_R(\Xi^2 + X H_n^2 + H_1 H_n) = 0.$$

Since $\{p = 0\}$ is the flow-out of $\{p = 0, x = 0\}$ under H_R we conclude that $p = 0$ on the surface $\Xi^2 + X H_n^2 + H_1 H_n = \chi^*(\xi^2 + x\eta_n^2 + \eta_1\eta_n) = 0$.

Thus, to prove (III.23) it only remains to show that χ is a boundary canonical transformation. To do this we translate (III.31) to a singular initial value problem:

(III.32) $$\begin{cases} WX = x,\quad WY_j = 0,\quad WH_k = 0 \\ WH_1 = - x H_n,\ WY_n = xY_1 \\ (X,Y,H) = (0,y,\eta) \quad \text{on } x = \lambda = 0 \end{cases}$$

where $W = xH_R$ expressed in terms of the coordinates (x,y,λ,η) on $\tilde{T}^*Z$. From (III.13) it follows that W is C^∞,

$$W = a(x\partial_x + \lambda\partial_\lambda) + x\,\Sigma\left(\frac{\partial R}{\partial\eta_j}\partial_{y_j} - \frac{\partial R}{\partial y_j}\partial_{\eta_j}\right) + O_2$$

where O_2 is a vector field vanishing to second order on $x = \lambda > 0$. The fact that the leading part of W is the radial vector field allows one to conclude easily that (III.32) has a unique solution, first in the sense of formal power series on $x = \lambda = 0$ and then in C^∞. This completes the proof of Proposition III.22.

IV. Transversal problems.

In this lecture we shall show, very briefly, how to quantize the boundary canonical transformations considered earlier; that is, discuss the definition and properties of the associated spaces of Fourier integral operators. Using these operators and the geometric normal forms obtained in Lecture III it will then be shown how the propagation of singularities for operators of real principal type in $\dot{L}_b$ can be analysed. The method in this application is precisely that of Duistermaat and Hörmander [7], adapted to the treatment of WF_b. It is interesting to see that WF_b involves "second microlocalization" at the boundary in that the residual problem, in $\dot{\Lambda}$ reduces to the analysis of local properties of pseudodifferential operators.

Suppose that M,N are C^∞ manifolds with boundary and

$$\text{(IV.1)}\qquad \tilde{\chi}\colon \tilde{T}^*M \supset \Gamma \longrightarrow \tilde{T}^*N$$

is a boundary canonical transformation defined in some open cone. To associate totally characteristic Fourier integral operators to $\tilde{\chi}$ introduce local coordinates (x,y) so that, by conjugation

$$\text{(IV.2)}\qquad \tilde{\chi}\colon \tilde{T}^*Z \supset \Gamma_1 \longrightarrow \tilde{T}^*Z.$$

As in the standard theory the first step is to parametrize $\tilde{\chi}$ by a phase function. Let

$$\Gamma_2 \subset \overline{\mathbb{R}^+_x} \times \mathbb{R}^n_y \times \mathbb{R}^n_{y'} \times (\mathbb{R}^{N+1}_{(\mu,\theta)} \setminus \{0\})$$

be an open cone, where $(\mu,\theta) \in \mathbb{R} \times \mathbb{R}^N$. Then $\phi \in C^\infty(Z \times \mathbb{R}^n \times (\mathbb{R}^{N+1} \setminus \{0\}))$ is a phase function in Γ_2 if it is real-valued, homogeneous of degree one in (μ,θ) and has

$$(\phi_y,\ \mu\phi_\mu,\ \phi_\theta) \neq 0.$$

Since boundary canonical transformations are special canonical diffeomorphisms we shall impose a strong non-degeneracy condition on ϕ:

$$\text{(IV.3)}\quad \begin{cases} \text{At } x = y \text{ if } \phi_\theta = 0 \text{ then } \phi_\mu \neq 0 \text{ and} \\ \det \begin{bmatrix} \phi_{yy'} & \phi_{y\theta} & \phi_{y\mu} \\ \phi_{\theta y'} & \phi_{\theta\theta} & \phi_{\theta\mu} \\ \mu\phi_{\mu y'} & \mu\phi_{\mu\theta} & (\mu\phi_\mu)_\mu \end{bmatrix} \neq 0 . \end{cases}$$

Now, if ϕ is such a phase function then the set

$$C_\phi = \{(x,y,y',\mu,\theta) \in \Gamma_2;\ \phi_\theta = 0\}$$

is a homogeneous submanifold of Γ_2 of codimension N. The parametrizing map

$$\text{(IV.4)}\quad p_\phi : C_\phi \ni (x,y,y',\mu,\theta) \longmapsto (x,y,x\phi_x+\mu\phi_\mu,\phi_y;x\phi_\mu,y',\mu\phi_\mu-\phi_{y'}) \in \tilde{C}_\phi$$

is a local diffeomorphism onto a homogeneous submanifold of $\tilde{T}^*Z \times \tilde{T}^*Z$.

(IV.5) Proposition. If ϕ is a phase function satisfying (IV.3) then near $x = 0$ $\tilde{C}_\phi$ is everywhere locally the graph of a boundary canonical transformation. Conversely, near any point in $\partial\tilde{T}^*Z \times \partial\tilde{T}^*Z$ of the graph of a boundary canonical transformation there is such a parametrization.

The construction of a parametrization follows the method of Caratheodory , Arnold and Hörmander once it is noted that (IV.4) is just the usual parametrization map lifted to $\tilde{T}^*Z$. Given a phase function ϕ the oscillatory integral

$$\text{(IV.6)}\quad K(z,z') = \int e^{i\phi(z,y,x\xi,\theta)-ix'\xi} a(z,y',x\xi,\theta)\, d\theta\, d\xi$$

can be used to define the Schwartz kernel of a Fourier integral operator associated to $\tilde{\chi}$. Naturally the amplitude a in (IV.6) should be a symbol as a function of z,y',μ,θ and should have essential support inside the cone Γ_2 in which ϕ in non-degenerate; let $S^m(\Gamma_2)$ denote the space of such symbols. As in the pseudodifferential operator case (IV.6) makes good sense in $x > 0$ but to extend it down to $x = 0$ in a useful manner the lacunary conditions, coming from the demand that the formal series at $x = 0$ should vanish, need to be satisfied. If

$$L_k(\phi)\ a = \int_a^\infty e^{i\phi(0,y',\mu,\theta)} a(0,y,y',\mu,\theta)\mu^k\, d_\mu$$

then:

(IV.7) Lemma. The map $L: S^{-\infty}(Z \times \mathbb{R}^n \times \mathbb{R}^{N+1}) \longrightarrow \prod_{k=0}^{\infty} S^{-\infty}$ is surjective.

This shows that the lacunary conditions are residual in $\dot{S}^m(\Gamma)$, so if $\dot{S}^m(\Gamma_2 \phi)$ is the space of symbols satisfying them,

$$\dot{S}^m(\Gamma_2,\phi) / \dot{S}^{-\infty}(\Gamma_2,\phi) \cong \dot{S}^m(\Gamma_2) / \dot{S}^{-\infty}(\Gamma_2) .$$

(IV.8) Proposition. If $a \in \dot{S}^m(\Gamma_2,\phi)$ the oscillatory intergral (IV.6) defines the Schwartz kernel of an operator

(IV.9) $$F: \dot{\mathcal{E}}'(Z) \longrightarrow \dot{\mathcal{D}}'(Z)$$

with restriction properties $F(\dot{C}_c^\infty(Z)) \subset \dot{C}^\infty(Z), F \subset C_c^\infty(Z)) \subset C^\infty(Z)$ and for $u \in C_c^\infty(Z)$

(IV.10) $$Fu|_{x=0} = F_0(u|_{x=0})$$

where F_0 is a Fourier integral operator on $\mathbb{R}^n$ associated to $\partial\tilde{\chi}$.

Naturally, to get a space of operators associated to (IV.1) it is necessary to show coordinate invariance. Since we also want a symbol calculus the essential independence of choice of parametrizing phase function, behaviour under composition and the taking of adjoints needs also to be proved. These technical details can be found in [3]. Here we just remark that a well-defined symbol map

$$\sigma: I_b^m(M,N_j\tilde{\chi}) \longrightarrow S^{m+\frac{n}{2}}(\text{graph } \tilde{\chi};\ L \otimes \Omega^{\frac{1}{2}} / S^{m-1+\frac{n}{2}}$$

with L, the Maslov and half-density bundle respectively is fixed by extending Hörmander's symbol map from the interior. It has the 'expected' properties, eg. Egorov's theorem holds.

Our objective here is to apply these operators. If $A \in L_b^m(M)$ has real principal symbol, and is of classical type, and $B \in L_b^{1-m}(M)$ is elliptic then the conditions (III.8),(III.9) on a point $\bar{\rho}$ as in (III.7), applied to BA, are independent of B. This is the (microlocal) real principal type condition.

The principal symbol $a \in C^\infty(\tilde{T}^*M \setminus 0)$ has Hamilton vector field on $\tilde{T}^*M \setminus 0$

$$H_a = x\frac{\partial a}{\partial \lambda}\partial_x - x\frac{\partial a}{\partial x}\partial_\lambda + \sum_{j=1}^{n}\left(\frac{\partial a}{\partial \eta_j}\partial_{y_j} - \frac{\partial a}{\partial y_j}\partial_{\eta_j}\right)$$

(see (III.13)). This is tangent to

$$\Sigma_b(A) = \{\rho \in \tilde{T}^*M\ \ 0;\ a(\rho) = 0\}$$

and is even tangent to the two invariantly defined pieces

(IV.11) $\Sigma_b(A) \cap \partial\tilde{T}^*M, \quad \Sigma_b(A) \cap T^*\partial M,$

i.e. $x = 0$ and $x = \lambda = 0$ in $\tilde{T}^*M$.

(IV.12) Theorem. If $A \in L_b^m(M)$ is of classical and real principal type then $\forall u \in \dot{\mathcal{D}}'(M)$

(IV.13) $WF_b(u) \cap WF_b(Au) \subset \Sigma_b(A)$

is a union of maximally extended bicharacteristics of H_a in $\Sigma_b(A)$.

Just as is the case away from the boundary this is really a microlocal theorem. Moreover it extends trivially to any operator in $L_b^m(M)$ with real principal symbol, the statement at points where A is not microlocally of real principal type being either elliptic theory or else trivial. In fact the proof is a review of the proof given by Duistermaat and Hormander [7] with standard Fourier integral operator calculus replaced by the totally characteristic calculus. Thus, in Proposition III.10 it is shown how to reduce the symbol of A to an elliptic multiple of that of D_{y_1} on Z. Application of the pseudodifferential calculus shows the existence of an elliptic Fourier integral operator F associated to $\tilde{\chi}$ such that

$$FA \equiv D_{y_1} F.$$

This reduces the proof to the case $A = D_{y_1}$ on Z, where it is trivial.

Now, since (IV.13) refers to $WF_b(u)$ it leaves open errors in $\dot{A}$. Before discussing these we consider the Cauchy problem for A. Suppose $N \subset M$ is a hypersurface in M meeting ∂M transversally. As an example of a composition result transcending those discussed above we have

(IV.14) Proposition. Let $u \in \dot{\mathcal{D}}'(M)$ and suppose that

(IV.15) $\tilde{N}^*N \cap WF_b(u) = \emptyset$

where $\tilde{N}^*N$ is the kernel of the natural projection

(IV.16) $\iota_N^*: \tilde{T}_N^*M \longrightarrow \tilde{T}^*N.$

Then the restriction $u|_N \in \dot{\mathcal{D}}'(N)$ is well-defined, by continuity, from $C_c^\infty(M)$.

We have not made the continuity statements precise here but see Hörmander [1] for the standard case - as there the topology comes from the pseudodifferential definition of (IV.15). Comparing (IV.15) and (IV.13) suggests the fact that for A of principal type the condition

$$\Sigma_b(A) \cap \tilde{N}^*N = \emptyset$$

is also the statement that H_a be transversal to $T^*_N M$ on $\Sigma_b(A)$. Working with the ideas of [1], [8] it is then easy to show that the microlocal problem

$$\iota_N^{*-1}(\bar{\sigma}) \cap WF_b(Au) = \emptyset$$

$$WF_b((u|_N) - u_0) \not\ni \bar{\sigma} \quad , \quad u_0 \in \dot{\mathcal{D}}'(N)$$

is easily solved whenever A is of real principal type, $\bar{\sigma} \in \tilde{T}^*N \setminus 0$ has

$$\iota_N^{*-1} \; \bar{\sigma} \; \cap \Sigma_b(A) \neq \emptyset$$

and H_a is transversal to $\tilde{T}^*_N M$ at one point of this set.

The microlocal solvability of the Cauchy problem implies the local solvability of $Au = f$, modulo $\dot{A}(M)$, provided A is of real principal type in the sense of [7], i.e., the intergral curves of H_a do not lie over compact sets, in M. These same conditions imply the local solvability of $Au = f$ modulo $\dot{C}^\infty(M)$. The last step from errors in $\dot{A}(M)$ to $\dot{C}^\infty(M)$ uses the standard calculus of Lagrangian distributions. The symbol equation to be solved can be reduced, using the transformation (II.34), to a real-principal equation in the usual sense.

To obtain some examples of such operators consider a C^∞ manifold with boundary M which carries a singular Riemann metric. Explicitely, g is a symmetric 2- cotensor which is positive, definite in the interior with boundary singularity

$$\text{(IV.17)} \quad g = \frac{a(x,y)}{x^2}\, dx^2 + 2x \sum_{i=1}^{n} \frac{b_i(x,y)}{x}\, dy^i . dx + \sum_{i,j=1}^{n} C_{ij}(x,y)\, dy^i . dy^j$$

where $a > 0$, $(C_{ij}) > 0$

(IV.18) Lemma. The Laplace-Beltrami operator, Δ_g, of a conic-type Riemannian manifold with boundary is an elliptic operator in $L^2_b(M)$.

Theorem IV.12 applies to the wave operator

$$A = (\partial_t^2 - \Delta g) \qquad \text{on } \mathbb{R}_t \times M,$$

or of course to the Laplace-Beltrami operator of an appropriate conic-type Lorentz metric.

V. Gliding rays

Using the Fourier integral operators of Lecture IV and the computation in symplectic geometry of Lecture III we shall show how boundary value problems for a non-characteristic differential (or pseudo-differential) operator of second order and real principal type can be reduced to normal form near any gliding point. The normal form itself can be taken as a modified version of Friedlander's example [9], see also [3].

$$P_0 = D_x^2 + x D_{y_n}^2 + D_{y_1} D_{y_n} \tag{V,1}$$

on $Z = \overline{\mathbb{R}^+} \times \mathbb{R}^n$.

The symbol of P_0 is

$$p_0 = \xi^2 + x\eta_n^2 + \eta_1 \eta_n \tag{V.2}$$

and we shall be considering boundary value problems microlocally near the point $x = \xi = 0,\ y = 0,\ \eta = \bar{\eta} = (0,\dots,0,1)$. The bicharacteristics of p_0 are solutions of the equations

$$\left\{\begin{array}{ll} \dfrac{dx}{ds} = 2\xi & \dfrac{d\xi}{ds} = -\eta_n^2 \\ \dfrac{dy_1}{ds} = \eta_n & \dfrac{d\eta_1}{ds} = 0 \\ \dfrac{dy_n}{ds} = \eta_1 + 2x\eta_n & \dfrac{d\eta_n}{ds} = 0 \\ \dfrac{dy_j}{ds} = 0 & \dfrac{d\eta_j}{ds} = 0 \quad j = 2,\dots,n-1 \end{array}\right. \tag{V.3}$$

i.e. the Hamilton vector field is:

$$H_{p_0} = 2\xi\, \partial_x - \eta_n^2 \partial_\xi + \eta_n \partial_{y_1} + (\eta_1 + 2x\,\eta_n)\, \partial_{y_n}\,. \tag{V.4}$$

Now the equations (V.3) are easily integrated and show that along the bicharacteristics of p_0 in $T^*\mathbb{R}^{n+1}$ through the base point $(0,0;,\bar{\eta})$

$$x = -s^2.$$

The same qualitative behaviour, non-degenerate second-order tangency of bicharacteristics to $x = 0$ from $x < 0$, is exhibited by a Hamiltonian of the form (III.21) exactly when

$$r(0,0,\bar{\eta}) > 0, \tag{V.5}$$

the <u>gliding condition</u>. Of course this can be stated rather more invariantly from (III,24) as

$$p = x = \{p,x\} = 0,\ \{x,\{p,x\}\} \neq 0, \{p,\{p,x\}\} < 0$$

at $\bar{\rho} \in \partial T^*Z$, with $x > 0$ inside Z.

In particular, Proposition (III.21) near such a gliding point. The main technical result that is needed to directly analyse boundary value problems for P is the 'quantization' of this geometric result. Since the proof is not yet to be found in the literature a sketch will be given here.

(V.6) <u>Theorem</u>. Let P be a second order differential operator on Z with principal symbol of the form (III.21), (V.4). Near $\bar{\rho} = (0,\pi) \in T^*\mathbb{R}^n \subset \tilde{T}^*Z$ there is a boundary canonical transformation $\tilde{\chi}$, fixing $\bar{\rho}$, and totally characteristic Fourier integral operators F,G associated to $\tilde{\chi}$ and elliptic at $\bar{\rho}$ such that

$$\text{(V.7)} \qquad \bar{\rho} \notin WF_b[(FP - P_0G)u] \quad \forall u \in \dot{\mathcal{E}}'(Z).$$

Before proceeding to outline the proof we shall make some remarks concerning a slight extension of the results of Lectures III & IV. The space $L_b^m(Z)$ of totally characteristic pseudodifferential operators is defined in Lecture IV. There are other invariant spaces of operators, for example the differential operators $\text{Diff}^k(Z)$. We can therefore consider, for each $k \in \mathbb{N}$

$$L_b^{m,k}(Z) = \sum_{0 \le p \le k} L_b^{m-p}(Z)\, \text{Diff}^p(Z),$$

the space of differential operators with totally characteristic pseudodifferential coefficients. This space is certainly coordinate independent and it is straightforward to show that, for instance, the transformation results of Lecture IV apply with only the obvious minor modifications. In particular the symbol calculus carries over, with the symbols being slightly more general functions namely functions $a \in C^\infty(T^*Z)$ such that $x^k a \in C^\infty(\tilde{T}^*Z)$. Boundary canonical transformations preserve this property. It is in fact no extra difficulty in proving Theorem V.6 for $P \in L_b^{2,2}(Z)$ (or indeed $L_b^{m,2}$) satisfying the appropriate invariant versions of (V.4).

Following these remarks we see that the canonical transformation $\tilde{\chi}$ of Proposition III.22 can be used to accomplish the first step in (V.7). Namely, if F_1 is elliptic at $\bar{\rho}$ and associated to $\tilde{\chi}$ then, with $G_1 = F_1$, from the transformation results of Lecture IV

$$\text{(V.8)} \qquad F_1 P \equiv P' F_1 \qquad \text{at } \bar{\rho}$$

in the sense of (V.7), with $P' \in L_b^{2,2}(Z)$ having principal symbol

$$\tilde{\sigma}_2(P) = a(x,y,\lambda,\eta)\ p_0.$$

The elliptic factor a can then be removed by replacing G_1 by BG_1 with B a microlocal parametrix for $A = a(x,y,xD_x,D_y)$. Thus, we obtain (V.8) where now:

$$\text{(V.9)} \qquad P_1 = P' - P_0 = C_1D_x^2 + C_2D_x + C_3 \in L_b^{1,2}(Z)$$

i.e. $C_i \in L_b^{i-2}(Z)$.

To try to remove these lower order terms we have two types of freedom of action. First, we can replace G_1 in (V.8) by $(Id - A_1)^{-1}G_1$, $A_1 \in L_b^{-1}$, which changes P' to $P' - A_1P'$ and so alters P_1 by subtracting A_1P'. Symbolically this changes $\tilde{\sigma}_1(P_1)$ to

$$\text{(V.10)} \qquad \tilde{\sigma}_1(P_1) - a_1p_0.$$

The second freedom in (V.8) is to replace F_1 by A_2F_1, A_2 elliptic at $\bar{\rho}$. This replaces P' by

$$P' - [P', A_2]\ A_2^{-1}$$

if G_1 is changed to A_2G_1. Again this leaves P_1 of first order with symbol changed to

$$\text{(V.11)} \qquad \tilde{\sigma}_1(P_1) - \{p_0, \log a_2\}.$$

Combining these two operations we wish to choose a_2 first so that,using (V.11), $\tilde{\sigma}_1(P_1)$ can be assumed to vanish on $p_0 = 0$. Then we choose a_1 so $\tilde{\sigma}_1(P_1) = 0$ using (V.10). If this can be done, the same construction works at all levels of the symbol filtration. Using the completeness property of the symbol calculus in a standard way this allows us to improve (V.9) to a remainder term in $L^{-\infty,2}(Z)$, and so gives (V.8) and the Proposition.

We leave the details to the reader and consider only the basic constructive problem:

(V.12) <u>Proposition</u>. If $f \in C^\infty(T^*Z)$, $f = c_1\ \xi^2 + c_2\ \xi + c_3$ has coefficients in $C^\infty(\tilde{T}^*Z)$ then there exists $a \in C^\infty(\tilde{T}^*Z)$ such that near $\bar{\rho}$

$$\text{(V.13)} \qquad H_{p_0}a = \{p_0, a\} = f \quad \text{on} \quad p_0 = 0.$$

A proof of this Proposition can be extracted, relatively easily, from [10]. Nevertheless we shall indicate briefly a slightly different approach. In (V.4) H_{p_0} is written out explicitly. Since we need to work with C^∞ functions of x,y,λ,η we change to these coordinates:

$$W' = x\,H_{p_0} = 2\lambda(\partial_x + \xi\,\partial_\lambda) - x^2\eta_n^2\,\partial_\lambda + x\,\eta_n\,\partial_{y_1}$$
$$+ x(\eta_n + 2x\,\eta_n)\,\partial_{y_n}\,.$$

This is certainly tangent to the surface $p_0 = 0$, on which $\xi^2 = -x\eta_n^2 - \eta_1\,\eta_n$. Since we also want to consider homogeneous solutions we see $\eta_n = 1$ and obtain

$$W' = W'' \equiv 2\lambda\partial_x - 3x^2\,\partial_\lambda - 2x\eta_1\,\partial_\lambda + x(\partial_{y_1} + (\eta_1 + 2x)\partial_{y_n})$$
$$\text{on } p_0 = 0.$$

Now, we wish to solve

(V.14) $$W''k = xf = g \quad \text{on } \lambda^2 = -x^2\eta_1 - x^3.$$

Since f_0 in (V.13) is quadractic in ξ it suffices to assume, by the Melgrange preparation theorem that f is the restriction to $p_0 = 0$ of a linear function in ξ. Correspondingly it is enough to consider (V.14) for

(V.15) $$g = a(x,y,\eta)\lambda + b(x,y,\eta)$$

with $b(0,y;\eta) = 0$. Similarly, by replacing the quadractic term in λ using $\lambda^2 = -x^2\eta_1 - x^3$ we can suppose $h = c\lambda + d$ and (V.14) becomes

$$W(c\lambda + d)$$

$$= [2\partial_x b + x(\partial y_1 + (\eta_1 + 2x)\,\partial y_n)c]\lambda +$$

(V.16) $$[2(-x^2\eta_1 - x^3)\partial_x c - 3x^2\,c - 2x\eta_1\,c + x(\partial_{y_1} + (\eta_1 + 2x)\partial_{y_n})d]$$

$$= \quad a\lambda + b.$$

We first investigate this in terms of formal power series in $x_1\eta_1$. In fact, if $(c,d) = (c_j,d_j)\eta_1^j$ then

$$W[c\lambda + d] = [2\partial_x d_j + (x\partial_{y_1} + 2x^2\partial_{y_n})c_j]\lambda$$

$$+ [-2x^3\partial_x\,c_j - 3x^2\,c_j + (x\partial_{y_1} + 2x\partial_{y_n})d_j)$$
$$+ O(\eta_1^{\,j+1}).$$

Proceeding by induction we can therefore suppose $(a,b) = (a(x,y,\eta'),\, b(x,y,\eta'))\eta_1^j$ and try to solve:

$$\text{(V.18)}\qquad \begin{aligned} &2\,\partial_x b + x\partial_{y_1} c + 2x^2\,\partial_{y_n} c = a \\ &-2x^3\,\partial_x c - 3x^2 c + x\,\partial_{y_1} d + 2x^2\,\partial_{y_n} d = b \end{aligned}$$

as formal power series in x, with $b = O(x)$.

To do this first solve

$$\partial_{y_1} d(y,\eta') = b_1(y,\eta')$$

so that we can assume $b = O(x^2)$. Then proceed by induction. At the kth stage the term $d_{k+1}\, x^{k+1}$ in d can be chosen to eliminate $a_k x^k$ and $c_k x^k$ in c chosen to eliminate $b_{k+2} x^{k+2}$. This process converges, in the sense of formal power series, to a solution of (V.16). Thus we can assume that a,b vanish to all orders at $x = \eta_1 = 0$.

A similar argument can then be used to solve (V.16) in the sense of formal power series in x, with coefficients vanishing to all orders at $\eta_1 = 0$. Thus, it only remains to show that (V.16) has an exact solution when a,b vanish to infinite order on $x = 0$.

Returning to the oringinal form (V.13) of the problem we can introduce $x,y,\xi,\eta' = (\eta_2,\ldots,\eta_n)$ as coordinates on $p_0 = 0$ and assume that f vanishes to infinite order on $x = 0$. The objective is to find a C^∞ solution a which, when restricted to $x = 0$ is even under the exchange $\xi \longmapsto -\xi$. Indeed, the extension of a off $p_0 = 0$ as a linear function of ξ can then be make in such a way that the coefficient of ξ vanishes at $x = 0$, i.e., is a function of x,y,λ,η, even linear in λ.

To see this we make a singular change of coordinates from $x \geq 0$ to new variables (r,s,y_1') from (x,ξ,y_1) keeping $y_2,\ldots,y_n$, $\eta_2,\ldots,\eta_n$ parameters (and $\eta_n = 1$). The idea is to arrange to have

$$\text{(V.19)}\qquad \partial_s = 2\xi\,\partial_x - \partial_\xi + 2(\eta_1 + 2x)\partial_{y_n}\,.$$

This can be accomplished by setting

$$\text{(V.20)}\qquad \begin{cases} x = r - s^2 \\ \xi = -s & |s| \leq r,\ r \geq 0 \\ y_n = y_n' + \eta_1 s + 2(rs - s^3/3)\,. \end{cases}$$

Setting $\theta = s/r^{\frac{1}{2}}$ transforms the vector field to

$$\text{(V.21)}\qquad V = H_{p_0} = r^{\frac{1}{2}}\,\partial_\theta + \partial_{y_1} \qquad |\phi| \leq 1$$

which is even C^∞, in $r > 0$, if the surfaces $\theta = \pm 1$ are identified to give the manifold

$$M = S^1_\theta \times \mathbb{R}^+_r \times \mathbb{R}_{y_1} \times (\mathbb{R}^{n-1}_{y'}\ \ \mathbb{R}^{n-1}_{\eta'}).$$

Now, the transformation from $x \geq 0$ to M is singular only at $x = 0$, and there only to finite order. If follows that f can be transferred to $f' \in C^\infty(\overline{M})$ which vanishes to infinite order at the boundary $r = 0$.

It follows easily that

$$\text{(V.22)} \qquad Va' = f' \qquad a' = 0 \qquad \text{on } y_1 = -\varepsilon$$

has a unique C^∞ solution on $\overline{M}$ and a' vanishes to infinite order on $r = 0$. Note that a', unlike f', need not vanish to all orders at $\theta = \pm 1$. However the inverse transformation to (V.20) is singular only at $r = 0$. Thus a' pulls back to a C^∞ function of x,y,ξ,η' in $x \geq 0$ which vanishes to all orders at $x = 0$ and has the same value at $\pm\xi$ on $x = 0$. This completes the proof of Proposition V.12 and hence of Theorem V.6.

The main consequence of (V.7) is that F and G can be used to intertwine solutions to problems for P and P_0. The operator P_0 has coefficients depending only on x so can be analysed easily. For example, the Dirichlet problem:

$$\text{(V.23)} \qquad \begin{cases} P_0 u = 0 & \text{in } x \geq 0 \\ u = u_0 \in \mathcal{E}'(\mathbb{R}^n) & \text{in } x = 0 \\ u = 0 & \text{in } y_1 \ll 0 \end{cases}$$

has a unique solution modulo C^∞, of u_0 has wavefront set in a small cone $|\eta_1| \leq \varepsilon\,\eta_n$. The behaviour of this solution is determined by the forward Neumann operator

$$\text{(V.24)} \qquad N_0 u_0 = D_x u\big|_{x=0}.$$

Explicitly N_0 is a convolution operator on $\mathbb{R}^n$,

$$\text{(V.25)} \qquad N_0 u_0\,(\eta) = -\eta_n^{2/3}\left[\frac{Ai'}{Ai}(-\zeta_o)\right]\hat{u}_0(\eta).$$

Here, Ai is the standard Airy function solving $Ai''(z_1 = zAi(z)$ and decreasing rapidly as $z \to \infty$. It is entire, but has zeros, of order one, on the negative real axis. The multiplier in (V.25) is the tempered distribution:

$$\lim_{\varepsilon\uparrow 0} \frac{Ai'}{Ai}(-(\eta_1 + i\varepsilon)\,\eta_n^{-1/3})\ ,\ \zeta_0 = \eta_1\eta_n^{-1/3}$$

A microlocally unique solution to the analogous problem for P.

$$\left.\begin{array}{l} Pv = 0 \\ v = v_0 \in \mathcal{E}'(\mathbb{R}^n) \quad \text{on} \quad x = 0 \\ v = 0 \quad \text{in} \quad y_1 < -\varepsilon \end{array}\right\} \quad \text{at} \quad \overline{\rho}$$

can be shown to exist when v_0 has wavefront set in a sufficiently small cone around $\overline{\rho}$. The microlocal Neumann operator for P:

$$Nv_0 = D_x v\big|_{x=0}$$

can be expressed in terms of N_0. Indeed, $v \equiv Gu$ (at $\overline{\rho}$) so using the calculus of totally characteristic Fourier integral operator

(V.26) $$N = F_1^{-1}(A \cdot N_0 + B) \cdot F_1 \quad \text{at} \quad \overline{\rho}$$

where F_1 is a classical Fourier integral operator elliptic at $\overline{\rho}$ and associated to $\partial\chi$, A and B are classical pseudo-differential operators and A is elliptic of order 0. Apriori B will be of order one but is easily shown to be of order zero.

The representation (V.26) can be used to analyse other boundary value problems and, for example, transmission problems near a gliding point.

VI. $\dot{A}$ - Hypoellipticity.

Elliptic operators in $L_b^m(M)$, for M a manifold with boundary, are $\dot{A}$-hypoelliptic; if $u \in \dot{\mathcal{D}}'(M)$

(VI.1) $$Au \in \dot{A}(M) \iff u \in \dot{A}(M) \quad \text{if} \quad a \text{ is elliptic.}$$

In this lecture we shall examine some classes of degenerate elliptic operators in $L_b^m(M)$ for which (VI.1) still holds.

So, suppose $A \in L_b^2(M)$ has real principal symbol $a \in C^\infty(\tilde{T}^*M \setminus 0)$, homogeneous of degree two. We shall assume that A is of classical type and that

(VI.2) $$a \geq 0 \quad \text{on} \quad \tilde{T}^*M \quad 0, \quad a = 0 \text{ precisely on } \tilde{T}^*\partial M\setminus 0.$$

More specifically it will be assumed that

(VI.3) $$a\big|\partial\tilde{T}^*M \text{ vanishes to precisely second order at } \tilde{T}^*\partial M$$

and that in the normal direction

(VI.4) $$da \neq 0 \text{ and inward pointing at } T^*\partial M \setminus 0.$$

To ensure hypoellipticity we shall need a further condition on A at $T^*\partial M$, such as

(VI.5) $$\operatorname{Im} \sigma_1(A_0) \neq 0 \text{ or } \operatorname{Re} \sigma_1(A_0) > 0.$$

Here A_0 is defined by $A_0 u_0 = Au|_{\partial M}$ if $u \in \dot{A}'(M)$ and $u_0 = u|\partial M$; the second part of (V1.2) shows that A_0 is of order one.

Although the hypoellipticity of A will be shown by the use of estimates, based as much on the material of Lecture I as on the calculus itself, and these can be carried out directly it is useful to see the extent to which A can be reduced to normal form. Since A is elliptic except on $T^*\partial M$ we know

(V1.6) $\quad Au \in \dot{A}(M) \Rightarrow WF_b(u) \subset T^*\partial M.$

It is therefore only necessary to examine A microlocally near points of $T^*\partial M$.

(V1.7) Proposition. If $A \in L_b^2(M)$ is classical and has real principal symbol satisfying (V1.2), (V1.3), (V1.4) near $\bar{\sigma} \in T^*\partial M \setminus 0$ there exists a boundary canonical transformation $\tilde{\chi}: \tilde{T}^*M, \bar{\sigma} \longrightarrow \tilde{T}^*Z, (0,0,0,\bar{\eta})$ and elliptic Fourier integral operators F, G associated to it such that

(V1.8) $\quad F \cdot A = [(xD_x)^2 + xD_{y_n}^2 + B(y, D_y)] \cdot G$ at σ.

Here $B \in L_\ell^1(\mathbb{R}^n)$ has symbol b_1 with

(V1.9) $\quad \sigma_1(A_0) = \alpha \cdot (\partial\chi)^* b_1, \quad \alpha > 0.$

The method of proof of such a proposition should, by now, be relatively clear. Details will not be given here. However, we make some remarks on the choice of $\tilde{\chi}$. Suppose that $\tilde{\chi}_1$ can be found with

(V1.10) $\quad (\tilde{\chi}_1^{-1})^* a = \alpha(\lambda^2 + x\beta(y,\eta) + O(x^\infty)) \quad \alpha, \beta > 0.$

A canonical transformation, taking β to η_n^2, then gives the principal symbol equation for (V1.8). Note that

$$\frac{\lambda^2 + x\lambda_n^2 + \gamma}{\lambda^2 + x\eta_n^2} \in C^\infty$$

if γ vanishes to infinite order at $x = 0$. This means that (V1.10) suffices to give the leading part of (v1.8). The construction of $\tilde{\chi}_1$ is therefore only a problem in fomal power series at $x = 0$, it is reasonably easy. Similar remarks apply to the simplification of the lower order terms. Observe that the invariance of the symbol of B, apart from a conformal factor, as stated in (V1.9) follows directly from (V1.8).

The basic result to be discussed is:

(V1.11) Theorem. If $A \in L_b^2(M)$ satisfies (V1.2), (V1.3), (V1.4) and (V1.5) near $\bar{\sigma} \in T^*\partial M \setminus 0$ and $u \in \mathcal{D}'(M)$ then

(V1.12) $\quad \bar{\sigma} \notin WF_b(Au) \Longleftrightarrow \bar{\sigma} \notin WF_b(u).$

To deduce (V1.12), even with the simplified from (V1.8) available, the basic idea is to show the existence of a fixed Sobolev space H_s such that for a fixed cone γ around $\bar{\sigma}$ in $\tilde{T}^*Z$

(V1.13) $\quad Du \in H^s_{loc}(Z) \quad \forall\, P \in L^k_b(Z)$ with ess.sup $P \subset \gamma$.

Naturally, proof is by induction over the order of P.

To start the induction observe that there exists s such that (V1.13) holds for $k \le 0$. This follows from the fact that $u \in H^s$ for some s and

(V1.14) Proposition. If $s \in \mathbb{Z}$ and $P \in L^0_b(Z)$ then

$$P : H^s_{comp}(Z) \longrightarrow H^s_{loc}(Z).$$

Proof. The L^2-boundedness of operators in $L^0_b(Z)$ has already been noted in Theorem II.23. If $s < 0$ then $u \in H^s_{comp}$ if and only if it can he written in the form

(V1.15) $$u = \sum_{|\alpha| < -s} D^\alpha v_\alpha \;, \quad v_\alpha \in L^2_{comp}(Z).$$

Since

(V1.16) $$P\, D^\alpha = \sum_{|\gamma| \le |\alpha|} D^\gamma P_\gamma \qquad P_\gamma \in L^0_b(Z)$$

for any $P \in L^0_b(Z)$ it follows that Pu is of the form (V1.15) locally, hence is in $H^s_{loc}(Z)$.

The same type of argument works for $s \in \mathbb{N}$. Indeed, $u \in H^s_{comp}(Z)$ if and only if

(V1.17) $$D^\alpha u = v^\alpha + \sum_{j \le s-1} \delta^{(j)}(x) \otimes u_j^{(\alpha)}(y) \qquad \forall |\alpha| \le s$$

with $v^\alpha \in L^2_{comp}$ and $u_j^{(\alpha)} \in H^{s-j-\frac{1}{2}}(\mathbb{R}^n)$, since u can then be recovered by solving an elliptic boundary value problem. Using (V1.16) and the behaviour of P on boundary terms it is clear that Pu is also locally of the form (V1.17). Hence the proposition is proved.

(V1.18) Corollary. If $P \in L^0_b(M)$ then

$$P : \dot{H}^s_{comp}(M) \longrightarrow \dot{H}^s_{loc}(M) \quad \forall\, s \in \mathbb{N}.$$

For simplicity we shall first make the global assumption that

(V1.19) $\quad [(xD_x)^2 + xD^2_{y_n} + B(y.D_y)]u = f \in \dot{A}(Z).$

Now from the calculus we have (V1.6), $WF_b(u) \subset T^*\partial Z$. First observe that if $f \in \dot{A}(Z)$ and $k \in \mathbb{N}$ is given, there is an integer $N_0 \in \mathbb{N}$ such that

$$x^N f \in \dot{H}^k_{loc}(Z) \qquad \forall\, N \geq N_0 \tag{V1.20}$$

provided the order of f is finite.

We therefore multiply (V1.19) by x^N. A short caculation shows that if $g = x^N f$ and $v = x^N u$,

$$[(D_x x)(xD_x) + \frac{i(N-1)}{2}(xD_x + D_x x) + \frac{(N-1)^2}{2} + xD^2_{y_n} + B]v = g \tag{V1.21}$$

Now, suppose that $P = P(y,Dy)$ is a compactly supported self-adjoint pseudodifferential operator with essential support near $\overline{\sigma} = (0,\overline{\eta}) \in T^*\mathbb{R}^n$ and

$$\text{order of } P = -k - \tfrac{1}{2}. \tag{V1.22}$$

Applying P^2 to both sides of (v1.21) and formally pairing in L^2 the real part gives:

$$\begin{cases} \|xD_x Pv\|^2 + \dfrac{(N-1)^2}{2}\|Pv\|^2 + \langle Pv, x\,PD^2_{y_n}v\rangle \\ + \mathbb{R}e\langle Pv, PBv\rangle = \mathbb{R}e\langle Pv, g\rangle. \end{cases} \tag{V1.23}$$

Assuming that the second condition in (V1.5) holds, so $\sigma_1(B) > 0$, this can be written:

$$\begin{cases} \|xD_x Pv\|^2 + \dfrac{(N-1)^2}{2}\|Pv\|^2 + \|x^{\frac{1}{2}}D_{y_n}Pv\|^2 \\ \quad + \mathbb{R}e\langle Pv, BPv\rangle \leq |\langle xG_1 v, v\rangle| + |\langle G_2 v, v\rangle| \\ \qquad\qquad + \|Pv\|\,.\,\|g\|\,. \end{cases} \tag{V1.24}$$

The commutator terms on the right in (V1.24) have $G_1, G_2 \in L^{\cdot}(\mathbb{R}^n)$ of order 2x order P and 2x order P-1 respectively. Since there are elliptic terms of the same type, but higher order, on the left it is clear that, at least at this formal level, induction over the order of P can be used to bound the left side of (V1.24), for any k in (V1.22). Natrually, the essential supports of successive test operators are nested and to justify (V1.24) it is necessary to approximate P by a sequence of smoothing operators. These technical details are however routine.

If the first condition in (V1.5) holds it is necessary to consider the imaginary part of the inner product of P^2v with (IV.21). In either case we deduce, not precisely (V1.23) but this condition on $v = x^N u$. Note also that this conclusion really only depends on the assumption (V1.12) rather than its global analogue (V1.19).

To complete the proof of the theorem it is necessary to show that

(V1.25) $$\bar{\sigma} \notin WF_b(x^N u) \Rightarrow \bar{\sigma} \in WF_b(u),$$

given (V1.12). If $P \in L^0(Z)$ has essential support near $\bar{\sigma}$ then (V1.25) implies

$$P'(x^N u) = x^N Pu \in \dot{A}(Z).$$

Replacing u by Pu we can assume that $x^N u \in \dot{A}(Z)$. Division by x^N is possible in $\dot{A}(Z)$, i.e., there exists $u_1 \in \dot{A}(Z)$ with $x^N u_1 = x^N u$. Subtracting u_1 it only remains to show that if u' is known, a priori, to be of the form

(V1.26) $$u' = \sum_{0 \leq j < N} \delta^{(j)}(x) \otimes u_j(y)$$

$\bar{\sigma} \notin WF_b(Au')$ implies $\bar{\sigma} \notin WF(u')$.

Substituting (V1.26) into (V1.19) directly gives

(V1.27) $$\sum_{0 \leq j < N} \delta^{(j)}(x) \otimes \{-(j+1)^2 u_j + Bu_j - (j+1) D^2_{y_n} u_{j+1}\} \in \dot{A}(Z).$$

Thus, for each j, the bracketed term in (V1.27) is C^∞, at least has no wavefront set at $\bar{\sigma}$. The top term satisfies

$$\bar{\sigma} \notin WF(Bu_{N-1} - N^2 u_{N-1})$$

and since B is elliptic of order 1, $\bar{\sigma} \notin WF(u_{N-1})$. The same argument applies to successive terms completing the proof of the theorem.

The type of argument outlined above can be applied more generally. For example, if (V1.5) is replaced by

(V1.28) $$\sigma_1(A_0) \geq 0$$

then the first part of the discussion above, but now using Gårding's inequality, carries over and (V1.12) implies $\bar{\sigma} \notin WF_b(x^N u)$ if N is large enough. The only possible micro-local solutions are of the form (V1.26) and result from the existence of non-trivial microlocal solutions to

(V1.29) $$\bar{\sigma} \notin WF(Bu_j - (j+1)^2 u_j) \qquad (j \geq 0).$$

For example

(V1.30) Proposition. If $A_0 \equiv 0$ and (V1.2), (V1.3), (V1.4) holds then P is $\dot{A}$-hypoelliptic at $\bar{\sigma}$.

In fact there are even more degenerate operators that can be handled in this way. In some circumstances it is possible to replace the condition (VI.4) by a weaker "tangential hypoellipticity" condition. This arises in the treatment of the Laplace-Beltrami operator for the complete Kähler metric

$$g_{i\bar{j}} = - \partial_j \alpha_i \log(- \phi)$$

where ϕ is a C^∞ strictly plurisubharmonic defining function for a bounded, C^∞, strictly pseudo convex domain $M \subset \mathbb{C}^n$. Details of such applications will appear elsewhere.

References

1. L. Hörmander. Fourier integral operators I Acta Math. 127 (1971) 79-183.

2. L. Hörmander. Linear partial differential operators. Springer 1963.

3. R.B. Melrose. Transformation of boundary value problems. to appear.

4. R. B. Melrose & J. Sjöstrand. Singularities of boundary value problems I. Comm. Pure Appl. Math. 31 (1978), 593-617.

5. R. B. Melrose. Differential boundary value problems of principal type. in Seminar on singularities of solutions of linear partial differential equations. Ed L. Hörmander P.U.P. 1979.

6. R. B. Melrose. Equivalence of glancing hypersurfaces. Invent. Math. 37 (1976), 165-191.

7. J. J. Duistermaat & L. Hörmander. Fourier integral operators II. Acta. Math. 128 (1972), 183-269.

8. V.W. Guillemin & R. B. Melrose. The Poisson summation formula for manifolds with boundary. Adv. in Math. 32 (1979) 204-232.

9. F.G. Friedlander. The wavefront set of a simple initial-boundary value problem with glancing rays. Math. Proc. Camb. Phil Soc. 79 (1976) 145-149.

10. K.G. Andersson & R. B. Melrose. The propagation of singularities along gliding rays. Invert. Math. 41 (1977) 197-232.

PROPAGATION OF SINGULARITIES AND THE SCATTERING MATRIX

James V. Ralston

University of California, Los Angeles

ABSTRACT.

In exterior domains in $\mathbb{R}^n$ the scattering phase for the laplacian often plays the role like that of the eigenvalue counting function in interior domains. These lectures give an introduction to the scattering matrix, and show how the basic results on propagation of singularities and parametrices for the wave equation apply here to give "spectral asymptotics" for exterior domains and information about the poles of the scattering matrix

0. INTRODUCTION.

The topic of these lectures will be the laplacian in an exterior domain Ω in $\mathbb{R}^n$. Our approach will be that of scattering theory in that we will study the spectral properties of the laplacian on Ω "relative to" the laplacian on all of $\mathbb{R}^n$. My goal is to present analogues for these exterior problems of the well-known spectral asymptotics for interior problems. The technical tools in this work are the known results on the propagation of singularities and parametrices for the wave equation in Ω. However, to be applied here these results need to be combined with explicit formulas from scattering theory. Hence the plan of these lectures will be to derive the needed results from scattering theory first, then present the "spectral asymptotics", and finally show how one of these leads to a new result on the position of the poles of the scattering matrix. There are presently more conjectures than theorems concerning the scattering matrix and I will discuss open problems as we come to them.

H. G. Garnir (ed.), Singularities in Boundary Value Problems, 169–184.

All of my own contributions to this are joint work with J.W. Helton, A. Majda and mostly recently with C. Bardos and J.C. Guillot. In particular, the two theorems in the later lectures are joint work with Bardos and Guillot. In addition I am grateful to R. Melrose and R. Redheffer for helpful discussions of trace formulas and the proof of Theorem 2.

I. SOME SCATTERING THEORY.

For our purposes here it is a little easier to begin with the wave equation in Ω — instead of the Schrödinger equation — and consider the boundary value problem:

$$u_{tt} - \Delta u = 0 \quad \text{in} \quad \Omega \times \mathbb{R}$$

$$u = 0 \quad \text{on} \quad \partial\Omega \times \mathbb{R} \quad .$$

We will use the Dirichlet boundary condition on $\partial\Omega$ in these lectures, but all these results can be extended to the Neumann condition as well.

In scattering theory one begins by studying the interaction of plane wave trains with $\partial\Omega$. The plane wave train $\exp(ik(t - \theta\cdot x))$, $|\theta| = 1$, $k > 0$, satisfies differential equation but not the boundary condition. Hence one looks for solutions in the form

$$v(t,x) = e^{ik(t-\theta\cdot x)} + e^{ikt}w_s(x)$$

where

$$\text{(i)} \quad \Delta w_s + k^2 w_s = 0$$

$$\text{(ii)} \quad e^{-ik\theta\cdot x} + w_s = 0 \quad \text{on} \quad \partial\Omega \quad .$$

To make w_s unique, one requires that it be the limit of the (unique) square-integrable solutions to this problem when $\operatorname{Im} k < 0$ — this is one form of the "radiation condition". With this additional requirement w_s exists and is unique (see [ST] or [HR, Appendix] for a proof of this). Moreover, as $|x| \to \infty$

$$w_s(x) = \frac{e^{-ik|x|}}{|x|^{\frac{n-1}{2}}} T(\theta',\theta,k) + O\left(\frac{1}{|x|^{\frac{n+1}{2}}}\right)$$

where $\theta' = \frac{x}{|x|}$. Note that this says that the reflected wave train $\exp(ikt)w_s$ is spherical for $|x|$ large, and propagating outward – as is intuitively correct. Note also that, given (i) and $T(\theta',\theta,k)$, w_s is uniquely determined, so it is natural to study $T(\theta',\theta,k)$ instead of w_s.

The "transmission coefficient", $T(\theta',\theta,k)$, satisfies a number of identities. The most surprising of these is the "reciprocity law": $T(-\theta,\omega,k) = T(-\omega,\theta,k)$. To prove this one proceeds as follows:

$$0 = \int_{|x|=R} w_s(x,\theta,k) \frac{\partial w_s}{\partial r}(x,\omega,k) - w_s(x,\omega,k) \frac{\partial w_s}{\partial r}(x,\theta,k)dS$$

$$= \int_{\partial\Omega} - e^{-ikx\cdot\theta} \frac{\partial w_s}{\partial \nu}(x,\omega,k) + e^{-ikx\cdot\omega} \frac{\partial w_s}{\partial \nu}(x,\theta,k)dS$$

$$= \int_{\partial\Omega} (w_s(x,\omega,k) \frac{\partial}{\partial \nu}(e^{-ikx\cdot\theta}) - e^{-ikx\cdot\theta} \frac{\partial w_s}{\partial \nu}(x,\omega,k))dS$$

$$- \int_{\partial\Omega} (w_s(x,\theta,k) \frac{\partial}{\partial \nu}(e^{-ikx\cdot\omega}) - e^{-ikx\cdot\omega} \frac{\partial w_s}{\partial \nu}(x,\theta,k))dS$$

$$= \int_{|x|=R} w_s(x,\omega,k) \frac{\partial e^{-ikx\cdot\theta}}{\partial r} - e^{-ikx\cdot\theta} \frac{\partial w_s}{\partial r}(x,\omega,k)dS$$

$$- \int_{|x|=R} w_s(x,\theta,k) \frac{\partial}{\partial r} e^{-ikx\cdot\omega} - e^{-ikx\cdot\omega} \frac{\partial w_s}{\partial r}(x,\theta,k)dS$$

The integrals in the last equation above can be evaluated by computing the leading term by stationary phase $R \to \infty$ (cf [LP 2, p. 744]). This yields $0 = a(k)\, T(-\theta,\omega,k) - a(k)\, T(-\omega,\theta,k)$ where $a(k) \neq 0$. This argument is typical of many computations in scattering theory: Green's formula plus a little stationary phase.

The functions $\varphi_+ = e^{-ikx\cdot\theta} + w_s(x,\theta,k)$ are known as (outgoing) distorted plane waves. Note that $\varphi_-(x,\theta,k) \equiv \varphi_+(x,-\theta,k)$ also satisfies (i) and (ii). However, $\varphi_- - e^{-ikx\cdot\theta}$ is the limit of the square-integrable solutions for $\operatorname{Im} k > 0$. In the picture given earlier $e^{ikt}\varphi_-(x,\theta,k)$ is a plane wave plus a reflected asymptotically spherical wave which is propagating <u>inward</u>. Hence one calls φ_- an incoming distorted plane wave.

The scattering matrix, $\mathcal{S}(k)$, can be defined as the operator on $L^2(S^{n-1})$ mapping $\bar{\varphi}_+(x,\cdot,k)$ to $\bar{\varphi}_-(x,\cdot,k)$. To see that there is such an operator and determine it explicitly, it suffices to find a smooth kernel on $S^{n-1} \times S^{n-1}$ such that

$$0 \equiv \bar{\varphi}_-(x,\theta,k) - \bar{\varphi}_+(x,\theta,k) - \int_{S^{n-1}} K(\theta,\omega)\bar{\varphi}_+(x,\omega,k)d\omega . \tag{1}$$

Letting $|x| \to \infty$, evaluating $\int_{S^{n-1}} K(\theta,\omega)e^{ikx\cdot\omega}d\omega$ by stationary phase, and substituting the asymptotic form for the spherical waves in the φ's, one sees that $K(\theta,\theta')$ is necessarily $c(k)\,T(-\theta',-\theta)$, which equals $c(k)\,T(\theta,\theta')$ by the reciprocity law. Moreover, with this choice of K the function on the right of (1) satisfies

$$\Delta u + k^2 u = 0 \qquad\qquad u = 0 \quad \text{on} \quad \partial\Omega$$

and

$$\frac{\partial u}{\partial r} - iku = O\left(|x|^{-\frac{n+1}{2}}\right), \qquad u = O\left(|x|^{-\frac{n-1}{2}}\right),$$

Hence, by Rellich's uniqueness theorem $u \equiv 0$ and (1) holds identically for $x \in \Omega$. Thus $\mathcal{S}(k) = I + \mathcal{K}(k)$, where $\mathcal{K}(k)$ is the integral operator with kernel $K(\theta,\omega,k)$. Finally, if one now returns to (1) and computes the leading term of the asymptotic development as $|x| \to \infty$ for the right hand side, one has

$$0 = -\bar{T}(\theta',\theta) - \frac{c(k)}{\bar{c}(k)}\,T(\theta,\theta') - \int_{S^{n-1}} c(k)\,T(\theta,\omega)\,\bar{T}(\theta',\omega)d\omega$$

which says that $\mathcal{S}(k)$ is unitary.

The derivation of the reciprocity law contains the formula:

$$\int_{\partial\Omega} e^{-ikx\cdot\omega}\frac{\partial}{\partial\nu}\left(e^{ikx\cdot\theta}\right) - e^{ikx\cdot\theta}\frac{\partial w_s}{\partial\nu}(x,\omega,k)dS$$
$$= a(k)\,T(\theta,\omega,k) \quad . \tag{A}$$

Since it is easy to show that $w_s(x,\omega,k)$ extends to a holomorphic function in $\mathrm{Im}\,k < 0$, formula (A) shows that $T(\theta,\omega,k)$ and hence $\mathcal{S}(k)$ extend to holomorphic functions $\mathrm{Im}\,k < 0$. (Formula (A) is also a starting point for determining the asymptotic behavior of $T(\theta,\omega,k)$ as $k \to \infty$.)

Since $\mathcal{S}(k)$ has the form $I + \mathcal{K}(k)$ where $\mathcal{K}(k)$ has a smooth kernel and hence is a trace class operator, it follows that $\det \mathcal{S}(k)$ exists and is holomorphic in $\operatorname{Im} k \leq 0$ (see [GK, Chapter IV]). Thus, defining $\mathcal{S}(k) = (\mathcal{S}^*(\bar{k}))^{-1}$ we have a meromorphic continuation of $\mathcal{S}(k)$ to $\mathbb{C} - \{k \leq 0\}$. In the case that n is odd $(\Omega \subset \mathbb{R}^n)$ this analytic continuation extends across the negative real axis and $\mathcal{S}(k)$ is meromorphic on $\mathbb{C}$, (see [LP 1], Chapter V).

The following identity ([GK, Chapter IV])

$$\frac{d}{dk} \log \det \mathcal{S}(k) = \operatorname{tr}\left(\mathcal{S}^* \frac{d\mathcal{S}}{dk}\right) = -\operatorname{tr}\left(\mathcal{S} \frac{d\mathcal{S}^*}{dk}\right) ,$$

and the same sort of manipulations with Green's formula used earlier (see [HR, Addendum]), yield the following formula for the kernel of $-\mathcal{S} \frac{d\mathcal{S}^*}{dk}$

$$\frac{i}{8\pi^2} \left(\frac{k}{2\pi}\right)^{n-3} \int_{\partial\Omega} \frac{\partial \varphi_+}{\partial \nu} (-\theta) \frac{\partial \bar{\varphi}_+}{\partial \nu} (-\omega)\, x \cdot \nu \, dS .$$

Thus we have

$$\frac{1}{2\pi i} \frac{d}{dk} \log \det \mathcal{S}(k) = \frac{1}{16\pi^3} \left(\frac{k}{2\pi}\right)^{n-3} \int_{\partial\Omega} dS \int_{|\theta|=1} (x \cdot \nu) \left|\frac{\partial \varphi_+}{\partial \nu} (\theta)\right|^2 d\theta \quad \text{(B)}$$

Formula (B) is the starting point for a proof that $\frac{d}{dk} \log \det \mathcal{S}(k)$ has an asymptotic expansion in inverse powers of k and logarithms as $k \to \infty$, when $\partial\Omega$ is strictly convex. This proof was outlined in [R2] for the Dirichlet boundary condition, and is really a special case of the results of Melrose [M] in the case of either Dirichlet or Neumann conditions.

The function $s(k) = \frac{1}{2\pi i} \log \det \mathcal{S}(k)$ is real since $|\det \mathcal{S}(k)| = 1$, and it is known as the scattering phase. In our next lecture we will describe two situations where it plays the same role for the exterior Dirichlet problem that the counting function $N(\lambda) = \{\#\text{ eigenvalues} \leq \lambda^2\}$ plays for the interior Dirichlet problem.

II. TRACE FORMULAS.

The spectral asymptotics for the laplacian in an interior domain Ω begin with two simple trace formulas:

$$\operatorname{tr}(e^{t\Delta}) = \Sigma\, e^{-\lambda_j^2 t}$$

and

$$\frac{1}{\sqrt{2\pi}} \operatorname{tr}\left(\int \rho(t) \cos t\sqrt{-\Delta}\, dt\right) = \frac{1}{2} \sum_j \hat{\rho}(\pm \lambda_j)\ , \quad \rho \in C_0^\infty(\mathbb{R})\ ,$$

where $\{-\lambda_j^2\}$ are the eigenvalues of the laplacian in Ω. To emphasize the analogy we want to draw here, we rewrite these as:

$$\operatorname{tr}(e^{t\Delta}) = t\int_0^\infty e^{-\mu t}\, N(\sqrt{\mu})d\mu \qquad \text{and}$$

$$\frac{1}{\sqrt{2\pi}} \left(\int \rho(t) \cos t\sqrt{-\Delta}\, dt\right) = \frac{1}{2} \int \frac{d\hat{\rho}}{d\lambda}(\lambda)\, N(\lambda)d\lambda$$

Here for $\lambda \geq 0$, $N(\lambda) = \{\# \lambda_j \leq \lambda\}$ and $N(-\lambda) = -N(\lambda)$.

When Δ is the laplacian on the exterior domain Ω, neither $e^{t\Delta}$ nor $\int \rho(t) \cos t\sqrt{-\Delta}\, dt$ is trace class, but, if we let Δ_0 denote the laplacian on $\mathbb{R}^n$ and extend operators on $L^2(\Omega)$ to be zero on $L^2(\mathbb{R}^n - \Omega)$, then we have

$$\operatorname{tr}(e^{t\Delta_0} - e^{t\Delta}) = t\int_0^\infty e^{-\mu t}\, s(\sqrt{\mu})d\mu \tag{C}$$

and

$$\frac{1}{\sqrt{2\pi}} \operatorname{tr}\left(\int \rho(t)(\cos t\sqrt{-\Delta} - \cos t\sqrt{-\Delta_0})\, dt\right) = \frac{1}{2} \int_{-\infty}^{\infty} \frac{d\hat{\rho}}{d\lambda}(\lambda)\, s(\lambda)d\lambda \tag{D}$$

where $s(-\lambda) = -s(\lambda)$. These formulas are consequences of Krein's theory of the spectral shift functions. Formula (C) was derived from that theory by Jensen and Kato in [JK], and (D) is easily derived by the same method. However, the analogue of (C) for the corresponding resolvents (multiply both sides of (C) by

$t^{\frac{n}{2}+\varepsilon} e^{-\sigma t}$ and integrate from 0 to ∞) was given by Buslaev in [B 1], and the analogue of (C) for the laplacian plus a potential was found independently by Colin de Verdière [CdV].

In [B 2] Buslaev gives a derivation of formulas of this sort which is fairly intuitive and goes as follows. The functions $\varphi_+(\cdot,\omega,k)$ actually form a complete set of generalized eigenfunctions for the laplacian in Ω [ST]. Hence given $\psi(\lambda) \in C_0^\infty(R)$, the operator $\psi(\Delta)$ in the functional calculus of Δ has the kernel

$$(2\pi)^{-n} \int_{S^{n-1}} d\omega \int_{\mathbb{R}_+} \varphi_+(x,\omega,k)\psi(-k^2)\,\bar{\varphi}_+(y,\omega,k)k^{n-1}dk$$

Hence, assuming $\psi(\Delta) - \psi(\Delta_0)$ is trace class and computing its trace as the integral of its kernel on the diagonal,

$$\operatorname{tr}(\psi(\Delta) - \psi(\Delta_0))$$

$$= \lim_{R\to\infty}(2\pi)^{-n}\int_{S^{n-1}} d\omega \int_{\mathbb{R}_+} \psi(-k^2)k^{n-1}dk \int_{|x|\le R} (|\varphi_+(x,\omega,k)|^2 - 1)dx,$$

where $\varphi_+ \equiv 0$ on the complement of Ω. Differentiating $\Delta\varphi_+ + k^2\varphi_+ = 0$ with respect to k, one has $(\Delta + k^2)\frac{\partial\varphi_+}{\partial k} = -2k\varphi_+$ and after substituting this and using Green's formula one gets

$$\operatorname{tr}(\psi(\Delta) - \psi(\Delta_0))$$

$$= \lim_{R\to\infty} (2\pi)^{-n}\int_{S^{n-1}} d\omega \int_{R^+} \psi(-k^2)k^{n-2}dk \int_{|x|=R} -\frac{1}{2}\left[\frac{2k|x|}{n} + \varphi_+\frac{\partial^2\varphi_+}{\partial|x|\partial k} - \frac{\partial\bar{\varphi}_+}{\partial|x|}\frac{\partial\varphi_+}{\partial k}\right]dS$$

At this point it is clear that $\operatorname{tr}(\psi(\Delta) - \psi(\Delta_0))$ depends only on the asymptotic behavior of φ_+ as $|x| \to \infty$. Substituting the asymptotic form for φ_+ and using stationary phase one eventually (the computation is actually rather lengthy) arrives at (for $n > 1$):

$$\operatorname{tr}(\psi(\Delta) - \psi(\Delta_0)) = -\int_0^\infty \psi(-k^2) \frac{ds}{dk}(k)\, dk.$$

Applied with $\psi(x) = e^{xt}$ and $\psi(x) = \hat{\rho}(\sqrt{-x}) + \hat{\rho}(-\sqrt{-x})$ this leads, at least formally, to (C) and (D), respectively. However, it is clear that to make this derivation rigorous one must know something about the behavior of $s(k)$ as $k \to \infty$. The advantage of Krein's theory is that it overcomes this difficulty: once one has shown that

$$(I - \Delta)^{-\frac{n}{2}-\varepsilon} - (I - \Delta_0)^{-\frac{n}{2}-\varepsilon}$$

is a trace class, it follows that (C) and (D) hold and

$$\int_0^\infty |s(\lambda)\lambda|(1 + \lambda^2)^{-\frac{n}{2}-1-\varepsilon} d\lambda < \infty .$$

Once one has (C) the computation of the asymptotic behavior of $s(k)$ as $k \to \infty$ — which we think of as analogous to the asymptotic behavior of $N(k)$ as $k \to \infty$ — is greatly simplified. The asymptotic behavior of $\operatorname{tr}(e^{t\Delta} - e^{t\Delta_0})$ as $t \to 0_+$ is easy to determine using the results of Pleyel, Minakshisundaram and McKean-Singer (see [MS]). So once one knows that $s(k)$ has an asymptotic expansion as $k \to \infty$, one has only to equate coefficients on the two sides of (1) to arrive at (cf. [B 1] and [MR])

$$s(k) = \frac{(4\pi)^{-n/2}}{\Gamma(1 + n/2)} k^n \text{ (Volume } \Omega^c)$$

$$- \frac{-(4\pi)^{-(n-1)/2}}{\Gamma\left(1 + \frac{n-1}{2}\right)} \frac{k^{n-1}}{4} \text{ (Area } \partial\Omega) + \cdots$$

The truly difficult step is showing that $s(k)$ has an asymptotic expansion. As was noted earlier for $\partial\Omega$ strictly convex one may prove this by substituting Melrose's parametrix into (B). However, by analogy with variable coefficient problems, $s(k)$ should have an asymptotic expansion in inverse powers of k as long as there are no trapped ray paths in Ω. Proving this looks like a formidable problem. Moreover, by analogy with Weyl's theorem for $N(\lambda)$, at least the leading term in the expansion of $s(k)$ should be valid for all obstacles regardless of geometry. The best result in this direction is in Jensen and Kato [JK] where it is shown that (C), combined with the monotonicity of $s(k)$ for starlike obstacles (see (B)) and a tauberian theorem of G. Freud implies that the leading term is valid for starlike obstacles.

Formula (D) does not have such direct consequences in spectral theory. However, since $s(k) = \frac{1}{2\pi i} \log \det \mathcal{S}(k)$ on $\mathbb{R}$ for n odd, if one could deform the contour of integration on the right hand side of (D) into the lower half-plane one would have

$$\frac{1}{\sqrt{2\pi}} \operatorname{tr} \int \rho(t)(\cos t\sqrt{-\Delta} - \cos t\sqrt{-\Delta_0})dt = \frac{1}{2}\sum_j \hat{\rho}(z_j) \tag{E}$$

where the z_j's are the zeros of $\det \mathcal{S}(z)$, counted by multiplicity. The conjugates of the z_j's are the poles of the scattering matrix and (E) can be interpreted as saying

$$\operatorname{tr}(\cos t\sqrt{-\Delta} - \cos t\sqrt{-\Delta_0}) = \frac{1}{2} \sum_{\text{poles}} e^{it\mu_j}$$

in the sense of distributions. Unfortunately, even assuming $\rho(t) \in C_0^\infty(a,\infty)$, $a \gg 0$, one does not yet know enough about $\det \mathcal{S}(z)$ to justify the contour deformation. Nonetheless, using results of Lax and Phillips [LP 1] one can prove (E) **by a simple argument**, provided n is odd and $\rho \in C_0^\infty(2R,\infty)$, where $\partial\Omega \subset \{|x| < R\}$. Since this argument has not appeared elsewhere, we will sketch it here. Showing that (E) holds for $\rho \in C_0^\infty(0,\infty)$ is an interesting unsolved problem.

To apply the results of Lax and Phillips we need to move from $L^2(\Omega)$ to $H_E(\Omega)$. The energy space H_E is defined as the closure of $C_0^\infty(\Omega) \times C_0^\infty(\Omega)$ in the norm

$$\|\{f_1,f_2\}\|_E^2 = \int_\Omega |\nabla f_1|^2 + |f_2|^2 dx$$

For solutions of $u_{tt} = \Delta u$ in $\Omega \times \mathbb{R}$, $u = 0$ on $\partial\Omega \times \mathbb{R}$ with initial data in $C_0^\infty(\Omega) \times C_0^\infty(\Omega)$ the mapping

$$U(t): \{u(0),\ u_t(0)\} \to \{u(t),\ u_t(t)\}$$

is isometric in $\| \ \|_E$ and $U(t)$ extends to a unitary group on $H_E(\Omega)$. The corresponding group in $H_E(\mathbb{R}^n)$ is denoted by $U_0(t)$. Finally one extends $U(t)$ to $H_E(\mathbb{R}^n)$ by defining it equal to zero on $H_E(\mathbb{R}^n) \ominus H_E(\Omega)$. With these definitions is is not hard to check that for $\rho \in C_0^\infty(0,\infty)$

$$\operatorname{tr}_{H_E}\left(\int \rho(t)(U(t) - U_0(t)\right)dt$$

$$= \operatorname{tr}_{L^2}\left(2\int \rho(t)(\cos t\sqrt{-\Delta} - \cos t\sqrt{-\Delta_0})dt\right).$$

The following subspaces of H_E play a central role in Lax and Phillips's theory:

$$D_\pm = \{f \in H_E : U_0(t)f = 0 \text{ for } |x| < R \pm t\} .$$

Let $P_\pm$ be the orthogonal projections on $H_E \ominus D_\pm$ respectively. Since ρ is supported in $(0,\infty)$, it follows that

$$\int \rho(t)\ (U(t) - U_0(t))f\ dt = 0 \quad \text{for} \quad f \in D_+$$

and

$$\left(\int \rho(t)(U(t) - U_0(t))dt\right)^* f = \int \rho(t)(U(-t) - U_0(-t))f\ dt = 0$$

for $f \in D_-$. Thus

$$\operatorname{tr} \int \rho(t)\ (U(t) - U_0(t))dt =$$

$$\operatorname{tr}\ P_- \int \rho(t)\ (U(t) - U_0(t))dt\ P_+$$

When n is odd, P_- and P_+ commute [LP, Chapter IV] and using $\operatorname{tr} AB = \operatorname{tr} BA$ we have

$$\operatorname{tr}\int \rho(t)(U(t) - U_0(t))dt$$

$$= \operatorname{tr}\int \rho(t)(P_+ U(t)P_- - P_+U_0(t)P_-)dt$$

If we now assume ρ is supported in $(2R,\infty)$, since $P_+ U_0(t)P_- \equiv 0$ for $t > 2R$ we have

$$\operatorname{tr}\int \rho(t)(U(t) - U_0(t))dt = \operatorname{tr}\int \rho(t) Z(t)dt$$

where $Z(t) = P_+ U(t)P_-$. The function $Z(t)$ is a semigroup on

$K = H_E \ominus (D_+ \oplus D_-)$ and the central result of the theory is that the spectrum of its infinitesimal generator is $\{-iz_j\}$ where the zeros of $\det S(z)$ (with the same multiplicities [LP 1, Chapter III]). Since $C = \int \rho(t) Z(t)dt$ is a compact operator commuting with $Z(t)$, it follows that the nonzero eigenvalues of C are $\sqrt{2\pi}\,\hat{\rho}(z_j)$ with the same multiplicities. Thus by Lidskiĭ's theorem [Ri, p. 139],

$$\operatorname{tr} C = \Sigma\sqrt{2\pi}\,\hat{\rho}(z_j)$$

and, combining this with (D), we have

$$\frac{1}{\sqrt{2\pi}}\operatorname{tr}\left(\int \rho(t)(\cos t\sqrt{-\Delta} - \cos t\sqrt{-\Delta_0}\right)dt$$

$$= \frac{1}{2}\int \frac{d\hat{\rho}}{d\lambda}(\lambda)\, s(\lambda)d\lambda \qquad (E')$$

$$= \frac{1}{2}\sum_j \hat{\rho}(z_j)$$

For $\partial\Omega$ starlike the final equality was established by Lax and Phillips [LP 4] **and by Melrose.**

III. POISSON FORMULA AND CONSEQUENCES

One way to state formula (E) is to say that the distribution

$$\ell(\rho) = \operatorname{tr}\left(\int \rho(t)(\cos t\sqrt{-\Delta} - \cos t\sqrt{-\Delta_0})dt\right)$$

in $\mathcal{D}'(2R,\infty)$ is equal to $\sum_{\text{poles}} e^{it\mu_j}$ where the sum is interpreted in the sense of distributions. This identity corresponds to the trivial first step in the derivation of the Poisson relation and formula for interior problems in the work of Chazarain [C] Duistermaat-Guillermin [DG], and Andersson-Melrose [AM]. By "Poisson relation" we mean the result that the singular support of ℓ is contained in the set of lengths of periodic ray paths in Ω, and by "Poisson formula" we mean a formula giving the singularities of ℓ — the classical Poisson formula is the case $\Omega = S^1$. After a simple reduction, the proof the Poisson relation here is precisely that used for interior problems.

To study the singularities of ℓ on $(2R,T)$ it suffices to assume that $\rho \in C_0^\infty(2R,T)$. Since $\cos t\sqrt{-\Delta}$ is the fundamental solution for the mixed problem $u_{tt} = \Delta u$, $u = 0$ on $\partial\Omega \times \mathbb{R}$ $u(0) = f$ $u_t(0) = 0$, it follows by standard domain of dependence arguments that, choosing cutoffs

$$\psi(x) = \begin{cases} 0 & |x| > R + T + 1 \\ 1 & |x| < R + T \end{cases} \quad \text{and} \quad \varphi(x) = \begin{cases} 0 & |x| > R + 2T + 2 \\ 1 & |x| < R + 2T + 1, \end{cases}$$

$$\int \rho(t)(\cos t\sqrt{-\Delta} - \cos t\sqrt{-\Delta_0})dt$$

$$= \int \rho(t)\varphi \cos t\sqrt{-\Delta}\,\psi dt - \int \rho(t)\,\varphi \cos t\sqrt{-\Delta_0}\,\psi dt$$

The two operators on the right are both in trace class, and by Huyghens' principle the (smooth) kernel of $\int \rho(t)\varphi \cos t\sqrt{-\Delta_0}\psi dt$ vanishes in a neighborhood of the diagonal. Thus for $\rho \in C_0^\infty(2R,T)$

$$\operatorname{tr} \int \rho(t)(\cos t\sqrt{-\Delta} - \cos t\sqrt{-\Delta_0})dt$$

$$= \operatorname{tr} \int \rho(t)\varphi \cos t\sqrt{-\Delta}\,\psi dt$$

The arguments of [AM, §8] can be applied directly to the right hand side above, and, combined with the propagation of singularities result of [M,Sj], they prove

Theorem 1: (Poisson relation, [BGR]). For an exterior domain Ω in $\mathbb{R}^n$, n odd, whose smooth boundary makes finite order contact with all straight line segments, the singular support of

$\sum_{\text{poles}} e^{it\mu_j}$ as a distribution in $\mathcal{D}'(2R,\infty)$ is contained in the set of lengths of periodic generalized ray paths in Ω.

As in the interior case one can compute the contributions to the singularities of ℓ from nongrazing periodic ray paths as in [GM]. Moreover, the exterior problems are sometimes so much better behaved than the interior problems that it is possible to use the Poisson formulas to get new information about the poles themselves. The final topic in these lectures will be one such case.

The main difficulty in using the Poisson formula for interior domains — or more generally compact manifolds with boundary — is that one has a great many periodic ray paths and the length spectrum is not only not discrete but also very wildly behaved at $t \to \infty$ (actually for manifolds without boundary it is discrete "generically" [DG] but the Zoll surfaces are the only exception I know to the second statement). However, if we consider an exterior domain Ω formed by removing two strictly convex bounded regions from $\mathbb{R}^n$, the set of lengths of periodic ray paths is just $\{2md : m \in \mathbb{Z}+\}$, where d is the minimal distance between the regions. This yields the Poisson formula:

$$\sum_{\text{poles}} e^{it\mu_j} = 2d \sum_{m=1} |\det(P^m - 1)|^{-\frac{1}{2}} \delta(t - 2md) + h(t) \qquad (F)$$

where P is the Poincaré map associated with the ray path going from one region to the other along the shortest path and back again, δ is the point mass at 0, and $h(t) \in L^2_{loc}(0,\infty)$. As always the equality here is to be interpreted in $\mathcal{D}'(2R,\infty)$.

To derive (F) one builds a parametrix for $\cos t\sqrt{-\Delta}$ in a micro-local neighborhood of the ray path running back and forth along the shortest path connecting the two regions. Actually in this case this construction can be carried out by classical geometrical optics as in Lax [L]. Then one computes the contributions to the singularities of ℓ from this parametrix directly. To show that the remaining part of $\cos t\sqrt{-\Delta}$ contributes nothing one must appeal to [AM, Propositions 8.17,8.20].

In the case considered here the eigenvalues of P are all real positive and different from 1. Since P is symplectic, they occur in reciprocal pairs and we denote the eigenvalues greater than 1 by $\lambda_1,\ldots,\lambda_{n-1}$. Then one can rewrite (F) in the suggestive form:

$$\left(\sum_{\text{poles}} e^{it\mu_j}\right) - h(t) = \tag{F}$$

$$= \sum_{(m_0, m)\in \mathbb{Z}\times\mathbb{Z}_+^{n-1}} e^{(2\pi i m_0 + (\frac{1}{2}+m_1)\beta_1 + \cdots + (\frac{1}{2}+m_{n-1})\beta_{n-1})\frac{t}{2d}}$$

where $\beta_j = -\log \lambda_j$. This formula is really the analogue for a hyperbolic periodic bicharacteristic of the formula for the elliptic case in [GW, Remark 1]. It raises the question: Is the arrangement of the poles μ_j in $\operatorname{Im} z > 0$ anything like the arrangement of the "pseudo-poles"

$$\frac{1}{2d}(2\pi i m_0 + (\tfrac{1}{2} + m_1)\beta_1 + \cdots + (\tfrac{1}{2} + m_{n-1})\beta_n) \quad ?$$

While the following theorem is only a very modest step toward an affirmative answer, it is quite unlike results known previously.

Theorem 2. Let Ω be the complement of two disjoint strictly convex bodies in $\mathbb{R}^n$, n odd. Then there are poles of the scattering matrix in all regions

$$S_\varepsilon = \{z : 0 \le \operatorname{Im} z \le \varepsilon \log |z|\}, \ \varepsilon > 0 \quad .$$

Remark: In [LP 3] Lax and Phillips prove the partially converse result that if Ω satisfies the "generalized Huyghen's principle" — which follows from [MSj] and the absence of trapped rays in Ω — then there is an $\varepsilon_0 > 0$ such that S_{ε_0} contains no poles.

Proof: To begin one must go back to the proof that $C = \int \rho(t)(\cos t\sqrt{-\Delta} - \cos t\sqrt{-\Delta_0})dt$ is trace class and check that in fact $\|C\|_{tr} \le K\|\rho\|_{H_0^{n+1}(2R,T)}$ where $\|\ \|_{tr}$ denotes the trace norm. Then one uses (E') with $\hat{\rho}$ equal to

$$\hat{\rho}_a(\xi) = e^{-2iR\xi}\, e^{-ia(n+2)\xi}\left(\frac{2\sin a\xi}{\xi}\right)^{n+2} \quad .$$

Since the trace norm dominates the sum of moduli of the eigenvalues, it follows that $\sum_{\text{poles}} |\hat{\rho}_a(\bar{\mu}_j)| < C$ uniformly for

$1 \le a \le 2$. Integrating this inequality from 1 to 2 in a, one concludes

$$\sum_{\text{poles}} e^{-2R \operatorname{Im} \mu_j} |\mu_j|^{-n-2} < \infty \quad . \tag{G}$$

Next one multiplies both sides of F') by $e^{-i(\gamma+i\delta)t}\sigma_a(t)$, where $\sigma_a(t) = \rho_a(t-a)$, and integrates. Since $\rho_a(t)$ is supported in $[2R, 2R + 2a(n + 2)]$, if we set $\delta = \frac{1}{4d}(\beta_1 + \cdots \beta_{n-1})$, it follows that

$$\sum_{\text{poles}} \hat{\sigma}_a(\gamma + i\delta - \mu_j) - \sum_{\text{pseudo-poles}} \hat{\sigma}_a(\gamma + i\delta - \nu_j)$$

is a square-integrable function of γ. However, if we assume that all the μ_j lie in $\{\operatorname{Im} z > \max\{2\delta, \varepsilon \log|z|\}\}$, it follows from (G) and the explicit form of $\ddot{\sigma}_a$, that for $a > a(\varepsilon)$ $\left|\sum_{\text{poles}} \hat{\sigma}_a(\gamma + i\nu - \mu_j)\right|$ is uniformly bounded for $\gamma \in \mathbb{R}$ independently of a. For the pseudo-poles the same sort of direct estimate shows for $\left|\sum_{\text{pseudo-poles}} \hat{\sigma}_a(i\delta - \nu_j)\right| > (2a)^{n+2} - C$ where C is independent of a. Since $\sum_{\text{pseudo-poles}} \hat{\sigma}_a(\gamma + i\delta - \nu_j)$ is periodic in γ (with period $\frac{\pi}{d}$) and continuous, we derive a contradiction to the assertion that the difference of the sums over the poles and pseudo-poles is square-integrable by taking "a" sufficiently large. Moreover, since there are only a finite number of poles in the region $\{z : \max\{0, \varepsilon \log|z|\} \le \operatorname{Im} z \le \max\{\delta, \varepsilon \log|z|\}\}$ and the contribution to the sum from a finite set of poles is always square-integrable, Theorem 2 follows.

The basic conjecture on poles and periodic ray paths is that such paths produce poles near the real axis and that the poles are closer for more stable ray paths (see [R 1] for examples). The case we have considered here is the "least" stable one, so one would expect that Theorem 2 holds whenever there are periodic ray paths. The difficulty with the argument given here is that as soon as the number of periodic ray paths becomes infinite the structure of the pseudo-poles becomes unwieldy.

REFERENCES.

[AM] K. Andersson and R. Melrose, Invent. Math. 41 (1977), pp. 23-95.

[B 1] V.S. Buslaev, Dokl. Akad. Nauk. SSSR, 197, (1971), pp. 999-1002.

[B 2] V.S. Buslaev, Dokl. Akad. Nauk USSR 143 (1962), pp. 1067-1070.

[C] J. Chazarain, Invent. Math. 24 (1974), pp. 65-82.

[DG] J.J. Duistermaat and V. Guillemin, Invent. Math. 29 (1975), pp. 39-79.

[GK] I. Gohberg and M.G. Krein, "Introduction to the Theory of Linear Non-self-adjoint Operators", AMS Translations, Vol. 18, Providence 1969.

[GM] V. Guillemin and R. Melrose, Adv. in Math. 32 (1979), pp. 204-232.

[GW] V. Guillemin and A. Weinstein, Bull. AMS 82, pp. 92-94 966.

[HR] J.W. Helton and J. Ralston, J. Diff. Eq. 21 (1976), pp. 378-399. Addendum, J. Diff. Eq. 28 (1978), pp. 155-162.

[BGR] C. Bardos, J.-C. Guillot and J. Ralston, C.R. Acad. Sc. Paris, t. 290 (1980).

[L] P.D. Lax, Duke Math. J. 24 (1957), pp. 627-646.

[LP 1] P.D. Lax and R.S. Phillips, "Scattering theory", Academic Press, New York, 1967.

[LP 2] P.D. Lax and R.S. Phillips, CPAM 29 (1969), pp. 737-787.

[LP 3] P.D. Lax and R.S. Phillips, Arch. Rat. Mech. and Anal. 40 (1971), pp. 268-280.

[LP 4] P.D. Lax and R.S. Phillips, "The time delay operator and a related trace formula", preprint, Stanford 1976) .

[M] R. Melrose, "Forward scattering by a convex obstacle", CPAM 33 (1980), 461-500.

[MS] H. McKean and I. Singer, J. Diff. Geom. 1 (1967), pp. 43-69.

[MSj] R. Melrose and J. Sjostrand, CPAM 31 (1978), pp. 593-617.

[MR] A. Majda and J. Ralston, Duke Math. J., 46 (1979), pp. 725-731.

[R 1] J. Ralston, CPAM 24 (1971), pp. 571-582.

[R 2] J. Ralston, "Diffraction by convex bodies" Seminaire Goulaouic-Schwartz 1978-1979 Expose 23,, May 1979.

[Ri] J.R. Ringrose, "Compact Non-self-adjoint Operators", Van Nostrand, New York, 1971.

[ST] N. Shenk and D. Thoe, J. Math. Anal. and Appl. 36 (1971), pp. 313-351.

[CdV] Y. Colin de Verdière, "Une formule de traces pour l'opérateur de Schrödinger dans R^3", preprint, Grenoble, 1980.

PROPAGATION AT THE BOUNDARY OF ANALYTIC SINGULARITIES

Pierre SCHAPIRA

Departement de Mathématiques
C.S.P. Université Paris-Nord
93430 - Villetaneuse France

We study non elliptic boundary value problems in the framework of microfunctions. We introduce the concept of N-regularity and apply it to the problem of propagation of analytic singularities at the boundary with some possible diffraction.

H. G. Garnir (ed.), Singularities in Boundary Value Problems, 185–212.

1. VARIOUS SHEAVES OF MICROFUNCTIONS.

Let X be a complex analytic manifold, T^*X its cotangent bundle, $T^*_X X \simeq X$ the zero section of T^*X, $\dot{T}^*X = T^*X - T^*_X X$, π the projection of T^*X on X, $\mathcal{O}_X$ the sheaf of holomorphic functions on X.

The space T^*X is endowed with the sheaf $\mathcal{E}_X$ of microdifferential operators (of finite orders) constructed in(20). Let us recall that if X is open in a complex vector space E, and U is open in $X \times E^*$, $\mathcal{E}_X(U)$ is isomorphic to the set of series $\{\sum_{j \in \mathbf{Z}} p_j(z,\zeta)\}$ with $p_j = 0$ for $j \gg 0$, p_j is holomorphic on U, homogeneous of degree j on ζ and such that for any compact set $K \subset U$ there exists $t > 0$ whith

$$\sum_{j \leq 0} |p_j|_K \; t^{-j}/(-j)! < \infty$$

The composite $P \circ Q$ of two microdifferential operators P and Q is given in a system of local coordinates, by the usual Leibnitz formula, and makes $\mathcal{E}_X$ a unitary non commutative sheaf of algebra, whose restriction to $T^*_X X$ is isomorphic with $\mathcal{D}_X$, the sheaf of holomorphic differential operators on X, (cf. (17) for a quick expository on this subject). If P is a section of $\mathcal{E}_X$, we denote by $\sigma(P)$ its principal symbol.

Let now M be a real analytic manifold of dimension n, X a complexification of M, N a real analytic submanifold of M, Y the complexification of N in X. Let $T^*_N X$ be the conormal bundle of N in X that is the kernel of the map $T^*X \times_X N \longrightarrow T^*N$

The sheaf $C_{N|X}$ on $T^*_N X$ is constructed in(20) in a intrinsic way but to understand it, let us assume that M is a real vector space, N a linear subspace of M, F a supplementary of N in M, so that $M = N \oplus F$, $X = N^{\mathbb{C}} \oplus F^{\mathbb{C}}$,

$$T_N^\star X = (N \times \{0\}) \times (i N^\star \times (F^{\mathbb{C}})^\star)$$

Let ω be an open set in N , I a conic open set in $i N^\star \times (F^{\mathbb{C}})^\star$. The sheaf $C_{N|X}$ is associated to the presheaf

$$\check{C}_{N|X}(\omega \times I) = \varinjlim_{U} H^n_{(\omega \times G)\cap U}(U, O_X)$$

where G is the closed cone of $i N \times F^{\mathbb{C}}$ polar of I , and U runs over the neighborhoods of ω in X .

We also define the sheaf $\Gamma_N(B_M)$ on N (the sheaf of hyperfunctions on M supported by N) :

$$\Gamma_N(B_M) = \mathcal{H}^n_N(O_X) .$$

The sheaf $C_{N|X}$ is locally constant on the orbits of the action of $\mathbb{R}^+$ on $T_N^\star X$, and the natural morphism

$$\Gamma_N(B_M) \longrightarrow \pi_\star C_{N|X} = C_{N|X} \mid T_N^\star N$$

associated to the inclusions $\omega \times \{0\} \subset \omega \times G$, is an isomorphism. When $N = M$, the sheaf $C_{M|X}$, just denoted C_M , is the sheaf of microfunctions introduced by M. Sato in 1969, and $\Gamma_M(B_M) = B_M$ the sheaf of hyperfunctions of M. Sato (introduced in 1959). The sheaves $C_{N|X}$ are naturally endowed with a structure of $\mathcal{E}_X | T_N^\star X$ - modules. (cf. (3), (8) for an explicit representation of the action of $\mathcal{E}_X$ on C_M).

Remark For the sake of simplicity we have systematically forgotten the various sheaves of relative orientation that should appear in the definitions of the sheaves $C_{N|X}$. If u is a microfuntion (that is a section of C_M), its support is also called its singular support and is denoted $SS(u)$. If u is an hyperfunction , $SS(u)$ is a closed conic set of $T_M^\star X$, and $SS(u)$ is contained in $T_M^\star M$ iff u is real analytic, and one says sometime that u is micro-analytic at $x^\star \in T_M^\star X$ if $x^\star \notin SS(u)$. The sheaves $C_{N|X}$ and C_M are related by the following theorem of Kashiwara (5), to appear in (10)(cf.also (9)).

Theorem 1.1. We have at, least outside of $T^{\star}_{Y}X$, natural injective morphisms of sheaves on $T^{\star}_{N}X \cap T^{\star}_{M}X$:

$$C_{N|X}|_{T^{\star}_{N}X \cap T^{\star}_{M}X} \rightarrowtail \Gamma_{T^{\star}_{N}X \cap T^{\star}_{M}X}(C_M)$$

$$\rightarrowtail C_M|_{T^{\star}_{N}X \cap T^{\star}_{M}X} \rightarrowtail \mathcal{H}^p_{T^{\star}_{N}X \cap T^{\star}_{M}X}(C_{N|X})$$

where p is the codimension of N in M .

Let us assume that N is an hypersurface of M, M_+ an open half space of M with N as boundary, and let Q_+ be the half space of $T^{\star}_{N}X$ with $T^{\star}_{N}X \cap T^{\star}_{M}X$ as boundary such that in a choice of local coordinates, if $(z^o, \zeta^o) \in Q_+$, the polar of a neighborhood of ζ^o contains $\overline{M}_+$ (cf. figure below). Then we have a natural morphism from $\Gamma_{\overline{M}_+}(M, B_M)$ to $\Gamma(Q_+, C_{N|X})$ and this morphism is injective.

We can describe the preceding morphisms by the following pictures, where C_M and $C_{N|X}$ are associated to the cohomology groups with support in the closed sets $M \times \Gamma$ and $N \times G$.

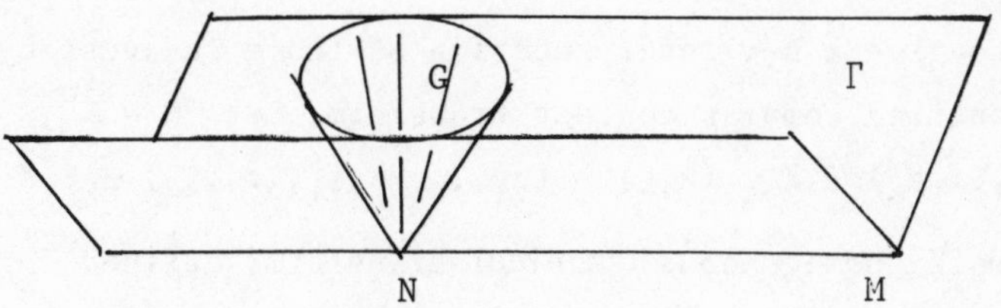

$$C_{N|X}|_{T^{\star}_N X \cap T^{\star}_M X} \longrightarrow \Gamma_{T^{\star}_N X \cap T^{\star}_N X}(C_M)$$

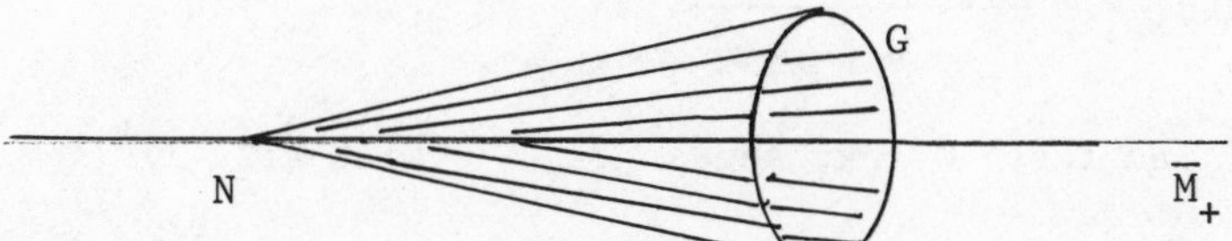

$$\Gamma_{\overline{M}_+}(B)|N \longrightarrow C_{N|X}|Q_+$$

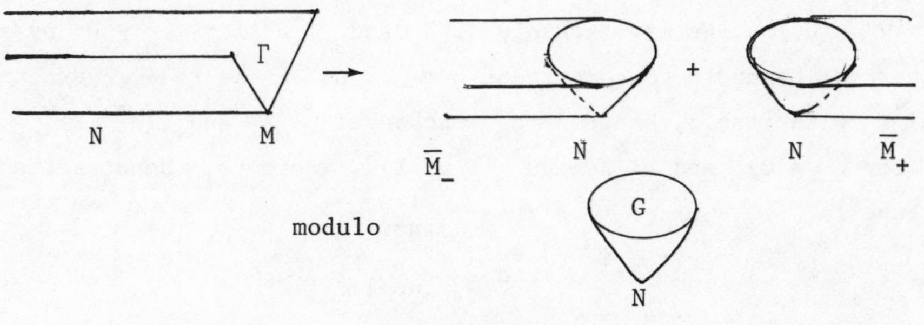

$$C_M|T^{\star}_M X \cap T^{\star}_N X \longrightarrow \mathcal{H}^p_{T^{\star}_M X \cap T^{\star}_N X}(C_{N|X}) \text{ when } p = 1 .$$

As these cohomology groups are rather non intuitive objects, it is satisfactory to give a new interpretation of those sheaves, by the aid of a quantized complex contact transform. Let $M = \mathbb{R}^n$, $X = \mathbb{C}^n$, $z = (z_1,\dots z_n) \in X$, $(z,\zeta) = (z_1,\dots z_n,\zeta_1,\dots,\zeta_n) \in T^\star X$.

Let Φ be the complex homogeneous canonical transform, defined for $\zeta_n \neq 0$ by :

$$\varphi : T^\star\mathbb{C}^n - \{\zeta_n = 0\} \longrightarrow T^\star\mathbb{C}^n$$

$$(z,\zeta) \longrightarrow (z + i\,\varphi'(\zeta),\zeta)$$

$$\text{where } \varphi(\zeta) = \frac{\zeta_1^2 + \dots + \zeta_{n-1}^2}{4\,\zeta_n}$$

It is easy to see that Φ exchanges $T^\star_{\mathbb{R}^n}\mathbb{C}^n$ with $(T^\star_{\partial\Omega_o}X)^+$,

the exterior conormal bundle to the boundary of the tube

$$\Omega_o = \{z \in \mathbb{C}^n \; ; \; y_n > \sum_{i=1}^{n-1} y_i^2\} .$$

Let $p < n$, and let $N = \{x \in M \; ; \; x_1 = \dots = x_p = 0\}$, $Y = N^{\mathbb{C}} = \{z \in X \; ; \; z_1 = \dots = z_p = 0\}$ and $L = \{z \in X; y_{p+1} = \dots = y_n = 0\}$. We may exchange $T^\star_N X$ and $T^\star_L X$ for $\zeta_n \neq 0$ by a partial Legendre contact transform Ψ. As Ψ is real, Ψ exchanges $T^\star_M X$ with itself, hence $\Phi \circ \Psi$ exchanges $T^\star_M X$ and $(T^\star_{\partial\Omega_o}X)^+$ (for $\zeta_n \neq 0$) and $T^\star_N X$ and $(T^\star_{\partial\Omega_1}X)^+$, where Ω_1 denotes the tube :

$$\Omega_1 = \{z \in \mathbb{C}^n \; ; \; y_n > \sum_{i=p+1}^{n-1} y_i^2\} .$$

Let $\mathcal{O}^+_{\partial\Omega_j}$ $(j = 0,1)$ be the sheaf on $\partial\Omega_j$ of boundary values of holomorphic functions on Ω_j :

$$\mathcal{O}^+_{\partial\Omega_j} = i_\star(\mathcal{O}_{\Omega_j})/\mathcal{O}_X$$

where $i_\star(O_{\Omega_j})(U) = O(U \cap \Omega_j)$ for any open set U in X. We denote by $C^+_{\partial\Omega_j}$ the inverse image of $O^+_{\partial\Omega_j}$ by the projection $\pi : (T^\star_{\partial\Omega_j} X)^+ \to \partial\Omega_j$. The sheaf $C^+_{\partial\Omega_j}$ is naturally endowed with a structure of $\mathcal{E}_X$ - module (cf. (11)) and it can be proved that the contact transform $\Phi \circ \Psi$ can be extended (locally) as an isomorphism of $\mathcal{E}_X$ onto itself, and an isomorphism of C_M on $C^+_{\partial\Omega_o}$ and $C_{N|X}$ on $C^+_{\partial\Omega_1}$, these last two isomorphism being compatible with the action of $\mathcal{E}_X$. Then theorem 11 can be re-interpreted with the sheaves $O^+_{\partial\Omega_o}$ and $O^+_{\partial\Omega_1}$, as shown in the following pictures, (Ω_o and Ω_1 are open, and $p = 1$) (cf. (16) for a construction analogous to the isomorphism of C_M onto $C^+_{\partial\Omega_o}$).

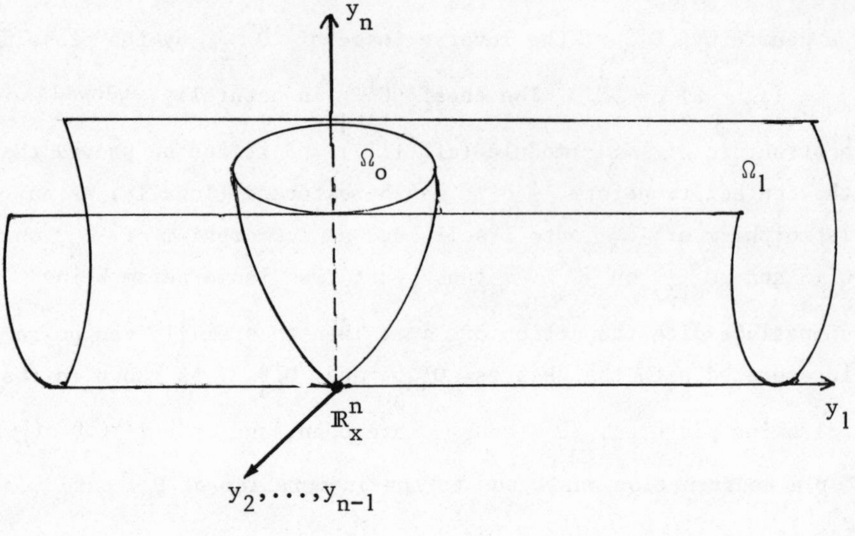

$$\mathcal{O}^+_{\partial\Omega_1} \,|\, \partial\Omega_o \cap \partial\Omega_1 \quad \longrightarrow \quad \Gamma_{\partial\Omega_o \cap \partial\Omega_1}(\mathcal{O}^+_{\partial\Omega_o})$$

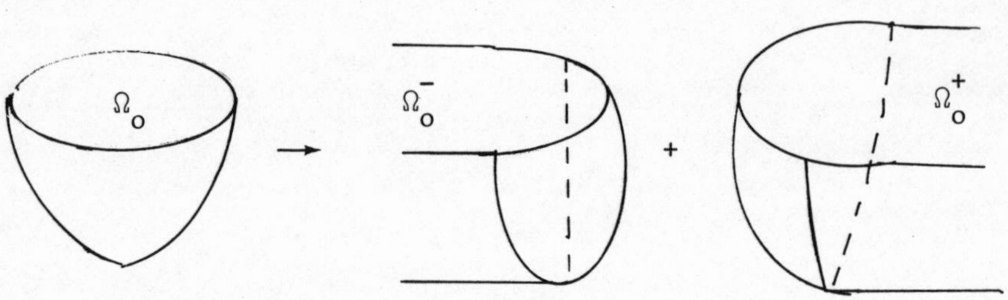

modulo

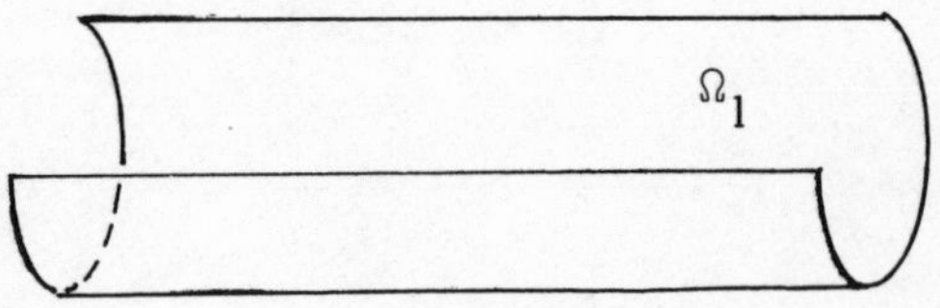

$$\mathcal{O}^+_{\partial\Omega_o} \,|\, \partial\Omega_o \cap \partial\Omega_1 \quad \longrightarrow \quad \mathcal{H}^1_{\partial\Omega_o \cap \partial\Omega_1}(\mathcal{O}^+_{\partial\Omega_1})$$

$\forall\, f \in \mathcal{O}(\Omega_o)$, $f = f_o^- + f_o^+ \mod h$

where $f_o^\pm \in \mathcal{O}(\Omega_o^\pm)$, $h \in \mathcal{O}(\Omega_1)$

2. MICRO LOCAL HOLMGREN THEOREM

Let $N \subset M \subset X$ be as in the preceding section, and Y the complexification of N in X.
The manifold $T_N^{\star}X$ admits a natural foliation by complex manifolds of dimension p, where p is the codimension of N in M: these are the bicharacteristic leaves of the complex involutive manifold $Y \times_X T^{\star}X \subset T^{\star}X$, $T_N^{\star}X$ being invariant by the Hamiltonian flow of $Y \times_X T^{\star}X$.

The following theorem was announced by Kashiwara and Kawai in (7).

<u>Theorem 2.1</u> Let u be a section of $C_{N|X}$ on an open set $U \subset T_N^{\star}X$. Then the support of u is a union of bicharacteristic leaves of $Y \times_X T^{\star}X$.

<u>Proof</u>

a) We first reduce the problem to the case where $U \cap T_Y^{\star}X = \emptyset$, by a trick introduced by M. Kashiwara. We set $\tilde{X} = X \times \mathbb{C}$, $\tilde{N} = N \times \mathbb{R}$, $\tilde{U} = \{ z,t \,;\, \zeta, \tau) \in T_{\tilde{N}}^{\star}\tilde{X} \,;\, \tau \neq 0 \,,\, \zeta/\tau \in U \}$ where t is a coordinate on $\mathbb{C}$.

If u belongs to $C_{N|X}$, $u \otimes \delta(t)$ belongs to $C_{\tilde{N}|\tilde{X}}$, and :

$$(z,\zeta) \in SS(u) \iff (z,o\,;\zeta,i) \in SS(u \otimes \delta(t))$$

it is thus enough to prove the theorem for $C_{\tilde{N}|\tilde{X}}$, at points $(z,t\,;\zeta,\tau)$ where $\tau \neq 0$.

b) We can exchange, by a quantized complex transform, the problem to the following.

Let $\Omega_1 = \{ z \in \mathbb{C}^n \,;\, y_n > \sum_{j=p+1}^{n-1} y_j^2 \}$, let W be an open set of $\mathbb{C}^n$ with $W \cap \partial\Omega_1 = U$, and set $W \cap \Omega_1 = W^+$. Let f be holomorphic in W^+, which extends holomorphically in a neighbourhood of the point $z^o = (z'^o_1, z''^o) \in U$ where $z' = (z_1, \ldots z_p)$ (that is : f gives 0 in $C^+_{\Omega_1}$ near that point).

Then f is holomorphic near any point (z'_1, z''^o) which belongs to the connected component of (z'^o, z''^o) in U. The proof is an immediat consequence of the well known Bochner local tube theorem cf. (6) or (14).

We assume now that N is an hypersurface of the real analytic manifold M, and let $\overline{M}_+$ be one of the closed half-space of M with N as boundary.

The manifold $T^*_M X \underset{M}{\times} N$ is involutive in $T^*_M X \simeq i\, T^* M$, and admits a foliation by the so-called bicharacteristic curves.

<u>Theorem 2.2</u> (Holmgren-Kashiwara) (5)

Let u be an hyperfunction on M supported by $\overline{M}_+$. Then $SS(u) \cap (T^*_M X \underset{M}{\times} N)$ is a union of bicharacteristic curves of $T^*_M X \underset{M}{\times} N$.

If we take a system of local coordinates $(x_1, \ldots, x_n)$ on M, such that $N = \{x \in M \,;\, x_1 = 0\}$, and if we set $z = x + iy$, $\zeta = \xi + i\eta$, where (z, ζ) are local coordinates on $T^* X$, the theorem asserts that if $(0, x'^o; i\eta_1^o, i\eta'^o) \in SS(u)$ then $(0, x'^o; i\eta_1, i\eta'^o) \in SS(u)$ for any $\eta_1 \in \mathbb{R}$.

If we take $\eta_1^o = 0$, $\eta'^o = 0$, we find in particular that $(0, x'^o) \in \operatorname{supp}(u)$ implies that $(0, x'^o; +i, 0 \ldots 0)$ and $(0, x'^o; -i, 0 \ldots 0)$ belong to $SS(u)$: this is the "classical" Holmgren theorem (20).

<u>Proof of theorem 2.2.</u>

a) By the same trick as in the proof of theorem 2.1 it is enough to prove the theorem for $i\eta'^o \neq 0$.

b) Let Q_+ be the open half space of $T^*_N X$ associated to $\overline{M}_+$ (cf. theorem 1.2) and let us consider the diagram :

$$\begin{array}{ccc} \Gamma_{\overline{M}_+}(M, B_M) & \longrightarrow & \Gamma(T^*_M X, C_M) \\ \downarrow & & \downarrow \\ \Gamma(Q_+, C_{N|X}) & \longrightarrow & H^1_{T^*_M X \cap T^*_N X}(T^*_N X, C_{N|X}) \end{array}$$

Let $u \in \Gamma_{\overline{M}_+}(B_M)$ and assume that $x^* = (0, x'^o ; i\eta_1^o, \eta'^o)$ does not belong to SS(u). Then $u = 0$ near x^* in Q_+ as a section of $C_{N|X}$, and theorem 2.1. implies that $u = 0$ near any point $(0, x'^o ; \zeta_1, i\eta'^o)$ of Q^+. Then the image of u in $H^1_{T^*_M X \cap T^*_N X}(T^*_N X, C_{N|X})$ will be 0 at any point $(0, x'^o ; i\eta_1, i\eta'_o)$, and this achieves the proof, because of the injectivity of the second vertical arrow (theorem 1.1).

3. DIVISION THEOREM FOR THE SHEAF $C_{N|X}$

We choose local coordinates $z = (z_1, \ldots, z_n)$ on X, with $z = x + iy$, $M = \{z \in X ; y = 0\}$, $N = \{x \in M ; x_1 = 0\}$. We say that a microdifferential operator P of order m defined in a neighborhood of $(z^o, \zeta^o) \in T^*X$ if of Weierstrass type in D_{z_1} if P is written :

$$P(z, D_z) = \sum_{j=o}^{m} A_j(z, D_{z'}) D_{z_1}^j$$

where $D_{z'} = (D_{z_2}, \ldots, D_{z_n})$, and A_j are microdifferential operators of order $\leqslant m - j$, with $A_m \equiv 1$.

Let $Y = \{z \in X ; z_1 = 0\}$ and let ρ denotes the projection $T^*X \underset{X}{\times} Y \longrightarrow T^*Y$.

Theorem 3.1 (7) Let P be a microdifferential operator defined in a neighborhood of $x^\star = (0, x'^o ; \tau^o, i\eta'^o) \in T^\star_N X$. Let μ be the multiplicity, that we assume to be finite, of the root at $\tau = \tau^o$ of the equation $\sigma(P)(0,x'^o ; \tau, i\eta'^o) = 0$. Then the map from $(C_{N|X})_{x^\star} \times (C^\mu_N)_{\rho(x^\star)}$ to $(C_{N|X})_{x^\star}$:

$$(v, (w_j)^{\mu-1}_{j=o}) \longrightarrow P\,v + \sum_{j=o}^{\mu-1} w_j \otimes \delta^j_1$$

is an isomorphism.

For a proof we refer to (11 § 6). We can get a semi-global version of the preceding theorem, by purely algebraic considerations (cf. (22) or (12)).

Theorem 3.2 - Let $y^\star = (x'^o ; i\eta'^o) \in T^\star_N Y$, and let P be a microdifferential operator of Weïerstrass type in D_{z_1}, defined in a neighborhood of $\rho^{-1}(y^\star)$. Let U be an open set in $T^\star_N X$ which contains exactly m' roots (counted with multiplicities) of the equation $\sigma(P)(0,x'^o ; \tau, i\eta'^o) = 0$. Then the map from $(C_{N|X})_{\rho^{-1}(y^\star) \cap U} \times (C^{m'}_N)_{y^\star}$ to $(C_{N|X})_{\rho^{-1}(y^\star) \cap U}$:

$$(v, (w_j)^{m'-1}_{j=o}) \longrightarrow P\,v + \sum_{j=o}^{m'-1} w_j \otimes \delta^j_1$$

is an isomorphism.

This theorem is in particular true all over $T^\star_N X$ and implies :

Corrollaire 3.3 (21) - Let P be a differential operator of order m for which N is non characteristic. Then the map from $\Gamma_N(B_M) \times B^m_N$ to $\Gamma_N(B_M)$:

$$(v, (w_j)^{m-1}_{j=o}) \longrightarrow P\,v + \sum_{j=o}^{m-1} w_j \otimes \delta^{(j)}_{x_1}$$

is an isomorphism.

Remark that this corollary is in fact equivalent to the Cauchy-Kowalewska theorem (21). We recall now the concept of N-regularity introduced in (22) which will allow us to study propagation of singularities at the boundary.

Definition 3.4. (22)- Let P be a microdifferential operator defined near $x^\star \in T^\star_M X \cap T^\star_N X$. We say that P is N-regular at $x^\star$ if for any $u \in \Gamma_{T^\star_M X \cap T^\star_N X}(C_M)_{x^\star}$, $P\,u \in (C_{N|X})_{x^\star}$ implies that $u \in (C_{N|X})_{x^\star}$.

This definition does not make any difference between $\overline{M}_+$ and $\overline{M}_-$ and is not precise enough for diffrative problems. It has been refined by K . Kataoka as follow. First, for a section $u \in C_M \mid T^\star_M X \cap T^\star_N X$, we will say that $SS^2_N(u)$ is contained in Q_+ if the image of u in $\mathcal{H}^1_{T^\star_M X \cap T^\star_N X}(C_{N|X})$ belongs to the image of the sheaf $C_{N|X}|Q_+$ in $\mathcal{H}^1_{T^\star_M X \cap T^\star_N X}(C_{N|X})$ (cf. theorem 1.1).
For example, if $u \in \Gamma_{\overline{M}_+}(B_M)$, $SS^2_N(u)$ is contained in Q_+ .

Definition 3.5 (13) - We say that P is N^+-regular at $x^\star$, if for any $u \in \Gamma_{T^\star_M X \cap T^\star_N X}(C_M)$, such that $SS^2_N(u)$ is contained in Q_+ , $P\,u \in (C_{N|X})_{x^\star}$ implies $u \in (C_{N|X})_{x^\star}$.

4. PROPAGATION AND REFLECTION

Let N, Y, M, X be as in the preceding section, and let P be a differential operator of order m for which N is non characteristic. Let M_+ be an open half space of M , with N as boundary (we assume of course that M_+ is locally on one side of N). If u is an hyperfunction on M_+ solution of the equation $P\,u = 0$, we have constructed in (21)(cf. also (15)) the m traces of u on N as follows : let $\overline{u} \in \Gamma_{\overline{M}_+}(M, B_M)$ be an extension of u (which always exists the sheaf B_M being flabby)

Then $P\,\overline{u} \in \Gamma_N(M,B_M)$ and $P\,\overline{u}$ can be written for a choice of coordinates $(x_1,\dots,x_n)$:

$$P\,\overline{u} = P\,v + \sum_{j=o}^{m-1} w_j \otimes \delta_{x_1}^{(m-1-j)}$$

with $v \in \Gamma_N(M,B_M)$, $w_j \in B_N(N)$ (cf. corollary 3.3). It is clear that $(w_j)_{j=o}^{m-1}$ only depend on u , and not on the choice of $\overline{u}$. We say that $(w_o,\dots,w_{m-1})$ are the traces of u on N . The traces depend on the choice of the coordinates, but to say for example, that the p-first traces, $(w_o,\dots,w_{p-1})$, are zero at some point $y^\star \in T_N^\star Y$, does not depend of this choice.

Theorem 4.1 (22) - In the preceding situation let u be an hyperfunction on M_+ , solution of $P\,u = 0$. Let $y^\star \in T_N^\star Y$, and let $Z = \rho^{-1}(y^\star) \cap \{\sigma^{-1}(P)(0)\}$, where ρ denotes the canonical projection of $T_N^\star X$ on $T_N^\star Y$. Let $Z^+ = Z \cap Q^+$, and $Z^o = Z \cap T_M^\star X$. Let $Z^o = Z^{o,1} \sqcup Z^{o,2}$ be a partition of Z^o.
We assume :

a) at any point of $Z^{o,1}$, P is N^+-regular ;

b) the closure in $T_M^\star X$ of $SS(u)$ does not intersect $Z^{o,1}$;

c) there are $m-p$ points (counted with their multiplicities) in $Z^+ \sqcup Z^{o,1}$;

d) the p-first traces of u are zero near $y^\star$.

Then :
- the m traces of u are zero near $y^\star$
- the closure in $T_M^\star X$ of $SS(u)$ does not intersect $\rho^{-1}(y^\star)$.

Proof

There are $m-p$ points in $Z^+ \sqcup Z^{o,1}$, and p points in $Z^- \sqcup Z^{o,2}$, if we set $Z^- = Z \setminus (Z^o \sqcup Z^+)$. Let $\overline{u} \in \Gamma_{\overline{M}_+}(M,B_M)$ be an extension of u , and let us consider $P\,\overline{u} \in \Gamma_N(M,B_M)$

as an element of $\Gamma(T^{\star}_{N}X, C_{N|X})$. By hypothesis d) we have, on $\rho^{-1}(y^{\star})$:

$$P\,\overline{u} = P\,v + \sum_{j=o}^{m-1-p} w_j \otimes \delta_1^{(j)}$$

where $v \in (C_{N|X})_{\rho^{-1}(y^{\star})}$, $w_j \in (C_N)_{y^{\star}}$.

At any point $x^{\star} \in Z^{+} \sqcup Z^{o,1}$, we can find $\tilde{v} \in (C_{N|X})_{x^{\star}}$ with $\overline{u} = \tilde{v}$: if $x^{\star} \in Z^{+}$ this is because $\Gamma_{\overline{M}_+}(M,B_M)$ is sent into $\Gamma(Q^{+},C_{N|X})$, and if $x^{\star} \in Z^{o,1}$ this follows from hypothesis a) and b) . Let U be an open set of $T^{\star}_{Z}X$, with

$$U \supset Z^{+} \sqcup Z^{o,1} \quad , \quad U \cap (Z^{-} \sqcup Z^{o,2}) = \emptyset .$$

Then $\overline{u} = \tilde{v}$ for $\tilde{v} \in (C_{N|X})_{\rho^{-1}(y^{\star}) \cap U}$. If we replace $\tilde{v}$ by $\tilde{v} - v$, we obtain :

$$P\,(\tilde{v} - v) = \sum_{j=o}^{m-1-p} w_j \otimes \delta_1^{(j)}$$

with $(\tilde{v} - v) \in (C_{N|X})_{\rho^{-1}(y^{\star}) \cap U}$, $w_j \in (C_N)_{y^{\star}}$.

We conclude by theorem 3.2. that all w_j, $j = 0,\dots,m-1-p$, are zero near $y^{\star}$. Then u admits an extension $\overline{u} \in \Gamma_{\overline{M}_+}(M,B_M)$ with $P\,\overline{u} = 0$ on $\rho^{-1}(y^{\star}) \cap T^{\star}_{M}X$. As P is invertible on $\rho^{-1}(y^{\star}) \cap T^{\star}_{M}X - Z^{o}$, $SS(\overline{u}) \cap \rho^{-1}(y^{\star}) = \emptyset$ by the Holmgren-Kashiwara theorem (theorem 2.2.).

We shall explain how to apply this theorem.

Example 1 - Assume $Z^{o} = \emptyset$. Then we get a micro-local version of the Morrey-Niremberg theorem (18) : if p traces of u are zero near $y^{\star}$, where $m-p = \# Z^{+}$, then the m traces of u are zero near $y^{\star}$.

Example 2 - Assume for simplicity that P has real simple characteristics, the bicharacteristic curves being transversal to N (that is the roots of the equation $\sigma(P)(0,x';\tau,i\eta')=0$ are all purely imaginary, and simple). Let $(b_i)_{i=1}^m$ be the bicharacteristic curves issued from $\rho^{-1}(y^\star)$, where $y^\star = (x'^o, i\,\eta'^o)$. Then if the p-first traces of u are zero at $y^\star$, and if u is zero along $m-p$ bicharacteristic curves, u will be zero along all the bicharacteristic curves. To apply theorem 4.1. one has to prove the N-regularity of P, but this will be a particular case of theorem 5.1. below. Thus we obtain the Lax - Nirenberg theorem on reflection on singularities, in the analytic case.

In fact theorem 5.1. allows to treat propagation along leaves of any dimension, as shown in the following.

Example 3 - Let us take $M = \mathbb{R}^n$,

$N = \{x \in M ; x_1 = 0\}$, $M_+ = \{x \in M ; x_1 > 0\}$,

$P = \sum_{j=1}^{n-1} D_{x_j}^2 - D_{x_n}$. Let $u \in B(M_+)$ be an hyperfunction solution of $P\,u = 0$. Then if u is micro-analytic in the $+i\,d\,x_n$ direction (resp. $-i\,d\,x_n$ direction), the same will be true for the traces of u on N.

More precisely let $x_n^o \in \mathbb{R}$ and let $b^+(x_n^o)$ be the leaf :

$b^+(x_n^o) = \{(x,i\eta) \in T^\star_{M_+} X ; x_n = x_n^o, i\eta = 0,\ldots,0, + i)\}$.

We know by the main theorem of (3) that $b^+(x_n^o)$ is contained or disjoined from $SS(u)$. As the operator P is N-regular at any point $(x ; 0,\ldots,0, \pm i)$ by theorem 5.1. we conclude that the two traces of u will be micro-analytic at any point

$y^\star = (x_2,\ldots,x_{n-1},x_n^o ; 0,\ldots,0, + i)$ of $T^\star_N Y$.

We now discuss some situations with diffraction.

Example 4 - Assume for simplicity $Z = Z^o$ reduced to a single point $x^\star \in T^\star_M X \times_M N$ (with multiplicity m). Then if P is N^+-regular at $x^\star$, and if at least one of the m traces of u is non zero at $\rho(x^\star)$, $x^\star$ belongs to the closure of $SS(u)$ in $T^\star_M X$.

This will be the case if P is micro-hyperbolic for N , or even N^+-semi-hyperbolic (cf. § 6 below), as for the operator $D_1^2 - x_1^k D_2^2$, where $k \in \mathbb{N}$, $M_+ = \{x \in M , x_1 > 0\}$ (cf. (2),(22), (13), (4)). An other very interestingsituation is the following. P is of order two, with simple and real characteristics, the equation $\sigma(P)\ (0,x'^o; \tau, i\,\eta'^o) = 0$ has a double root at $\tau = \tau^o \in i\,\mathbb{R}$, and the bicharacteristic b through $x^\star = (0, x'^o; \tau^o, i\,\eta'^o)$ is contained in $\pi^{-1}(M_+) \cup \{x^\star\}$, and tangent at the first order to $\pi^{-1}(N)$ at $x^\star$. Then P is N^+-regular at $x^\star$ by a theorem of K. Kataoka (13) (cf. also J. Sjöstrand (24) who obtain the same conclusions by a very different approach). This means that if the traces are not both zero at $\rho(x^\star)$, at least one of the half bicharacteristic b^+ or b^- issued from $(0,x'^o ; \tau^o , i\,\eta'^o)$ is contained in $SS(u)$.

5. A FIRST CLASS OF N-REGULAR OPERATORS

Let Λ be a conic involutive complex submanifold of $T^\star X$. Let us recall (cf. (1), (3)), that a vector $(x^\star, \theta) \in (T_\Lambda T^\star X)_{x^\star}$ is non micro-characteristic on Λ for a microdifferential operator P , if $\sigma(P)$ being zero to the order r on Λ , $\sigma(P)$ is not zero to an order $r' > r$ on Λ in the θ-direction at $x^\star$. If S is an hypersurface of $T^\star X$ and $f = 0$ is an equation of S , we say that S is non micro-characteristic for P at $x^\star$ if it is the case for $\theta = H_f$, where H_f denotes the Hamiltonian field of f .

We have proved in (23) the following.

Theorem 5.1. (23) Let N be an hypersurface of M, Y the complexification of N in X, and let P be a microdifferential operator defined near $x^\star \in T^\star_N X \cap T^\star_M X$. We assume that there exists an involutive conic analytic manifold $\Lambda^{\mathbb{R}}$ of $T^\star_M X$ with $x^\star \in \Lambda$, such that $T^\star X \underset{X}{\times} Y$ is non micro-characteristic at $x^\star$ for P on Λ, the complexification of $\Lambda^{\mathbb{R}}$ in $T^\star X$. Then P is N-regular at $x^\star$.

Sketch of the proof

a) By the trick of the adjunction of an auxillary variable, we may assume that Λ is involutive and regular (the fundamental 1-form ω on $T^\star X$ does not vanish on Λ). If $\sigma(P)(x^\star) = 0$, Λ and $T^\star X \underset{X}{\times} Y$ are orthogonal (there exists f which is zero on $T^\star X \underset{X}{\times} Y$ and g which is zero on Λ such that the Poisson bracket $\{f,g\}$ is not zero at $x^\star$) and we may find a real contact transform such that for new coordinates (z,ζ) on $T^\star X$, we have :

$$T^\star X \underset{X}{\times} Y = \{(z,\zeta) \in T^\star X \; ; \; z_1 = 0\}$$

$$\Lambda = \{(z,\zeta) \in T^\star X \; ; \; \zeta_1 = \ldots = \zeta_p = 0\} .$$

As the hypothesis of the theorem remains true if we replace Λ by another submanifold $\Lambda' \subset \Lambda$, we may even assume :

$\Lambda = \{(z,\zeta) \in T^\star X \; ; \; \zeta_1 = \ldots = \zeta_{n-1} = 0\}$. Let

$\tilde{\Lambda} = \{(z,\zeta) \in T^\star X \; ; \; \zeta_1 = \ldots = \zeta_{n-1} = y_n = \zeta_n = 0\}$ be the complexification of the bicharactéristic leaves of $\Lambda \cap T^\star_M X$, and let C_Λ be the sheaf on $\tilde{\Lambda}$ of microfunction in the $(z_1,\ldots,z_{n-1}, x_n)$ variables, holomorphic in $z_1,\ldots,z_{n-1}$. If we restrict our study to the set of $\tilde{\Lambda}$ where η_n is positive, C_Λ is isomorphic to the inverse image by the projection π of $T^\star X$ on X of the sheaf O_+ on $\mathbb{C}^{n-1} \times \mathbb{R}$ of boundary values of holomorphic functions of $\mathbb{C}^{n-1} \times \mathbb{C}_+$, where $\mathbb{C}_+ = \{z_n \in \mathbb{C} ; y_n > 0\}$.

Then one can developp a theory analogous to the theory of hyperfunctions but with the sheaf C_Λ instead of O_X, ((5), (10), (9)), and we get :

$$\mathcal{H}^p_{T^\star_M X \cap \Lambda}(C_\Lambda) = 0 \qquad p \neq n-1 .$$

The sheaf $B_\Lambda =_{\text{def.}} \mathcal{H}^{n-1}_{T^\star_M X \cap \Lambda}(C_\Lambda)$ is flabby.

There exists a natural injective morphism of $\mathcal{E}_X$-modules

$$C_M \mid T^\star_M X \cap \Lambda \longrightarrow \check{B}_\Lambda$$

(this last point is a particular case of theorem 1.1.).

If we replace Λ by $\Lambda_Y = \rho(T^\star X \underset{X}{\times} Y \cap \Lambda)$ and if we denote by C_{Λ_Y} the sheaf on $\tilde{\Lambda}_Y$ of microfunctions of the $(z_2,\ldots,z_{n-1},x_n)$ variables, holomorphic in the $z_2,\ldots,z_{n-1}$ variables, we construct in the same way the sheaf $\check{B}_{\Lambda_Y}$ on $T^\star_N Y \cap \Lambda_Y$.

b) Let $x^\star = (0,x'^o ; o,\ldots,o,i) \in T^\star_M X \cap T^\star_N X \cap \Lambda$ and let m be the order of the root of the equation $\sigma(P)(0,x'^o,\tau,o,\ldots,o,i)$ at $\tau = 0$. By the hypothesis that $T^\star X \underset{X}{\times} Y$ is non micro-characteristic for P on Λ we may write, after dividing P by an invertible operator :

$$P(z,D_z) = \sum_{|\alpha| \leqslant m} A_\alpha(z,D_{z'})\, D^\alpha_z$$

where $A_\alpha(z,D_{z'})$ are micro-differential operators not depending on D_{z_1}, of order 0 defined near $x^\star$, and where :

$$A_{(m,0,\ldots,0)} \equiv 1 \text{ , and } \alpha = (\alpha_1,\ldots,\alpha_{n-1},0) \text{ when}$$

$|\alpha| = m$.

c) Let us write N for $T^\star_M X \underset{N}{\times} N$. We want to prove that :

$$u \in \Gamma_N (C_M)_{x^\star} \text{ , } P u = \sum_{j=o}^{m-1} w_j \otimes \delta_1^{(j)}$$

where $w_j \in (C_N)_{\rho(x^\star)}$, implies $u = 0$. By the injectivity of the morphism from $C_M | \Lambda \cap T^\star_M X$ to B_Λ , it is enough to prove the following stronger result :

<u>Lemma 5.2</u> Under the previous hypothesis on the operator P , the map from $\Gamma_N(\check{B}_\Lambda) \times \check{B}^m_{\Lambda_Y}$ to $\Gamma_N(\check{B}_\Lambda)$:

$$(v,(w_j)_{j=o}^{m-1}) \longrightarrow P v + \sum_{j=o}^{m-1} w_j \otimes \delta_1^{(j)} \quad \text{is an}$$

isomorphism.

(Compare with corollary 3.3.).

<u>Proof of the lemma</u>

We identify C_Λ with the sheaf $\mathcal{O}_+$ on $\mathbb{C}^{n-1} \times \mathbb{R}$, and C_{Λ_Y} with the corresponding sheaf on $(\mathbb{C}^{n-1} \times \mathbb{R}) \cap Y$. At this stage of the proof it is convenient to use the language of derived category.

Let $\mathcal{M}$ be the coherent $\mathcal{E}_X$-module $\mathcal{E}_X/\mathcal{E}_X P$, and $\mathcal{M}_Y$ the induced system by $\mathcal{M}$ on Y (then $\mathcal{M}_Y$ is locally isomorphic to $\mathcal{E}^m_Y$) . We have the two following isomorphisms which are easy consequences of the results of (3) or (11), which traduce the precise Cauchy-Kowalewska theorem for P in C_Λ :

$$(\star)\quad \mathbb{R}\mathcal{H}om_{\mathcal{E}_X}(\mathcal{M}, C_\Lambda)|\, Y \simeq \mathbb{R}\mathcal{H}om_{\mathcal{E}_Y}(\mathcal{M}_Y, C_{\Lambda_Y})$$

$$(\star\star)\quad \mathbb{R}\mathcal{H}om_{\mathcal{E}_X}(\mathcal{M}, C_\Lambda)|\, Y \simeq \mathbb{R}\Gamma_Y\ \mathbb{R}\mathcal{H}om_{\mathcal{E}_X}(\mathcal{M}, C_\Lambda)\ [+2]\ .$$

Then we get

$$(\star\star\star)\ \mathbb{R}\Gamma_Y\ \mathbb{R}\mathcal{H}om_{\mathcal{E}_X}(\mathcal{M}, C_\Lambda)\ [+2] \simeq \mathbb{R}\mathcal{H}om_{\mathcal{E}_Y}(\mathcal{M}_Y, C_{\Lambda_Y})$$

and one apply the functor

$\mathbb{R}\Gamma_N \circ [n-1]$ to this isomorphism.

As $\mathbb{R}\Gamma_N = \mathbb{R}\Gamma_N \circ \mathbb{R}\Gamma_M$, we get :

$$\mathbb{R}\Gamma_N\ \mathbb{R}\mathcal{H}om_{\mathcal{E}_X}(\mathcal{M}, \mathbb{R}\Gamma_M(C_\Lambda))\ [n+1]$$

$$\simeq \mathbb{R}\mathcal{H}om_{\mathcal{E}_Y}(\mathcal{M}_Y, \mathbb{R}\Gamma_N(C_{\Lambda_Y}))\ [n-1]\ .$$

But we have :

$$\mathbb{R}\Gamma_N\ \mathbb{R}\Gamma_M(C_\Lambda)\ [n] = \Gamma_N(\check{B}_\Lambda)$$

$$\mathbb{R}\Gamma_N(C_{\Lambda_Y})\ [n-1] = \check{B}_{\Lambda_Y}$$

we obtain :

$$\mathcal{E}xt^1_{\mathcal{E}_X}(\mathcal{M}, \Gamma_N(\check{B}_\Lambda)) \simeq \mathcal{H}om_{\mathcal{E}_Y}(\mathcal{M}_Y, \check{B}_{\Lambda_Y})$$

or $\Gamma_N(\check{B}_\Lambda) / P\Gamma_N(\check{B}_\Lambda) \simeq B^m_{\Lambda_Y}$

Of course one should verify that this last isomorphism is equivalent to that of the lemma, but there is no difficulty in it.

Corollary 5.2 In the situation of theorem 5.1 we assume moreover that P is a differential operator of order m, and that any $\theta \in T_\Lambda T^\star X$ is non micro-characteristic for P on Λ. Let $\Lambda_Y = \rho(\Lambda \cap T^\star X \times_X Y)$ and $\Lambda^{\mathbb{R}}_Y = \Lambda_Y \cap T^\star_N Y$, and assume that $\rho^{-1}(\Lambda_Y) \cap \sigma^{-1}(P)(0) = \Lambda$. Then if u is an hyperfunction on M^+ solution of $Pu = 0$, and if $\gamma(u)$ denotes the m traces of u, we have: $SS(\gamma(u))$ is a union of bicharacteristic leaves of $\Lambda^{\mathbb{R}}_Y$.

Proof

Let $y^\star \in \Lambda^{\mathbb{R}}_Y$, $\{x^\star_i, i \in I\} = \rho^{-1}(y^\star) \cap \sigma^{-1}(P)(0)$. Let $b(y^\star)$ in the bicharacteristic leaf through $y^\star$ in $\Lambda^{\mathbb{R}}_Y$ and $b(x^\star_i)$ the bicharacteristic leaves through $x^\star_i$ in $\Lambda^{\mathbb{R}}$.

Assume that $y^\star \notin SS(\gamma(u))$. Then $\rho^{-1}(y^\star) \cap \overline{SS(u)} = \emptyset$ by the Holmgren-Kashiwara theorem (or by theorem 4.1 where we take $Z^{0,1} = \emptyset$, $m = p$). It follows from the main theorem of [3] that $b(x^\star_i) \cap SS(u) = \emptyset$ $\forall i \in I$. As the operator P is N-regular at each point of $\Lambda^{\mathbb{R}} \cap T^\star_M X \times_M N$ by theorem 5.1 we get by theorem 4.1 that $b(y^\star) \cap SS(\gamma(u)) = \emptyset$.

Example Let $f(x,t)$ be an hyperfunction on $\mathbb{R}^2$ which is the boundary value of $g(x + iy, t)$, hyperfunction defined for $y > 0$, solution of $(D_x + i D_y)g = 0$. Then if $(x_o, t_o; 0, idt) \notin SS(f)$, we have : $(x, t_o; 0, idt) \notin SS(f)$ $\forall x \in \mathbb{R}$.

6. HYPERBOLICITY AND N-REGULARITY

Let M be a real analytic manifold, X a complexification of M, P a microdifferential operator defined in a neighborhood of $x^\star \in T^\star_M X$. Let $\theta \in T_{T^\star_M X} T^\star X$ be a vector of the normal bundle to $T^\star_M X$ in $T^\star X$ at $x^\star$. Recall the definition of Kashiwara - Kawai (8) : P is micro-hyperbolic at $x^\star$ in the θ-direction if, for a choice of local coordinates $z = (z_1,..,z_n)$ on X, (z,ζ) on $T^\star X$, with $z = x + iy$, $\zeta = \xi + i\eta$, $T^\star_M X = \{ (z,\zeta) \in T^\star X \; ; \; y = \xi = 0 \}$, $x^\star = (x^o, i\eta^o)$, we have for an $\varepsilon_o > 0$:

$$\begin{cases} \sigma(P) \; ((x,i\eta) + \varepsilon\, \theta) \neq 0 \quad \text{for} \\ |(x,i\eta) - (x^o,i\eta^o)| \leqslant \varepsilon_o \,, \quad 0 < \varepsilon \leqslant \varepsilon_o \,. \end{cases}$$

Thanks to the local Bochner tube theorem this definition is free from coordinates (and can be extended to overdetermined systems (11)).

We denote by H the symplectic isomorphism from $T^\star T^\star X$ to $T T^\star X$ defined for $v \in T T^\star X$, $\theta \in T^\star T^\star X$ by :

$$< \theta, v > \; = \; < d\omega \,, v \wedge H(\theta) >$$

where ω is the canonical 1-form on $T^\star X$. For local coordinates (z,ζ) on $T^\star X$, with $\omega = \sum_j \zeta_j \, dz_j$,

$$H(dz_j) = - \frac{\partial}{\partial \zeta} \quad , \quad H(d\,\zeta_j) = \frac{\partial}{\partial z_j} \quad .$$

Let $H^{\mathbb{R}}$ be the real symplectic isomorphism associated to $\mathrm{Re}\, d\omega$ on the real analytic manifold $(T^\star X)^{\mathbb{R}}$. Then $H^{\mathbb{R}}$ defines an isomorphism from $T^\star(T^\star_M X)$ on $T_{T^\star_M X}(T^\star X)$. Moreover, if f is an holomorphic function on $T^\star X$, we have $\mathrm{Re}\, H_f = H^{\mathbb{R}}_{\mathrm{Re} f}$.

If f is a real analytic function on M, we still write f to denote the complexification of f in X, or even to denote $f \circ \pi$, the inverse image of f on $T^{\star}_{M}X$, or on $T^{\star}X$. Let now N be an analytic hypersurface of M defined by $f = 0$, with $df \neq 0$ on N, and let $M_{+} = \{ x \in M \,;\, f(x) > 0 \}$. If $x \in N$, $df(x)$ is called the conormal to $\overline{M}_{+}$ at x and we say (cf. (11)) that $df(x)$ is micro-hyperbolic for P at $x^{\star} \in T^{\star}_{M}X \times_{M} N$, with $\pi(x^{\star}) = x$, if $H(df(x))$ is micro-hyperbolic for P at that point.

Theorem 6.1 (22) In the preceding situation, assume that the conormal to $\overline{M}_{+}$ at $x^{\star}$ is micro-hyperbolic for P at $x^{\star}$. Then P is N^{+}-regular at $x^{\star}$.

Proof

Let $\Omega_{o} = \{ z \in \mathbb{C}^{n} \,;\, y_{n} > \sum_{j=1}^{n-1} y_{j}^{2} \}$,

an $\Omega_{1} = \{ z \in \mathbb{C}^{n} \,;\, y_{n} > \sum_{j=2}^{n-1} y_{j}^{2} \}$,

be the open sets of $\mathbb{C}^{n}$ already considered in § 1, $\mathcal{O}^{+}_{\partial\Omega_j}$ the sheaves on $\partial\Omega_{j}$ $(j = 0,1)$ of boundary values of holomorphic functions of Ω_{j}, and $\mathcal{C}^{+}_{\partial\Omega_{j}}$ the inverse image by π on $(T^{\star}_{\partial\Omega_{j}}X)^{+}$, the exterior conormal bundles to $\partial\Omega_{j}$, of the preceding sheaves (cf. (11)). A quantized complex transform Φ allows us to replace $\mathcal{C}_{M}$ by $\mathcal{C}^{+}_{\partial\Omega_{o}}$ and $\mathcal{C}_{N|X}$ by $\mathcal{C}^{+}_{\partial\Omega_{1}}$. We introduce also the sets :

$$\Omega^{+}_{o} = \Omega_{o} \cup \{ z \in \Omega_{1} \;,\; x_{1} > 0 \}$$

$$\Omega^{-}_{o} = \Omega_{o} \cup \{ z \in \Omega_{1} \;,\; x_{1} > 0 \}$$

and the corresponding sheaves $\mathcal{C}^{+}_{\partial\Omega^{+}_{o}}$ and $\mathcal{C}^{+}_{\partial\Omega^{-}_{o}}$. Then we have exact sequence of sheaves on $(T^{\star}_{\partial\Omega_o})^{+} \cap (T^{\star}_{\partial\Omega_1}X)^{+}$

$$0 \longrightarrow C^{+}_{\partial\Omega_1} \longrightarrow C^{+}_{\partial\Omega_o^{+}} \oplus C^{+}_{\partial\Omega_o^{-}} \longrightarrow C^{+}_{\partial\Omega_o} \longrightarrow 0$$

we still denote by P the image of P by the quantized contact transform Φ . The hypothesis of micro-hyperbolicity implies that P is an isomorphism from $\Gamma_{\{x_1 \leqslant o\}}(C^{+}_{\partial\Omega_o})$ onto itsef at $\varphi(x^{\star})$ ((11), theorem 5.1.2), and an isomorphism from $C^{+}_{\partial\Omega_o^{+}}$ onto itself at $\varphi(x^{\star})$ ((11), proposition 5.5.1) . Let u be a section of $C^{+}_{\partial\Omega_o^{-}} \cap \Gamma_{\{x_1 \leqslant o\}}(C^{+}_{\partial\Omega_o}$ at $\Phi(x^{\star})$ with $P\,u \in C^{+}_{\partial\Omega_o^{+}}$. Let $v \in C^{+}_{\partial\Omega_o^{+}}$ be a solution of $P\,v = P\,u$. Then $u - v = 0$ in $C^{+}_{\partial\Omega_o}$ because $u - v \in \Gamma_{\{x_1 \leqslant o\}}(C^{+}_{\partial\Omega_o})$ and u comes from a section of $C^{+}_{\partial\Omega_1}$, which achieves the proof.

Remark

In fact, we had only proved theorem 6.1. in (22) under the stronger assumption that the conormal to $\overline{M}_{+}$ and its opposite, are micro-hyperbolic.

It is clear that theorem 6.1. extends to (overdetermined) systems.

Theorem 6.1 as been refined by K. Kataoka as follow (13).
We choose local coordinates $z = (z_1,\ldots,z_n)$ on X , with $z = x+i\,y$, $M = \{z \in X \; ; \; y = 0\}$, $M_{+} = \{x \in M \; ; \; x_1 > 0\}$.
Let P be a microdifferential operator defined near $(x^{o}\,;\, i\eta^{o}) \in T^{\star}_{M}X \times_{M} N$. One says, with A. Kaneko (4) and K. Kataoka (13) that P is semi-hyperbolic with respect to $d\,x_1$ at $(x^{o}, i\eta^{o})$ if there exists $\varepsilon_o > 0$ such that :

$$\begin{cases} |x - x^o| \leqslant \varepsilon_o \ , \ |\eta - \eta_o| \leqslant \varepsilon_o \ , \ x_1 \geqslant 0 \ , \ 0 < \varepsilon \leqslant \varepsilon_o \\ \text{implies} \quad \sigma(P)(x, i\eta + \varepsilon\theta) \neq 0 \quad \text{where} \quad \theta = (-1, 0, \ldots, 0). \end{cases}$$

Then Kataoka has proved :

Theorem 6.2 (13) If P is semi-hyperbolic with respect to dx_1 at $(x^o, i\eta^o)$, P is N_+-regular at that point.

We do not give the proof here wich is based on an inequality obtained by applying the local Bochner tube theorem to $\sigma(P)\ (z_1^2, z' ; \zeta)$ (cf. also (2) where the same method was already used to obtain a local (non micro-local) version of theorem 6.2).

Let Γ be an open convex cone of $\pi^{-1}(M_+)$ in $T^*_M X$, with vertex at x^* . We may extend the definition of semi-hyperbolicity and say that P is Γ-hyperbolic with respect to dx_1 at x^* if $\sigma(P)(x, i\eta + \varepsilon\theta) \neq 0$ for $|x - x^o| \leqslant \varepsilon_o$, $|\eta - \eta_o| \leqslant \varepsilon_o$, $0 < \varepsilon \leqslant \varepsilon_o$, $(x, i\eta) \in \overline{\Gamma}$, with $\theta = (-1, 0 \ldots 0)$.

It would be very interesting to prove the N_+-regularity of P under the assumption of Γ-hyperbolicity for a non void convex cone Γ such that $\langle dx_1, v \rangle \geqslant 0 \ \ \forall\, v \in \Gamma$, to cover the following example to which generic diffraction problems can be reduced.

Example Assume that

$\sigma(P)\ (x, i\eta) = (i\eta_1)^2 - (x_1 - x_2)^k\ \ a(x, i\eta')$ where a is an homogeneous symbol of order two not depending on η_1 , real and positive on $T^*_M X$, defined near $x^* = (0 ; i\, dx_n)$, with $n \geqslant 3$.

For $k = 1$, K. Kataoka (13) has proved the N_+-regularity of P at $x^\star$, and J. Sjöstrand (24) has obtained in this case similar results.

The case where k is odd, $k > 1$, seems to remain open, but we remark that P is "Γ-hyperbolic" at $x^\star$ in the dx_1-codirection for $\Gamma = \{(x, i\eta) \in T^\star_M X \; ; \; x_1 > |x_2|\}$.

To end this section, let us mention the very last work of J. Sjöstrand (25) who studies various situations where propagation of analytic singularities at the boundary occurs.

References

(1) BONY J.M. : 1975-76, Sem. Goulaouic-Schwartz, exposé 17.

(2) BONY J.M., SCHAPIRA P. : 1973, Astérique 2-3, pp. 108-116.

(3) BONY J.M., SCHAPIRA P. : 1976, Ann. Inst. Fourier, 26, pp. 81-140.

(4) KANEKO A. : 1975, Sc. Pap. Coll. Gen. Educ. Univ. Tokyo 25, pp. 59-68.

(5) KASHIWARA M. : 1972, Talks in Nice.

(6) KASHIWARA M. : 1970, Suyaku no Ayumi, pp. 9-70, (in Japanese)

(7) KASHIWARA M., KAWAI T. : 1971, Proc. Japan Acad. 48, pp. 712-715, 1972, ibid, 49, pp. 164-168

(8) KASHIWARA M., KAWAI T. : 1975, J. Math. Soc. Japan 27, pp. 359-404.

(9) KASHIWARA M., KAWAI T. : 1980, Proc. of Les Houches colloquium, Lecture Notes in Physics , n° 126, Springer.

(10) KASHIWARA M., LAURENT Y. : to appear.

(11) KASHIWARA M., SCHAPIRA P. : 1979, Acta Math. 142, pp. 1 - 55.

(12) KATAOKA K. : 1977, Publ. R.I.M.S., Kyoto Univ. 12 suppl. pp. 147 - 153.

(13) KATAOKA K. : 1979, Micro-local theory of boundary value problem II, preprint.

(14) KOMATSU K. : 1972, J. Fac. Sc. Univ. Tokyo, I A, 19, n° 2, pp. 201 - 214.

(15) KOMATSU K. : 1970, Proc. Intern. Conf. Funct. Analysis and Relat. Topics, Univ. of Tokyo Press, pp. 107 - 121.

(16) LEBEAU G. : 1980, Ann. Sc. Ec. Norm. Sup., pp. 269-297.

(17) MALGRANGE B. : 1977-78, Sém. Bourbaki n° 522.

(18) MORREY C.B., NIRENBERG L. : 1957, Comm. Pure Appl. Math.10, pp. 261-290.

(19) NIRENBERG L. : 1973, Regional Conferences Series in Math 17, Providence.

(20) SATO M., KAWAI T., KASHIWARA M. : 1973, Lecture Notes in Math. 287, Springer, pp. 265 - 529.

(21) SCHAPIRA P. : 1969, Bull. U.M.I. n° 3, pp. 367-372, or 1971, Bull. Soc. Math. France 99, pp.113-141.

(22) SCHAPIRA P. : 1977, Publ. R.I.M.S., Kyoto Univ. 12 Suppl. pp. 441-453.

(23) SCHAPIRA P. : 1976-77, Sem. Goulaouic-Schwartz, exposé 9.

(24) SJÖSTRAND J. : 1979, Comm. Part. Dif. Eq. 5, n° 2, pp. 187-207.

(25) SJÖSTRAND J. : 1980, Analytic singularities and micro-hyperbolic boundary value problems, prepint.

LOWER BOUNDS AT INFINITY FOR SOLUTIONS OF DIFFERENTIAL EQUATIONS WITH CONSTANT COEFFICIENTS IN UNBOUNDED DOMAINS

Y. Shibata

Department of Mathematics, University of Tsukuba,
Ibaraki, Japan

1. INTRODUCTION

Let Ω be an unbounded domain in R^n and P be a differential operator. In this note, for the equation:

$$Pu = 0 \text{ in } \Omega, \tag{1.1}$$

we consider the following problems in the case that P is an operator with constant coefficients.

Problem 1. Determine the lower bound of the growth order at infinity of solutions of the equation (1.1).

Problem 2. What conditions should we impose on P in order that the lower bound is finite in Problem 1?

We shall explain the background of Problems 1 and 2 by giving examples. According to a well-known result in the theory of analytic functions of one complex variable, what is called Liouville's theorem, if u satisfies the equation: $(\partial/\partial x_1 + i\partial/\partial x_2)u(x_1,x_2) = 0$ in R^2 ($i = \sqrt{-1}$) and condition:

$$\lim_{|(x_1,x_2)| \to \infty} |u(x_1,x_2)| = 0$$

then $u \equiv 0$. On the other hand, for any given N, if we choose $\phi \in C_0^\infty(R^2)$ so that $\hat{\phi}(\xi_1,i\xi_1) \not\equiv 0$ and put

$$u(x_1,x_2) = \int_R e^{ix_1\xi_1+ix_2\xi_2} \frac{\hat{\phi}(\xi_1,\xi_2)(i\xi_1+\xi_2)^{N+1}}{i\xi_1 - \xi_2} d\xi_1\, d\xi_2$$

H. G. Garnir (ed.), Singularities in Boundary Value Problems, 213–234.

then $u(x_1,x_2) = o(|(x_1,x_2)|^{-N})$ and $(\partial/\partial x_1 + i\partial/\partial x_2)u \in \mathcal{E}'(R^2)$ but $u \notin \mathcal{E}'(R^2)$. Here $\hat{\phi}$ is the Fourier-Laplace transform of ϕ. These show that in the case that $P = \partial/\partial x_1 + i\,\partial/\partial x_2$ we can conclude as follows: if $\Omega = R^2$ then the lower bound of the growth order at infinity is finite, but if Ω = the exterior of a compact set then it is $-\infty$. When $P = \Delta + k$ where Δ is a Laplacian and k is a positive number, the result is different from the case that $P = \partial/\partial x_1 + i\partial/\partial x_2$. The following follows from a result due to Rellich [20].

Theorem 1.1. Let u be a C^2 function defined in an unbounded domain Ω in R^n whose boundary is compact, and $\Omega \supset \{x;\ |x| \geqslant R_0\}$. Suppose that

$$\Delta u + ku = 0 \text{ in } \Omega \ (k > 0). \tag{1.2}$$

Then if u satisfies the condition:

$$\varliminf_{R\to\infty} R^{-1} \int_{R<|x|<2R} |u(x)|^2 \, dx = 0 \tag{1.3}$$

then the support of u is compact.

Theorem 1.1 shows that in the case that $P = \Delta + k$ the lower bound of the growth order at infinity is finite when Ω = the exterior of a compact set and $= R^n$, respectively.

The problems mentioned previously were studied by many authors (see [1]-[24] and further references in these papers). In particular, after Littman's work [10,11,12], Hörmander [4] and Murata [15,16] completed independently the study of the case that P is an operator with constant coefficients, $\Omega = R^n$ and Ω is the exterior of a compact set. When the boundary of Ω is unbounded; Ω contains a half space and P is a Schrödinger type, the results for Problems 1 and 2 were obtained by Agmon [1], Konno [7] and Tayoshi [23]. When Ω is the complement of a closed proper cone in R^n and P is a general operator with constant coefficients, Murata and Shibata [17] and Littman [13] independently got the results for Problems 1 and 2. When Ω is unbounded and contained in a half space, a situation is changed, which is showed by a following example.

Example 1.2. Put

$$u(x) = \int \exp i(x'\cdot\xi' + x_n\sqrt{k - |\xi'|^2})\, \phi(\xi')\, d\xi'$$

where $\phi(\xi') \in C_0^\infty(\{\xi' = (\xi_1,\dots,\xi_{n-1}) \in R^{n-1};\ |\xi'|^2 > k\})$, $x' = (x_1,\dots,x_{n-1})$, $x'\cdot\xi' = \Sigma_{j=1}^{n-1} x_j\cdot\xi_j$. We see easily that $u(x) \in \mathcal{S}(\bar{H})$ $(H = \{x \in R^n;\ x_n > 0\})$ and u(x) satisfies the equation: $(\Delta + k)u = 0$ but $u \notin \mathcal{E}'(\bar{H})$.

It is natural that such a solution should be excluded by boundary conditions. Agmon [1] studied the case that the operator is a Schrödinger type with Dirichlet boundary condition and that Ω is unbounded and contained in a half space (see also [9,18,19,20,21, 22]). In particular, Shibata [21] completed the study of the case of general homogeneous boundary value problem for differential operators with constant coefficients in a half space ($\Omega = H$), and Shibata [22] almost completed the study of the case that $\Omega = \bar{H}$ - { a compact set in $\bar{H}$} and that P is a partially hypoelliptic operator with respect to $\partial/\partial x_n$ with constant coefficients.

In this note, we give results for Problems 1 and 2 and outlines of their proof in the case that operators have <u>constant coefficients</u> and that $\Omega = R^n$, the exterior of a compact set, the exterior of a closed proper cone, a half space and the exterior of a compact set in a half space, respectively.

When the author was writing this note, the results due to Agmon, Hörmander and Murata were very helpful. All results stated in sections 2.1 and 2.2 of Chapter 2 are due to Hörmander [4]. The author wishes to express his gratitude to Prof. Agmon, Prof. Hörmander and Prof. Murata.

Finally, the interest of the results stated in this note also lies in its applications. For example, if Ω = the exterior of a compact set and we consider $-\Delta$ as a symmetric operator in $L^2(\Omega)$ with domain $C_0^\infty(\Omega)$, then it follows from Theorem 1.1 that any self-adjoint extension of $-\Delta$ has no positive eigenvalues (see Agmon [1], Lax-Phillips [30] and so on). And also, the problems stated in this note are closely related to Sommerfeld radiation condition (see [20,25,26,27] and further references in these papers). But, their applications and so on are not studied in this note.

2. LOWER BOUNDS AT INFINITY FOR SOLUTIONS OF DIFFERENTIAL EQUATIONS WITH CONSTANT COEFFICIENTS IN UNBOUNDED DOMAINS WHICH CONTAIN A HALF SPACE

In this chapter, we shall consider the equation $P(D)u = 0$ in an unbounded domain Ω in R^n which contains a half space. Here and in the rest of this chapter, R^n denotes n-dimensional Euclidean space and points of R^n are written as $x = (x_1,\dots,x_n)$. For differentiation we use the symbol $D = -i(\partial/\partial x_1,\dots,\partial/\partial x_n)$. For the real dual space of R^n we use the same notation R^n and their points are written as $\xi = (\xi_1,\dots,\xi_n)$. C^n denotes n-dimensional unitary space and points of C^n are written as $\zeta = (\zeta_1,\dots,\zeta_n)$ ($\zeta_j \in C^1$) or $\xi + i\eta$ with $\xi, \eta \in R^n$. The results given in sections 2.1 and 2.2 are all due to Hörmander [4] and the results given in section 2.3 are due to Murata and Shibata [17].

2.1. The case that $\Omega = R^n$

If $u \in \mathcal{S}'(R^n)$ and $P(D)u = 0$ in R^n, then the support of the Fourier transform $\hat{u}$ is contained in the set $A = \{\xi \in R^n;\ P(\xi) = 0\}$. First, we shall consider the asymptotic behavior of the Fourier transform $\hat{v}$ of a distribution v supported by a smooth manifold in R^n. When v is tangentially smooth, it is known that the wave front set is contained in the normal bundle of the manifold at the support of v (see [29]). So, noting this point and using Parseval's identity, we have

Theorem 2.1 ([4, 12]). If v is a smooth density with compact support on a C^∞ submanifold M of R^n of codimension k, then

$$\int_{|x|\leq R} |\hat{v}(x)|^2 dx \leq CR^k,\ R > 0.$$

If Γ is a closed cone in R^n which contains no element $\neq 0$ which is normal to M at a point in supp v, then

$$|\hat{v}(x)| \leq C_N(1 + |x|)^{-N},\ x \in \Gamma$$

for every integer N.

The following is essential in this section.

Theorem 2.2 ([4]). Let $v \in \mathcal{S}'$, $v \in L^2_{loc}$ and assume that there is a point $x_o \in$ supp v such that supp v in a neighborhood of x_o is contained in a C^∞ manifold M of codimension $k > 0$. If $\theta \in R^n$ is a normal of M at x_o and if $\varepsilon > 0$, then

$$\varliminf_{R\to\infty} R^{-k} \int_{|x/R - \theta| < \varepsilon} |v(x)|^2\ dx > 0.$$

The following is a main result in this section.

Theorem 2.3 ([4]). Let $u \in \mathcal{S}' \cap L^2_{loc}$ satisfy a differential equation: $P(D)u = 0$. Assume that the set $A = \{\xi \in R^n;\ P(\xi) = 0\}$ is not empty and of codimension k. Set

$$\Gamma_R = \{x \in \Gamma;\ R < |x| < 2R\}$$

where Γ is an open cone in R^n which for every analytic manifold $M \subset A$ and $x_o \in M$ contains some normal of M at x_o. If

$$\varliminf_{R\to\infty} R^{-k} \int_{\Gamma_R} |u(x)|^2\ dx = 0$$

it follows that $u = 0$.

Remark. (I) The codimension of A is the minimal codimension of analytic manifold M with $M \subset A$.
(II) If A is empty, it follows easily that $u \in \mathcal{S}'$ implies $u = 0$.

Sketch of a proof of Theorem 2.3. Since A is the union of analytic manifolds (see [34]), the theorem follows from Theorem 2.2 and the following lemma:

Lemma 2.4 ([4]). Let $v \in \mathcal{S}'(R^n)$, $v \in L^2_{loc}$, $\theta \in R^n$ and $\varepsilon > 0$. If $\chi \in C_0^\infty(R^n)$, and $w = \chi v$, it follows that for every $k \in R^1$

$$\underline{\lim}_{R \to \infty} R^{-k} \int_{|x/R - \theta|<\varepsilon} |\hat{w}(x)|^2 \, dx \lesssim \underline{\lim}_{R \to \infty} \int_{|x/R - \theta|<2\varepsilon} |\hat{v}(x)|^2 \, dx$$

where $C = (2\pi)^{-n}\int|\hat{\chi}|dx$.

Q.E.D.

The following theorem shows that Theorem 2.3 is very precise.

Theorem 2.5 ([4]). Assume that Γ is an open cone in R^n and N an integer such that every $u \in \mathcal{S}' \cap C^\infty$ with $P(D)u = 0$ and

$$\underline{\lim}_{R \to \infty} R^{-N} \int_{\Gamma_R} |u(x)|^2 = 0$$

is equal to zero. If $M \subset R^n$ is a C^∞ manifold where P vanishes and if $x_0 \in M$, it follows that the closure of Γ contains some normal $\neq 0$ of M at x_0 and that $N \lesssim \text{codim } M$.

Proof. Let u be a C_0^∞ density on M. Then $P(D)\hat{u} = 0$. So, the theorem follows easily from Theorem 2.1.

Q.E.D.

2.2 The case that Ω is the exterior of a compact set

Let Ω be the exterior domain of a compact set in R^n. We shall consider the equation:

$$P(D)u = 0 \text{ in } \Omega \tag{2.1}$$

in this section. Since we shall consider Problems 1 and 2, we may consider the equation:

$$P(D)u = f \text{ in } R^n \tag{2.2}$$

where $f \in \mathcal{E}'$ with supp $f \subset R^n - \Omega$ instead of the equation (2.1).

Theorem 2.6 ([4]). Assume that $P = cP_1^{m^1} \cdots P_k^{m^k}$ where c is a constant and for every j

$$P_j \text{ is real and irreducible,} \tag{2.3}$$

$$P_j(\xi^j) = 0 \text{ and } N^j = \text{grad } P_j(\xi^j) \neq 0 \text{ for some } \xi^j \in R^n. \tag{2.4}$$

Let Γ^1 be an open cone containing N^j and $-N^j$ for every j and Γ be the same as in Theorem 2.3. Put

$$\Gamma^2 = \Gamma \cup \Gamma^1, \ \Gamma^2_R = \{x \in \Gamma^2; \ R < |x| < 2R\}.$$

If $u \in \mathcal{S}' \cap L^2_{loc}$ satisfies the equation (2.2) and the condition:

$$\underline{\lim}_{R \to \infty} R^{-1} \int_{\Gamma^2_R} |u(x)|^2 \, dx = 0, \tag{2.5}$$

then supp u $\subset$ ch supp f (ch A denotes the convex hull of A).

Sketch of a proof. First, we shall show that the conditions (2.3), (2.4) and (2.5) imply that there exists a solution $v \in \mathcal{E}' \cap L^2$ of (2.2). For this purpose it suffices to show that $\hat{f}/P$ is an entire function (see [28, Theorems 3.2.2 and 3.4.2]). With suitable coordinate the real zeros of P_j near ξ^j are of the form:

$$\xi_n = s(\xi'), \ \xi' = (\xi_1, \ldots, \xi_{n-1}) \in \omega \subset R^{n-1}$$

where $s \in C^\infty(\omega)$ is real values, $\xi^j = (\xi'_0, s(\xi'_0))$ and N^j is proportional to $(s'(\xi'_0), -1)$ (s' = grad s). Note that $\hat{f}$ is an entire function. The condition (2.5) implies the following fact:

$$[(\partial/\partial\xi_n)^k \hat{f}](\xi', s(\xi')) \text{ vanishes identically in } \omega \tag{2.6}$$

for $0 \leqslant k \leqslant m^j - 1$ (see the proof of Theorem 3.1 in [4]). According to the fact (2.6), we see that there exists a small open set $\omega' \ni \xi^j$ in C^n such that $\hat{f}$ vanishes of order m^j in $\omega' \cap \Sigma_j$ where $\Sigma_j = \{\zeta \in C^n; \ P_j(\zeta) = 0, \text{ grad } P_j(\zeta) \neq 0\}$. Since P_j is irreducible, Σ_j is connected. So, $\hat{f}/P_j$ is holomorphic in $\Sigma_j \cup \{\zeta \in C^n; \ P_j(\zeta) \neq 0\}$. According the fact that the set of poles of a meromorphic function has pure complex codimension 1 or is empty, we see that $\hat{f}/P_j$ is an entire function (see [32]). By repeating the argument, we can conclude that $\hat{f}/P$ is an entire function. Putting v = the inverse Fourier transformation of $\hat{f}/P$, we have that $v \in \mathcal{E}' \cap L^2$ and $P(D)v = f$. Therefore, applying Theorem 2.3 to u-v, we have that u=v, which completes the proof.

The following two theorems shows that the hypotheses made in Theorem 2.6 cannot be much relaxed.

Theorem 2.7 ([4]). Assume that P has an irreducible factor p which is not proportional to a real polynomial or has no simple real zero. For any integer N one can then find $u \in L^\infty \cap C^\infty$ so that $P(D)u \in \mathcal{E}'$ and $u(x) = o(|x|^{-N})$ but $u \notin \mathcal{E}'$.

Theorem 2.8. ([4]). Let p be an irreducible real polynomial and Γ a closed semi-algebraic cone in R^n such that grad $p(\xi)$ or -grad $p(\xi)$ is in Γ for every real zero of p. Assume that there is no real ξ with $p(\xi) = 0$ and grad $p(\xi) \neq 0$ such that grad $p(\xi)$ and -grad $p(\xi)$ are both in Γ. Then, for every integer N one can find $u \in \mathcal{S}' \cap C^\infty$ such that $P(D)u \in \mathcal{E}'$ and $u(x) = o(|x|^{-N})$ in Γ but $u \notin \mathcal{E}'$.

2.3. The case that Ω is the exterior of a proper cone

Let C_o be a closed proper cone and $\Omega = R^n - C_o$. We consider the equation:

$$P(D)u = 0 \text{ in } \Omega. \tag{2.7}$$

Since P(D) is invariant under the translation, we may assume that the vertex of C_o is the origin.

Theorem 2.9 ([13, 17]). Let C be the convex hull of C_o and

$$P(\zeta) = Q(\zeta)\Pi_{j=1}^{p} (P_j(\zeta))^{m_j},$$

where $Q(\zeta)$ is a polynomial such that $Q(\zeta) \neq 0$ for any $\zeta \in R^n - iC'$ and $P_j(\zeta)$ $(j = 1,\dots,p)$ are irreducible polynomials with real coefficients, where $C' = \{\xi \in R^n;\ x\cdot\xi > 0 \text{ for any } x \in C - \{0\}\}$. Set

$$A_j = \{\zeta \in R^n - iC';\ P_j(\zeta) = 0,\ \text{grad } P_j(\zeta) \neq 0\} = \bigcup_k A_{j,k},$$

$$B_j = \{\xi \in R^n;\ P_j(\xi) = 0,\ \text{grad } P_j(\xi) \notin C \cup (-C)\},$$

where $A_{j,k}$ is a connected component of A_j. Assume that the closure $\overline{A_{j,k}}$ of each $A_{j,k}$ intersects B_j. Let Γ^3 be an open cone in $R^n - C$ such that

(i) for some $\xi^{jk} \in \overline{A_{j,k}} \cap B_j$, $\Gamma^3 \ni \pm\text{grad } P_j(\xi^{jk})$,

(ii) for every real analytic manifold $M \subset \{\xi \in R^n;\ P(\xi) = 0\}$ and $\xi^o \in M$, Γ^3 contains some normal of M at ξ^o.

Set $\Gamma^3_R = \{x \in \Gamma^3;\ R < |x| < 2R\}$. If $u \in \mathcal{S}' \cap L^2_{loc}(R^n - C)$ satisfies the conditions:

$$P(D)u = f,\ \ \text{supp } f \subset C \tag{2.8}$$

$$\underline{\lim}_{R \to \infty} R^{-1} \int_{\Gamma_R^3} |u(x)|^2 \, dx = 0 \tag{2.9}$$

then supp u $\subset$ C.

Sketch of a proof. It is sufficient to show that the conditions (2.8) and (2.9) and the fact that Γ^3 satisfies the condition (i) imply that $\hat{f}/P$ is holomorphic in a tubular domain $R^n - iC'$ (see [33]). Without loss of generality, we may assume that $C \ni (x_1, 0,\ldots,0)$ $(x_1 > 0)$ and $(\partial P_j/\partial\xi_n)(\xi^{jk}) \neq 0$. By the implicit function theorem, the root of P_j near ξ^{jk} is of the form:

$$\zeta_n = s(\zeta'), \quad \zeta' = \xi'+i\eta'$$

where s is a holomorphic function in an open set $\omega \subset C^{n-1}$, $s(\xi')$ is real valued for $\xi' \in \omega \cap R^{n-1} + i\{0\}$, $\xi^{jk} = ((\xi^{jk})', s((\xi^{jk})'))$, grad $P_j(\xi^{jk})$ is proportional to (grad $s((\xi^{jk})', -1) \notin C \cup (-C)$. By Taylor expansion, we have $\text{Im}\, s(\xi'+i\eta') = s'(\xi')\cdot\eta' + O(|\eta'|^2)$, where s' = grad s. Since $(\xi', s'(\xi')\cdot\eta')$ is proportional to the tangent plane of $\{(\xi', s(\xi')), \xi' \in \omega \cap R^{n-1} + i\{0\}\}$ at $(\xi', s(\xi'))$ and $(s'((\xi^{jk})'), -1) \notin C \cup (-C)$, there exist a small compact subcone S of $\{\eta' \in R^{n-1}; (\eta', s((\xi^{jk})')\cdot\eta') \in -C'\}$, a small open ball B with center at the origin of R^{n-1}, and a neighborhood E of $(\xi^{jk})'$ such that the set $\{(\xi'+i\eta'; s(\xi'+i\eta')); \xi' \in E, \eta' \in V = S \cap B\}$ is contained in $R^n - iC'$. The condition (2.9) and the fact that Γ satisfies the condition (i) imply

$$\lim_{\substack{\eta' \to \infty \\ \eta' \in V}} \int (\partial/\partial\zeta_n)^\nu \hat{f}(\xi'+i\eta', s(\xi'+i\eta'))\phi(\xi')d\xi' = 0 \tag{2.10}$$

for any $\phi(\xi') \in C_0^\infty(E)$ and $0 \leqslant \nu \leqslant m^j - 1$ (see [17]). The fact (2.10) and the well-known "edge-of-wedge" theorem (see [31,33]) imply that $[(\partial/\partial\zeta_n)^\nu \hat{f}](\zeta', s(\zeta'))$ vanishes identically in ω for any ν with $0 \leqslant \nu \leqslant m^j - 1$. Since $A_{j,k}$ is connected, $\hat{f}/P_j$ is holomorphic in $A_{j,k} \cup \{\zeta \in R^n - iC'; P_j(\zeta) \neq 0\}$. So, $\hat{f}/P_j$ is holomorphic in $R^n - iC'$. By repeating the argument, we can conclude $\hat{f}/P$ is holomorphic in $R^n - iC'$, which completes the proof.

The following theorems and an example show that the hypotheses made cannot be much relaxed.

Theorem 2.10 ([17]). Let C be a closed convex proper cone. Assume that P has an irreducible factor p with $\{\zeta \in R^n - iC'; p(\zeta) = 0\} \neq$ empty which is not proportional to a real polynomial or has no simple real zeros. For any integer N, one can then find $u \in L^\infty \cap C^\infty$ so that supp $P(D)u \subset C$ and $u(x) = o(|x|^{-N})$ in $R^n - C$ but supp $u \not\subset C$.

Theorem 2.11 ([17]). Let C be the interior of a closed convex proper semi-algebraic cone and p be an irreducible real polynomial with $\{\zeta \in R^n - iC';\ p(\zeta) = 0\} \neq$ empty. Assume that there is no real ξ with $p(\xi) = 0$ and grad $p(\xi) \neq 0$ such that grad $p(\xi)$ and $-$ grad $p(\xi)$ are both in $R^n - C$. Then one can find $u \in \mathcal{S}' \cap C^\infty$ for every integer N such that supp $P(D)u \subset C$ and $u(x) = o(|x|^{-N})$ in $R^n - C$ but supp $u \not\subset C$.

Example 2.12 ([17]). Let $P(D) = (-\Delta + 1)^2 - D_n$ and $C = \{x \in R^n;\ -x_n > |x'|\}$ where $x' = (x_1, \ldots, x_{n-1})$. Choose $f \in \mathcal{S}$ so that supp $f \subset C$ and $\hat{f}(\zeta^o) \neq 0$ where ζ^o satisfies the equation $\Sigma_{j=1}^n \zeta_j^2 + 1 + \sqrt{\zeta_n} = 0$ and $\zeta^o \in R^n - iC'$. For any given integer N, we set

$$\hat{u}(\xi) = \xi_n^N \hat{f}(\xi)(|\xi|^2 + 1 + \sqrt{\xi_n + i0})^{-1},\ \xi \in R^n.$$

Let u(x) be the inverse Fourier transform of $\hat{u}$. We see easily that $u \in \mathcal{S}' \cap C^\infty$, supp $P(D)u \subset C$ and $u(x) = o(|x|^{-N})$ in $R^n - C$ but supp $u \not\subset C$.

3. LOWER BOUNDS AT INFINITY FOR SOLUTIONS OF GENERAL BOUNDARY VALUE PROBLEMS FOR A SYSTEM OF PARTIAL DIFFERENTIAL OPERATORS WITH CONSTANT COEFFICIENTS IN A HALF SPACE

In this chapter, we change notations a little to emphasize the half space. R^{n+1} denotes the n+1-dimensional Euclidean space and points of R^{n+1} are written as $(x,y) = (x_1, \ldots, x_n, y)$. For the real dual space of R^{n+1} we use the same notation R^{n+1} and their points are written as $(\xi, \lambda) = (\xi_1, \ldots, \xi_n, \lambda)$. C^n denotes n-dimensional unitary space and points of C^n are written as $\zeta = (\zeta_1, \ldots, \zeta_n)$ ($\zeta_j \in C^1$) or $\xi + i\eta$ with $\xi, \eta \in R^n$. For differentiation we use the symbol $D = (D_x, D_y) = -i(\partial/\partial x_1, \ldots, \partial/\partial x_n, \partial/\partial y)$. We denote by H the half space $\{(x,y) \in R^{n+1};\ y > 0\}$ and $H' = \{(x,0) \in R^{n+1}\}$. For a positive number δ we put $C^\infty([0,\delta);\ \mathcal{S}'(\mathcal{D}')(R^n)) = \{u \in \mathcal{S}'(\bar{H})(\mathcal{D}'(\bar{H}));\ \langle u(\cdot,y), \phi(\cdot)\rangle$ is a C^∞ function of y in $[0,\delta)$ for any $\phi \in \mathcal{S}(R^n)$ $(\mathcal{D}(R^n))\}$, respectively. Let

$$P(D) = \Sigma_{j=0}^m p_{m-j}(D_x)D_y^j$$

be a differential operator with constant coefficients and $B_j(D)$, $j = 1, \ldots, p$, be some other differential operators with constant coefficients. Since the case that $m = 0$ is reduced to the case of Chapter 2, we assume that $m \geqslant 1$. We consider the following boundary value problem :

$$P(D)u = f \text{ in } H,\ B_j(D)u = g_j,\ j = 1, \ldots, p, \text{ in } H'.$$

3.1. The case that $f = 0$, $g_j = 0$, $j = 1, \ldots, p$

First, we shall consider the conditions imposed on boundary conditions.

(A.1) The number of roots with positive imaginary part of the equation $P(\xi,\lambda) = 0$ in λ is less than or equal to p whenever $\xi \in R^n$.

Put $P(\xi,\lambda) = \Pi_{j=1}^{m'} P_j(\xi,\lambda)^{m^j}$ and $\tilde{P}(\xi,\lambda) = \Pi_{j=1}^{m'} P_j(\xi,\lambda)$ where all P_j are irreducible polynomials. Let us denote by $Q(\xi)$ the resultant of $\tilde{P}$ and $\partial\tilde{P}/\partial\lambda$. Put $A = \{\xi \in R^n;\ p_o(\xi) = 0 \text{ or } Q(\xi) = 0\}$. We break up $R^n - A$ into the open connected component V_j, whose number is finite, that is,

$$R^n - A = \bigcup_{\text{finite}} V_j,\quad V_j \cap V_{j'} = \text{empty if } j \neq j'.$$

Write $V_j = V$ for the sake of brevity. Let us denote by $\lambda_j(\xi)$, $j = 1,\ldots,m$, the roots of the equation $P(\xi,\lambda) = 0$ in λ when $\xi \in V$. We have the imaginary part of $\lambda_j(\xi)$ (denoting them by $\text{Im}\ \lambda_j(\xi)$) is a real analytic function of ξ. Without loss of generality, we may assume that $\text{Im}\ \lambda_j$, $j = 1,..,\mu(\mu\geq 0)$, do not vanish identically in V and $\text{Im}\ \lambda_j$, $j = \mu+1,..,m$, vanish identically in V. Put $A_V = \{\xi \in V;\ \text{Im}\ \lambda_t(\xi) = 0 \text{ for some } t \in \{1,\ldots,\mu\}\}$. We break up $V - A_V$ into the open connected components $W_{V,j}$, that is

$$V - A_V = \bigcup_{\text{finite}} W_{V,j},\ W_{V,j} \cap W_{V,j'} = \text{empty if } j \neq j'.$$

For the sake of brevity, we write $W_{V,j} = W$. When $\xi \in W$, the roots of the equation $P(\xi,\lambda) = 0$ in λ have constant multiplicity and split into three classes; real roots, those with positive imaginary part and those with negative imaginary part. We denote those by $\lambda_j^o(\xi)$, $j = 1,\ldots,a$, with multiplicity α^j; $\lambda_j^+(\xi)$, $j = 1,\ldots,b$, with multiplicity β^j; $\lambda_j^-(\xi)$, $j = 1,\ldots,c$, with multiplicity γ^j, where $\text{Im}\ \lambda_j^o \equiv 0$, $\text{Im}\ \lambda_j^+ > 0$, and $\text{Im}\ \lambda_j^- < 0$, respectively. Put

$$P^+(\xi,\lambda) = \Pi_{j=1}^{\tilde{b}} (\lambda-\lambda_j^+(\xi))^{\beta^j},$$

$$L_{W,\sigma}(\xi) = \det((2\pi i)^{-1} \oint B_{\sigma_j}(\xi,\lambda)\lambda^{k-1}(P^+(\xi,\lambda))^{-1}d\lambda)_{j,k=1,..,\tilde{b}},$$

where $\tilde{b} = \Sigma_{j=1}^{b} \beta^j$ and σ_j, $j = 1,\ldots,b$, are integers with $1 \leq \sigma_j \leq p$. Here and in the rest of this note, for an analytic function $a(\lambda)$ and a polynomial $b(\lambda)$ we put

$$(2\pi i)^{-1} \oint a(\lambda)b(\lambda)^{-1}\ d\lambda = (2\pi i)^{-1} \int_\kappa a(\lambda)b(\lambda)^{-1}\ d\lambda$$

where κ is a simple closed curve in the complex λ-plane which surrounds all roots of the equation $b(\lambda) = 0$. We consider the following assumption:

(A.2) For any W such that $\tilde{b} > 0$, there exist σ_j, $j = 1,\ldots,\tilde{b}$, such that $L_{W,\sigma}(\xi)$ does not vanish identically in W.

Theorem 3.1 ([21]). Let a system $\{P(D), B_j(D), j = 1,\ldots,p\}$ satisfy the assumptions (A.1) and (A.2). Then there exist an open cone Γ in R^{n+1} and a natural number N depending only on n and $\{P(D), B_j(D), j = 1,\ldots,p\}$ such that if $u \in C^\infty([0,\delta); \mathcal{S}'(R^n)) \cap L^2_{loc}(\bar{H})$ for some positive number δ satisfies the equations:

$$P(D)u = 0 \text{ in } H,\ B_j(D)u = 0,\ j = 1,\ldots,p, \text{ in } H', \tag{3.1}$$

and u satisfies the condition:

$$\underline{\lim}_{R \to \infty} R^{-N} \int_{\Gamma_R} |u(x,y)|^2\, dxdy = 0 \tag{3.2}$$

then u = 0. Here $\Gamma_R = \{(x,y) \in \Gamma;\ y \geqslant 0,\ R < |(x,y)| < 2R\}$.

Theorem 3.2 ([21]). If at least one of the assumptions (A.1) and (A.2) is not fulfilled, there exists a solution $u \not\equiv 0$ of the equations (3.1) which belongs to $\mathcal{S}(\bar{H})$.

Sketch of a proof of Theorem 3.1. Let U be any sufficiently small open set contained in W. According to the assumptions (A.1) and (A.2), we have there exist an integer p' with $\tilde{b} \leqslant p' \leqslant p$, integers δ^j with $0 \leqslant \delta^j \leqslant \alpha^j$, $j = 1,\ldots,a$ and integers σ_j with $1 \leqslant \sigma_j \leqslant p$, $j = 1,\ldots,p'$, such that $\Sigma_{j=1}^a(\alpha^j-\delta^j) + \tilde{b} = p'$ and $L(\xi)$ does not vanish identically in U where

$$L(\xi) =$$

$$\det\left(\frac{1}{2\pi i}\oint B_{\sigma_j}(\xi,\lambda)\lambda^{k-1}\left(\Pi_{j=1}^a(\lambda-\lambda_j^o(\xi))^{\alpha^j-\delta^j}P^+(\xi,\lambda)\right)^{-1}d\lambda\right)_{j,k=1.p'}.$$

Let $\phi(\xi) \in C_o^\infty(U)$ and $v(\xi,y) = \phi(\xi)\hat{u}(\xi,y)$, where $\hat{u}$ is the partial Fourier transform of u with respect to x. First of all, we shall show the following assertion: there exist an open cone Γ in R^{n+1} and a natural number N depending only on $\{P(D), B_j(D), j = 1,\ldots,p\}$ and n such that if a solution $u \in C^\infty([0,\delta); \mathcal{S}'(R^n)) \cap L^2_{loc}(\bar{H})$ of the equation (3.1) satisfies the condition (3.2) then supp $v \subset \{\xi \in U;\ L(\xi) = 0\} \times [0,\infty)$.

To show the above assertion, we need the following lemma whose proof is not stated at all in this note.

Lemma 3.3 ([21, Lemma 2.7]). Put $M_j = \{(\xi,\lambda_j^o(\xi));\ \xi \in U\}$, $j = 1,\ldots,a$. Assume that $\theta_j = (\theta_1^j,\ldots,\theta_n^j,\theta_{n+1}^j) \in H$ is a nomal of M_j at $(\xi_o,\lambda_j^o(\xi_o))$ $(\xi_o \in U,\ |\theta_j| = 1)$ and $\varepsilon > 0$. If u satisfies the condition: for $s = 1,\ldots,a$, with $\delta^s \geqslant 1$

$$\underline{\lim}_{R\to\infty} R^{-(2(\alpha^s-\delta^s)+1)}\int_{|(x,y)/R-\theta|<\varepsilon;\,y\geqslant 0}|u(x,y)|^2dxdy = 0,$$

then there exists a small neighborhood ω of ξ_o contained in U such that

$$<\Sigma_{j=1+\nu}^{d} \Sigma_{k=0}^{j-1-\nu} \frac{(j-1-k)!}{(j-1-k-\nu)!} q_j(\xi)(\lambda_s^{o(-)}(\xi))^{j-1-k-\nu} \times$$

$$D_y^k Q(\xi,D_y)v(\xi,0),\ \chi(\xi)> = 0 \tag{3.4}$$

for any $\chi(\xi) \in C_o^\infty(\omega)$ and $0 \leqslant \nu \leqslant \delta^s-1(0 \leqslant \nu \leqslant \gamma^s-1)$, respectively. Here

$$d = \Sigma_{j=1}^{a} \delta^j + \Sigma_{j=1}^{c} \gamma^j,$$

$$p_o(\xi)\Pi_{j=1}^{a}(\lambda-\lambda_j^o(\xi))^{\delta^j} \cdot \Pi_{j=1}^{c}(\lambda-\lambda_j^-(\xi))^{\gamma^j} = \Sigma_{j=0}^{d} q_j(\xi)\lambda^j,$$

$$Q(\xi,\lambda) = \Pi_{j=1}^{a}(\lambda-\lambda_j^o(\xi))^{\alpha^j-\delta^j} \cdot P^+(\xi,\lambda),\ \xi \in U.$$

So if we put $N = \min\{2(\alpha^j-\delta^j)+1;\ 1 \leqslant j \leqslant a\}$ and choose an open cone Γ in R^{n+1} so that Γ is a conic neighborhood of the set $\bigcup_{j=1}^{a} \{(\xi,\ \lambda_j^o(\xi));\ \xi \in U\}$, the assertion mentioned previously follows from the formulae (3.5) and the fact that u satisfies the equations: $B_j(D)u = 0$, $1,\ldots,p$, in H'.

The set $A_L = \{\xi \in U;\ L(\xi) = 0\}$ is an algebraic variety. Without loss of generality, we may assume that A_L is of codimension e. Let $\xi^o = (\xi_1^o,\ldots,\xi_n^o) \in A_L$ and assume that there is an open set U' $\ni \xi^o$ such that $U' \cap A_L$ is an analytic manifold of codimension e. Without loss of generality, we assume that $U' \cap A_L$ is defined by the equation $\xi'' = \mu(\xi')$, $\xi' \in \omega'$, where $\xi' = (\xi_1,\ldots,\xi_{n-e})$, $\xi'' = (\xi_{n-e-1},\ldots,\xi_n)$, $\xi^o = ((\xi^o)',\ \mu((\xi^o)'))$, ω' is a small neighborhood of $(\xi^o)'$ and $\mu(\xi')$ is an analytic function in ω'. Let $\phi'(\xi) \in C_o^\infty(U')$ and $w(\xi,y)$ be the composition of $\phi'\phi\hat{u}$ and the map $\xi \longmapsto (\xi',\ \xi''+\mu(\xi'))$ (defined arbitrarily for $\xi' \notin \omega'$). The support of $w(\xi,y)$ is contained in the plane $\{\xi'' = 0\} \times [0,\infty)$.

Lemma 3.4 ([21, Lemma 2.1]). $w(\xi,y)$ has the form:

$$w(\xi,y) = \Sigma_{|\alpha|\leqslant q}\ w_\alpha(\xi',y)\otimes D_{\xi''}^{\alpha}\delta(\xi'')$$

where $w(\xi',y) \in C^\infty([0,\delta);\ \mathcal{S}'(R_{\xi'}^{n-e}))$ and $\delta(\xi'')$ is the Dirac measure at the origin in $R_{\xi''}^{e}$.

It follows from Lemma 3.4 that the formulae

$$P(\xi',\mu(\xi'),\ D_y)w_\alpha(\xi',y) = 0 \text{ in } \omega' \times [0,\infty)$$
$$B_j(\xi',\ \mu(\xi'),D_y)w_\alpha(\xi',0) = 0,\ j = 1,\ldots,p \text{ in } \omega' \tag{3.5}$$

hold in distribution sense. If $L_{W,\sigma}(\xi',\ \mu(\xi'))$ vanishes identically in ω' for any σ, it follows from Theorem 2.2 that we can find an open cone Γ' in R^{n+1} such that the condition:

$$\underline{\lim}_{R\to\infty} \quad R^{-e} \int_{\Gamma'_R} |u(x,y)|^2 \, dxdy = 0$$

implies that $w(\xi,y) = 0$ where $\Gamma'_R = \{(x,y) \in \Gamma';\ y \geqslant 0,\ R < |(x,y)| < 2R\}$. If $L_{W,\sigma}(\xi',\mu(\xi'))$ does not vanish identically in ω' for some σ, repeating the same argument as that mentioned previously, we can conclude that there exists an algebraic variety B of co-dimension more than e such that supp $v(\xi,y) \subset B \times [0,\infty) \subset (U' \cap A_L) \times [0,\infty)$.

By repeating the argument, we can conclude that we can choose an open cone Γ and a natural number N which are independent o u so that the condition (3.2) implies that $u \equiv 0$.

3.2. The case that $f = 0$ and $g_j \in \mathcal{E}'(R^n)$

In this section and next section, we assume that

(A.3) P(D) is partially hypoelliptic with respect to y, that is, P(D) has the form:

$$P(D) = D_y^m + \Sigma_{j=1}^m \, p_j(D_x) D_y^{m-j}.$$

For the sake of brevity, we shall say that a system $\{P(D),\ B_j(D),\ j = 1,\dots,p\}$ of differential operators with constant coefficients is an <u>R-system</u> if it satisfies the hypotheses (A.1), (A.2) and (A.3).

We begin with the following example.

Example 3.5. Let $n = 1$ and a and b be positive numbers. Put

$$P(D) = (D_y^2 + a^2(D_x^2 + 1))(D_y^2 + b^2(D_x^2 + 1)),\ B_j(D) = D_y^{j-1},$$

for $j = 1,2,3$. A system $\{P(D),\ B_j(D),\ j = 1,2,3\}$ is an R-system. Let $\phi \in C_0^\infty(R^n)$ and put

$$v(x,y) = \frac{i}{2\pi}\int (b-a)^{-1}(be^{-a\sqrt{\xi^2+1}\cdot y} - ae^{-b\sqrt{\xi^2+1}\cdot y})\phi(\xi)e^{ix\cdot\xi}\, d\xi.$$

We see easily that $v \in \mathcal{S}(\overline{R^2_+})$ and that $P(D)v = 0$ in R^2_+ and $v(x,0) = \phi(x)$, $D_y v(x,0) = 0$, $D_y^2\, v(x,0) = iab(D_x+1)\phi(x)$ in R^1 but $v \notin \mathcal{E}'$.

Thus, to consider Problem 1, we need to find algebraic conditions which an R-system $\{P(D),\ B_j(D),\ j = 1,\dots,p\}$ satisfies. For a moment, we assume that $u \in C^\infty([0,\delta);\ \mathcal{S}'(R^n)) \cap L^2_{loc}(\bar{H})$ satisfies the equations:

$$P(D)u = f \text{ in } H; \; B_j(D)u = g_j, \; j = 1,\dots,p, \text{ in } H' \tag{3.6}$$

where $f \in \mathcal{E}'(\bar{H})$, $g_j = g_j(x) \in \mathcal{E}'(R^n)$, $j = 1,\dots,p$.

First of all, we get

Lemma 3.6 ([22]). Let ω be a small open set contained in W. If u satisfies the equations (3.6) and the condition:

$$\varliminf_{R \to \infty} R^{-1} \int_{\Gamma_R^\omega} |u(x,y)|^2 \, dxdy = 0$$

then the following formulae hold for any $\phi \in C_0^\infty(\omega)$:

$$\langle -i\Sigma_{j=\nu}^{d-1}[\Sigma_{k=0}^{j-\nu} q_k(\xi)\frac{(j-k)!}{(j-k-\nu)!}(\lambda_s^{o(-)}(\xi))^{j-k-\nu}]D_y^{d-1-j} \; P^+(\xi,D_y)\times$$
$$\hat{u}(\xi,0), \phi(\xi)\rangle = \langle[(\partial/\partial\lambda)^\nu \hat{f}_o](\xi,\lambda_s^{o(-)}(\xi)), \phi(\xi)\rangle, \tag{3.7}$$

for any integer ν with $0 \leqslant \nu \leqslant \alpha^s-1$ and $s=1,\dots,a$ (with $0 \leqslant \nu \leqslant \gamma^s-1$ and $s=1,\dots,c$), respectively. Here we put $f_o = f$ when $y \geqslant 0$ and $= 0$ when $y < 0$, $\hat{f}_o$ is the Fourier transform of f_o,

$$d = \Sigma_{j=1}^a \alpha^j + \Sigma_{j=1}^c \gamma^j,$$
$$\Pi_{j=1}^a(\lambda-\lambda_j^o(\xi))^{\alpha^j}\cdot\Pi_{j=1}^c(\lambda-\lambda_j^-(\xi))^{\gamma^j} = \Sigma_{j=0}^d q_j(\xi)\lambda^{d-j} \text{ when } \xi \in \omega,$$

Γ^ω is an open conic neighborhood of the set $\bigcup_{j=1}^a \{(\xi,\lambda_j^o(\xi));\ \xi \in \bar{\omega}\}$ and $\Gamma_R^\omega = \{(x,y) \in \Gamma^\omega;\ y \geqslant 0,\ R < |(x,y)| < 2R\}$.

Let $g(\lambda)$ be an entire function in one complex variable λ and put

$$\tilde{H}_{m-1-j}(g;\lambda_1,\dots,\lambda_m) = \frac{\det\begin{pmatrix}\sigma_0(1),\dots,\sigma_{j-1}(1),g(\lambda_1),\sigma_{j+1}(1),\dots,\sigma_{m-1}(1)\\ \vdots \quad \vdots \quad \vdots \quad \vdots \quad \vdots \\ \sigma_0(m),\dots,\sigma_{j-1}(m),g(\lambda_m),\sigma_{j+1}(m),\dots,\sigma_{m-1}(m)\end{pmatrix}}{\Pi_{j<k}(\lambda_j-\lambda_k)}$$

where the $\sigma_j(k)$ are defined as the following equations:

$$\Pi_{s\neq k}(\lambda-\lambda_s) = \Sigma_{j=0}^{m-1}\sigma_j(k)\lambda^{m-1-j},\ 1 \leqslant k \leqslant m.$$

Since $p_o(\zeta) = 1$, all roots of the equation $P(\zeta,\lambda) = 0$ in λ are continuous multi-valued functions of $\zeta \in C^n$. So we count roots of the equation: $P(\zeta,\lambda) = 0$ in λ according to their multiplicity, and then we denote them by $\lambda_j(\zeta)$, $j = 1,\ldots,m$. Put

$$H_{m-1-j}(\zeta) = \tilde{H}_{m-1-j}(f_o(\zeta,\cdot);\lambda_1(\zeta),\ldots,\lambda_m(\zeta)),\ \zeta \in C^n.$$

We see easily that all $H_{m-1-j}(\zeta)$ are entire functions of ζ and satisfies the equations:

$$-i\Sigma_{j=\nu}^{d-1}[\Sigma_{k=0}^{j-\nu} q_k(\xi)\frac{(j-k)!}{(j-k-\nu)!}(\lambda_s^{o(-)}(\xi))^{j-k-\nu}]\times$$

$$[\Sigma_{k=0}^{\tilde{\beta}} p_k^+(\xi)H_{d-1-j+\tilde{\beta}-k}(\xi)] = [(\partial/\partial\lambda)^\nu\ \hat{f}_o](\xi,\lambda_s^{o(-)}(\xi)),\ \xi \in \omega, \tag{3.8}$$

for any ν with $0 \leqslant \nu \leqslant \alpha^s$, $s = 1,\ldots,a$ (with $0 \leqslant \nu \leqslant \gamma^s$, $s = 1,\ldots,c$), respectively. Here $\tilde{\beta} = \Sigma_{j=1}^{b} \beta^j$ and

$$P^+(\xi,\lambda) = \Sigma_{j=0}^{\tilde{\beta}} p_j^+(\xi)\lambda^{\tilde{\beta}-j}.$$

Therefore, if the assumption:

(A.4) for some W, $\tilde{\beta} = 0$

holds, then we have

$$D_y^j\ \hat{u}(\xi,0) = H_j(\xi),\ j = 1,\ldots,m-1, \tag{3.9}$$

in ω in distribution sense. Since u satisfies the equation (3.7) and all $\hat{g}_j(\zeta)$ are entire functions (in fact $g_j \in \mathcal{E}'$), the formulae (3.9) imply that

$$\Sigma_{k=0}^{m-1}\ b_k^j(\zeta)H_{m-1-k}(\zeta) = \hat{g}_j(\zeta),\ j = 1,\ldots,p,\ \zeta \in C^n. \tag{3.10}$$

Here and in the rest of this note, in view of the assumption (A.3), we may assume that the order of $B_j(D)$ with respect to D_y is less than m, and then we put

$$B_j(\zeta,\lambda) = \Sigma_{k=0}^{m-1}\ b_k^j(\zeta)\lambda^{m-1-k},\ j = 1,\ldots,p.$$

In particular, the fact that $f = 0$ implies that $g_j(x) = 0$, $j = 1,\ldots,p$.

Summing up, we have showed

Theorem 3.7 ([22, Theorem 6.2]). Let a system $\{P(D), B_j(D), j = 1,\ldots,p\}$ be an R-system and satisfy the assumption (A.4). Put

$$\Gamma^1 = \Gamma \cup \Gamma^\omega \text{ and } \Gamma_R^1 = \{(x,y) \in \Gamma^1;\ y \geqslant 0,\ R < |(x,y)| < 2R\}$$

where Γ is the same as in Theorem 3.1, ω is a small open set contained in W with $\tilde{\beta} = 0$, and Γ^ω is the same as in Lemma 3.6. If $u \in C^\infty([0,\delta); \mathcal{S}'(R^n)) \cap L^2_{loc}(\bar{H})$ satisfies the equations: $P(D)u = 0$ in H and $B_j(D)u(x,0) \in \mathcal{E}'(R^n)$, $j = 1,\ldots,p$, and the condition:

$$\varliminf_{R \to \infty} R^{-1}\int_{\Gamma_R^1} |u(x,y)|^2 \, dxdy = 0$$

then $u = 0$.

Next, we shall consider the case that the assumption (A.4) does not hold. So, $\tilde{\beta} > 0$ for any W. Let an R-system $\{P(D), B_j(D), j = 1,\ldots,p\}$ satisfy the following two assumptions:

(A.5) for each j, there exists at least one point $\xi^j \in R^n$ such that at least one root of the equation $P_j(\xi^j,\lambda) = 0$ in λ has negative imaginary part,

(A.6) $\tilde{\beta} < p$ for some W.

The formulae (3.7) and (3.8) and the assumptions (A.5) and (A.6) imply that

$$<\Sigma_{k=0}^{\tilde{\beta}} \; p_k^+(\xi)[D_y^{m-1-j-k} \hat{u}(\xi,0) - H_{m-1-j-k}(\xi)], \phi(\xi)> = 0 \tag{3.11}$$

for any $\phi \in C_0^\infty(\omega)$ and for any integer j with $0 \leq j \leq m-1-\tilde{\beta}$ where ω is an open set contained in W with $\tilde{\beta} < p$. Put

$$G_j(\zeta) = \hat{g}_j(\zeta) - \Sigma_{k=0}^{m-1} b_k^j(\zeta)H_{m-1-k}(\zeta), \; \zeta \in C^n.$$

Since $L_{W,\sigma}(\xi) \not\equiv 0$ in ω for some σ, it follows from (3.11) that there exist algebraic functions $a_{j,\omega,\sigma}(\xi)$, $j = 1,\ldots,p$, such that

$$\Sigma_{j=1}^p a_{j,\omega,\sigma}(\xi)G_j(\xi) = 0, \; \xi \in \omega. \tag{3.12}$$

By analytic continuation along any closed path in $\{\zeta \in C^n; Q(\zeta) \neq 0\}$ begining and terminating with ξ, we see that there exist a finite number $s(\omega)$ and functional elements $a_{j,k,\omega,\sigma}(\xi)$, $j = 1,..,p$, $k = 1,\ldots,s(\omega)$, which arise at ξ by continuing $a_{j,\omega,\sigma}(\xi)$. Obviously, the formula (3.12) implies

$$\Sigma_{j=1}^p \; a_{j,k,\omega,\sigma}(\zeta)G_j(\zeta) = 0, \; \zeta \in \omega, \; k = 1,\ldots,s(\omega) \tag{3.13}$$

Since the number of W is finite and the assumptions (A.2) and (A.6) hold, we can conclude that there exist a positive constant number s and algebraic functions $A_{j,k}(\zeta)$, $j = 1,\ldots,p$, $k = 1,\ldots,s$, satisfying the following three conditions:

(i) $\Sigma_{j=1}^p \; A_{j,k}(\zeta)G_j(\zeta) = 0, \; \zeta \in C^n, \; k = 1,\ldots,s,$ (3.14)

(ii) the set $A_{j,1}(\zeta),\dots,A_{j,s}(\zeta)$ permutes among themselves by analytic continuation along any closed path in $\{\zeta \in C^n; Q(\zeta) \neq 0\}$ begining and terminating with any ζ with $Q(\zeta) \neq 0$,

(iii) for any ω contained in some W with $\tilde{\beta} < p$; for any index σ with $L_{W,\sigma} \not\equiv 0$; for any integer k with $1 \leq k \leq s(\omega)$ there exists a integer ℓ such that $a_{j,k,\omega,\sigma}(\xi) = A_{j,\ell}(\xi)$ for $\xi \in \omega$.

Now, we consider the following assumption:

(A.7) there exists a point $\zeta_o \in C^n$ such that the rank of the matrix $(A_{j.k}(\zeta_o))_{\substack{j=1,\dots,p\\ k=1,\dots,s}}$ is equal to p.

Since $G_j(\zeta)$ is an entire function, the assumption (A.7) implies that $G_j(\zeta) \equiv 0$, $j = 1,\dots,p$.

Summing up, we have showed

Theorem 3.8 ([22, Theorem 6.2]). Let an R-system $\{P(D), B_j(D), j = 1,\dots,p\}$ satisfy the assumptions (A.5), (A.6) and (A.7). Put

$$\Gamma^3 = \Gamma \cup \Gamma^2,\ \Gamma^3_R = \{(x,y) \in \Gamma^3;\ y \geqslant 0,\ R < |(x,y)| < 2R\}.$$

Here Γ is the same as in Theorem 3.1 and $\Gamma^2 = \bigcup_{\omega} \Gamma^\omega$ where ω is a small open set contained in W with $\tilde{\beta} < p$ and Γ^ω is the same as in Lemma 3.6. If $u \in C^\infty([0,\delta); \mathcal{S}'(R^n)) \cap L^2_{loc}(\bar{H})$ satisfies equations: $P(D)u = 0$ in H and $B_j(D)u(x,0) \in \mathcal{E}'$ $j = 1,\dots,p$, and the condition:

$$\underline{\lim}_{R \to \infty} R^{-1} \int_{\Gamma^3_R} |u(x,y)|^2\, dxdy = 0$$

then $u = 0$.

The following theorem shows that the assumptions (A.4)-(A.7) cannot be much relaxed.

Theorem 3.9 ([22, Theorem 6.5]). Let $\{P(D), B_j(D), j = 1,\dots,p\}$ be an R-system which does not satisfy the assumption (A.4). Suppose that it is not a system which satisfies the assumptions (A.5), (A.6) and (A.7). Then, for any given integer N we can find $u \in \mathcal{S}'(\bar{H}) \cap C^\infty(\bar{H})$ such that u satisfies the equations: $P(D)u = 0$ in H, $B_j(D)u(x,0) \in \mathcal{E}'$, $j = 1,\dots,p$, and $u(x,y) = o(|(x,y)|^{-N})$ in H but $u \notin \mathcal{E}'(\bar{H})$.

3.3. The case that $f \in \mathcal{E}'(\bar{H})$ and $g_j \in \mathcal{E}'(R^n)$, $j = 1,\dots,p$

First of all we give the following example:

Example 3.10. Let $n = 1$ and $P(D) = (D_y - (D_x^2 + 1))^2$. $\{P(D)\}$ is an R-system and satisfies the condition (A.4). So, we can apply Theorem 3.8 to the equation: $P(D)u = 0$ in H. But now we put

$$u(x,y) = \int_{R^1} e^{ix\xi}\, \hat{\phi}(\xi) \int_{R^1} e^{i(y-1)(\lambda+i)} ((\lambda+i)-\lambda(\xi))^{-2}\, d\lambda\, d\xi$$

where $\lambda(\xi) = \xi^2 + 1$ and $\hat{\phi}(\xi)$ is the Fourier transform of $\phi(x) \in C_0^\infty(R^1)$. Obviously, $u \notin \mathcal{E}'(R_+^2)$ but $u(x,y) \in \mathcal{S}(\bar{H})$ and $P(D)u = \phi(x)\otimes\delta(y-1) \in \mathcal{E}'(R_+^2)$.

This example shows that if $f \not\equiv 0$, it is not sufficient to consider an R-system in order to solve Problem 1. Let $P(D)$ be partially hypoelliptic with respect to D_y. Let $S(\xi,\eta)$ be a polynomial in (ξ,η) which is obtained by eliminating λ from the equations $P(\xi+i\eta,\lambda) = 0$ and $\overline{P(\xi+i\eta,\lambda)} = 0$. Put $A_S = \{\xi+i\eta \in C^n;\ S(\xi,\eta) = 0\}$. We break up $C^n - A_S$ into the open connected components V whose number is finite. Since the roots of the equation $P(\zeta,\lambda) = 0$ in λ for $\zeta \in V$ are not real or constant real numbers, there exists a non negative constant μ depending only on V such that μ roots of the equation $P(\zeta,\lambda) = 0$ in λ for $\zeta \in V$ have positive imaginary parts and $m-\mu$ roots of this equation have non positive imaginary parts (of course, a root is counted according to its multiplicity). We denote the former (the latter) be $\lambda_j^+(\xi)$, $j = 1,\dots,\mu$, $(\lambda_j^-(\xi)$, $j = 1,\dots,m-\mu)$, respectively and put

$$P^+(\zeta,\lambda) = \Pi_{j=1}^{\mu} (\lambda-\lambda_j^+(\zeta)),\ \zeta \in V.$$

Let us consider the following assumptions:

(A.8) for any V, $p \geq \mu$,

(A.9) for V with $\mu > 0$, there exist natural numbers $\sigma_1,\dots,\sigma_\mu$ $(1 \leq \sigma_j \leq p)$ such that $L_{V,\sigma}(\zeta)$ does not vanish identically in V, where

$$L_{V,\sigma}(\zeta) = \det\left(\frac{1}{2\pi i}\oint B_{\sigma_j}(\zeta,\lambda)\lambda^{k-1}(P^+(\zeta,\lambda))^{-1}d\lambda\right)_{j,k=1,..,\mu}.$$

For the sake of brevity, we shall say that a system $\{P(D), B_j(D), j = 1,\dots,p\}$ is a C-system if it satisfies the assumptions (A.3), (A.8) and (A.9).

Proposition 3.11 ([22, Theorem 2.1]). A C-system is also an R system.

The consideration made in section 3.2 implies that if a C-system $\{P(D), B_j(D), j = 1,\dots,p\}$ satisfies the assumption (A.4) or the assumptions (A.5), (A.6) and (A.7), then for $\zeta \in V$

$$\begin{aligned}&\Sigma_{k=0}^{m-1} b_k^j(\zeta)H_{m-1-k}(\zeta) = \hat{g}_j(\zeta),\ j = 1,\dots,p,\\ &\Sigma_{k=0}^{m-1} \sigma_k(\lambda_s^-(\zeta))H_{m-1-k}(\zeta) = \hat{f}_o(\zeta,\lambda_s^-(\zeta)),\ s = 1,\dots,m-\mu,\end{aligned} \tag{3.15}$$

where the $\sigma_j(\lambda_k)$ are defined by the following equations:

$$\Pi_{s\neq k}(\lambda-\lambda_s) = \Sigma_{j=0}^{m-1} \sigma_j(\lambda_k)\lambda^{m-1-j},\ 1 \leqslant k \leqslant m.$$

The formula (3.15) and the assumptions (A.8) and (A.9) imply that for some positive constants C, M and M'

$$|H_j(\zeta)| \leqslant C(1 + |\zeta|)^{M'} \cdot e^{M|\mathrm{Im}\ \zeta'|},\ \zeta \in C^n, \tag{3.16}$$

where M is a positive constant number such that

$$\text{supp } f \subset \{(x,y) \in \bar{H};\ |(x,y)| < M\},$$

$$\text{supp } g_j \subset \{x \in R^n;\ |x| < M\}.$$

Put h_j = the inverse Fourier transform of H_j. From the formula (3.16) and Payley-Wiener-Schwartz's theorem we have $h_j \in \mathcal{E}'(R^n)$. Choose a $C_o^\infty(R^1)$ function $\psi(y)$ such that $\psi(y) = 1$ near $y = 0$. Put

$$v(x,y) = [\Sigma_{j=0}^{m-1} \frac{(iy)^j}{j!} h_j(x)]\psi(y),\quad w(x,y) = u(x,y)-v(x,y),$$

$$F(x,y) = f(x,y)-P(D)v(x,y) = P(D)w(x,y).$$

Note that $w \in C^\infty([0,\delta); \mathcal{S}'(R^n)) \cap L^2_{loc}(\{(x,y) \in \bar{H};\ |(x,y)| > M\})$ and that $F(x,y) \in \mathcal{E}'(\bar{H}) \cap C^\infty([0,\delta); \mathcal{S}'(R^n))$. The consideration made previously implies that $\hat{F}_o/P$ is an entire function. Putting $W(x,y)$ = the inverse Fourier transform of F_o/P, we have $W \in \mathcal{E}'(R^{n+1})$ and $P(D)W = F$ in H. Since P is partially hypoelliptic, $W \in C^\infty([0,\delta); \mathcal{D}'(R^n))$. Moreover, we have

$$P(D)(v+W) = f \text{ in } H,\ B_j(D)(v+W) = g_j,\ j = 1,\dots,p, \text{ in } H',$$

$$v+W \in \mathcal{E}'(\bar{H}) \cap C^\infty([0,\delta); \mathcal{S}'(R^n)).$$

Summing up, we have showed

Theorem 3.12 ([22, Corollary 6.4]). Let a C-system $\{P(D), B_j(D), j = 1,\dots,p\}$ satisfy the assumption (A.4) or the assumptions (A.5), (A.6) and (A.7). Let Γ^3 and Γ^3_R be the same as in Theorem 3.8. If $u \in C^\infty([0,\delta); \mathcal{S}'(R^n)) \cap L^2_{loc}(\bar{H})$ satisfies the equations: $P(D)u \in \mathcal{E}'(\bar{H})$, $B_j(D)u(x,0) \in \mathcal{E}'(R^n)$, $j = 1,\dots,p$, and the condition:

$$\underline{\lim}_{R \to \infty} \int_{\Gamma_R^3} |u(x,y)|^2 \, dxdy = 0$$

then $u \in \mathcal{E}'(\bar{H})$.

REFERENCES

1. Agmon, S., Lower bounds for solutions of Schrödinger-type equations in unbounded domains, Proc. Int. Conf. Functional Analysis and Related Topics, Tokyo, (1969), 216-224.
2. Agmon, S., Lower bounds for solutions of Schrödinger equations, J. d'Anal. Math., 23 (1970), 1-25.
3. Goodman, R. W., One sided invariant subspaces and domains of uniqueness for hyperbolic equations, Proc A.M.S., 15 (1964), 653-660.
4. Hörmander, L., Lower bounds at infinity for solutions of differential equations with constant coefficients, Israel J. Math., 16 (1973), 103-116.
5. Ikebe, T. and Uchiyama, J., On the asymptotic behavior of eigenfunctions of second order elliptic operators, J. Math Kyoto Univ., 11 (1971), 425-448.
6. Kato, T., Growth properties of solutions of the reduced wave equation with a variable coefficients, Comm. Pure Appl. Math., 12 (1959), 403-425.
7. Konno, R., Non-existence of positive eigenvalues of Schrödinger operators in infinite domains, J. Fac. Sci Univ. Tokyo Sec. IA, 19 (1972), 393-402.
8. Landis, E.M., Some problems of the qualitative theory of second order elliptic equations (case of several independent variables), Russian Math. Surveys, 18 : 1 (1963), 1-62.
9. Landis, E.M., On the behavior of solutions of higher order elliptic equations in unbounded domains, Trans. Moscow Math. Soc., 31 (1974), 30-54.
10. Littman, W., Decay at infinity of solutions to partial differential equations with constant coefficients, Trans. Amer. Math. Soc., 123 (1966), 449-459.
11. Littman, W., Maximal rates of decay of solutions of partial differential equations, Arch. Rat. Mech. Anal., 37 (1970), 11-20.
12. Littman, W., Decay at infinity of solutions to partial differential equations; removal of the curvature assumption, Israel J., Math., 8 (1970), 403-407.
13. Littman, W., Decroissance a l'infini des solutions, a l'exterieur d'un cône, d'equations aux derivees partielles a coefficients constants, C.R.Acad. Sc. Paris, t., 287 (3 juillet 1978) Serie A 15-17.
14. Mochizuki, K. and Uchiyama, J., On eigenvalues in the continuum

of 2-body or many body Schrödinger operators, Nagoya Math. J., 70 (1978), 125-141.
15. Murata, M., A theorem of Liouville type for partial differntial equations with constant coefficients, J. Fac. Sci. Univ. Tokyo Sec. IA, 21 (1974), 395-404.
16 Murata, M., Asymptotic behaviors at infinity of solutions of certain linear partial differential equations, J. Fac. Sci. Univ. Tokyo Sec. IA, 23 (1976), 107-148.
17. Murata, M. and Shibata, Y., Lower bounds at infinity of solutions of partial differential equations in the exterior of a proper cone, Israel J. Math., 31 (1978), 193-203.
18. Oleĭnik, O.A. and Radkevič, E.V., Analiticity and theorems of Liouville and Phragmén-Lindelöf type for general elliptic systems of differential equations, Math, USSR Sbornik, 24:1 (1974), 127-143.
19. Pavlov, A.L., On general boundary value problems for differential equations with constant coefficients in a half-space, Math. USSR. Sbornik, 32:3, (1977), 313-334.
20. Rellich, F., Über das asymptotische Verhalten der Lösungen von $\Delta u + \lambda u = 0$ in unendlichen Gebieten, Jber. Deutsch. Math. Verein., 53 (1943), 57-65.
21. Shibata, Y., Liouville type theorem for a system $\{P(D), B_j(D), j = 1,\ldots,p\}$ of differential operators with constant coefficients in a half-space, Publ. RIMS Kyoto, Univ., 16 (1980), 61-104.
22. Shibata, Y., Lower bounds of solutions of general boundary value problems for differential equations with constant coefficients in a half space, to appear.
23. Tayoshi, T., The asymptotic behavior of the solutions of $(\Delta+\lambda)u = 0$ in a domain with unbounded boundary, Publ.RIMS Kyoto Univ., 8 (1972), 375-391.
24. Trèves, F., Differential polynomials and decay at infinity, Bull. Amer. Math. Soc., 66 (1960), 183-186.
25. Agmon, S. and Hörmander, L., Asymptotic properties of solutions of differential equations with simple characteristics, J. d'Anal. Math., 30 (1976), 1-38.
26. Grusin, V.V., On Sommerfeld-type conditions for a certain class of partial differential equations, Mat. Sbornik, 61:2 (1963), 147-174; A.M.S. Transl. Ser. 2, 51 (1966), 82-112.
27. Vainberg, B.R., Principle of radiation, limit absorption and limit amplitude in the general theory of partial differential equations, Russian Math. Surveys, 21:3 (1966), 115-193.
28. Hörmander, L., Linear partial differential operators, Springer verlag, Berlin-Göttingen-Heidelberg, 1963.
29. Hörmander, L., Fourier integral operators I, Acta. Math., 127 (1971), 79-183.
30. Lax, P.D. and Phillips, R.S., Scattering theory, Pure and Applied Math., 26 Academic Press, New York (1967).
31. Rudin, W., Lectures on the edge-of-wedge theorem, CBMS Regional Conf., 6 (1970).

32. Trèves, F., Lectures on partial differential equations with constant coefficients, Notas de Math., 27, Instituto de Matemática Pura e Aplicada, Rio de Janeiro, (1961).
33. Vladimirov, V.S., Methods of the theory of functions of several complex variables, "nauka", Moscow, (1964); English transl., M.I.T. Press, Cambridge, Mass., (1966).
34. Whitney, H., Elementary structure of real algebraic variety, Ann. Math., 66 (1957), 545-556.

ANALYTIC SINGULARITIES OF SOLUTIONS OF BOUNDARY VALUE PROBLEMS

Johannes Sjöstrand

Université de Paris Sud

Abstract: We give a survey of some recent results concerning analytic singularities of solutions to linear differential equations and in particular to boundary value problems.

0. Introduction.

In these notes we present some recent work on analytic singularities . Since boundary value problems are difficult to isolate from the general theory of linear partial differential equations , we have found it suitable to spend a few sections on general problems away from the boundary . Most of the material is published or will be published elsewhere , so the proofs are indicated only very briefly . The only exception is the proof of Theorem 7.2 and 7.1 , that we have no plans of publishing in a separate article. Also in section 2 the exposition is slightly more detailed, in order to give an example of the classical and very important

H. G. Garnir (ed.), Singularities in Boundary Value Problems, 235–269.

method of weights, which takes a particularly simple form in connection with analytic singularities.

1. Pseudodifferential operators, analytic wave front sets and resolutions of the identity.

We discuss here very briefly some of the basic notions that are used in the proofs of many of the results later on. More details can be found in [22], [23], [25] and in lecture notes that are in preparation.

If $\Omega \subset C^n$ is open, we say that the function $u(z,\lambda)$ on $\Omega \times R_+$ is an analytic symbol, if u is holomorphic in z and $u(z,\lambda) = \mathcal{O}(1)\, e^{\varepsilon\lambda}$ for every $\varepsilon > 0$, locally uniformly with respect to z. Very roughly, we shall often ignore symbols that are exponentially decreasing and often consider formal symbols, which are analytic symbols modulo exponentially decreasing ones. A classical analytic symbol of order 0 is a symbol

$$u(z,\lambda) = \sum_{0 \le k \le \lambda/eC} \lambda^{-k} u_k(z) \quad ,$$

where $|u_k(z)| \le C^{k+1} k^k$. (See Boutet de Monvel - Kree [3].)

Using analytic symbols, one can consider formal pseudodifferential operators of the form $P(x,\tilde{D}_x,\lambda)$, $\tilde{D}_x = \lambda^{-1} D_x$, and their various realizations, acting on distributions, symbols et c. The main technical ingredient is here the analytic version of the method of stationary phase with remainder estimate, see [23].

Let $\varphi(x,y,\xi)$ be analytic and defined near (x_0,x_0,ξ_0), where $(x_0,\xi_0)\in T^*R^n\setminus 0$, such that

(1.1) $\quad \operatorname{Im}\varphi(x,y,\xi) \geq C\,|x-y|^2 \quad ,$

(1.2) $\quad \partial\varphi/\partial x = -\partial\varphi/\partial y = \xi \;, \varphi = 0$ when $x=y$.

Let $a(x,y,\xi,\lambda)$ be a classical analytic symbol, elliptic and of order 0. If $u\in \mathcal{D}'(X)$, where X is an open set in R^n containing x_0, we say that $(x_0,\xi_0)\notin WF_a(u)$ if

$$\int e^{i\lambda\varphi(x,y,\xi)}\,\chi(y)\,a(x,y,\xi,\lambda)\,u(y)\,dy$$

is uniformly exponentially decreasing for (x,ξ) in a neighbourhood of (x_0,ξ_0). Here $\chi\in C_0^\infty$ is equal to 1 near x_0. This particular definition of $WF_a(u)$ is essentially due to Bros-Iagolnitzer[4]. Using a suitable (simple) machinery of analytic pseudodifferential operators it is easy to show that the definition is independent of φ, a and that $WF_a(u)$ is a closed conic set whose projection in X is the analytic singular support of u.

In this connection we mention a closely related notion of resolution of the identity. Let $\alpha=(\alpha_x,\alpha_\xi)$ and let $\varphi(x,y,\alpha)$ be a phase function, defined near (x_0,x_0,x_0,ξ_0) such that

(1.3) $\quad \operatorname{Im}\varphi \geq C(|x-\alpha_x|^2+|y-\alpha_x|^2)\quad ,$

(1.4) $\quad \varphi = 0$ and $\varphi'_x = -\varphi'_y = \alpha_\xi$ when $x=y=\alpha_x$.

Then if $a(x,y,\alpha,\lambda)$ is a classical analytic symbol of degree 0, we can introduce the formal pseudodifferential operator

$$(1.5)\qquad Au(x,\lambda)=\lambda^{3n/2}\iint e^{i\lambda\varphi(x,y,\alpha)}\,a(x,y,\alpha,\lambda)u(y)dyd\alpha\,.$$

If A is formally the identity, we say that

$$(1.6)\qquad \Pi_{\alpha,\lambda}(x,y) = \lambda^{3n/2}\,a\,e^{i\lambda\varphi}$$

is a <u>resolution of the identity</u>. If $P(x,\tilde{D}_x,\lambda)$ is a pseudodifferential operator with classical analytic symbol one can show that there exists a classical formal analytic symbol $\tilde{a}(x,y,\alpha,\lambda)$ with values in the $(2n-1)$-forms in α such that $\tilde{\Pi}_{\alpha,\lambda} = \tilde{a}\,e^{i\lambda\varphi}$ satisfies

$$(1.7)\qquad \left[P, \Pi_{\alpha,\lambda}\right]d\alpha = d_{\alpha}\tilde{\Pi}_{\alpha,\lambda}\,.$$

This identity is technically quite useful.

2. Micro-non-characteristic operators and Hanges' conjecture.

The notion of micro-non-characteristic operators was introduced by Bony-Schapira[2] and Bony[1]. We present here some joint work with N.Hanges, which generalizes that notion slightly. Let $P(x,D_x)$ be a differential operator on an open set $X\subset R^n$. Let $p(x,\xi)$ be the principal symbol and fix a point $(x_0,\xi_0)\in T^*X\setminus 0$, where p vanishes.

We assume that near x_0, we can find analytic functions $\varphi(x)$, $\psi(x)$ and an integer $m \geq 1$ with the following properties :

(2.1) $\operatorname{Im}\varphi \geq 0$, $\operatorname{Im}\varphi(x_0) = 0$, $\varphi'(x_0) = \xi_0$,

(2.2) p vanishes to the order m on

$\Lambda_\varphi = \{(x, \varphi'_x)$; x is in some complex neighborhood of $x_0\}$,

(2.3) ψ is real valued and $(\frac{\partial}{\partial t})^m p(x, \varphi'_x + t\psi'_x) \neq 0$ at $x = x_0$, $t=0$.

It is possible to show that (2.2)and (2.3) are equivalent to the seemingly weaker assumption that ψ is real valued and and $p(x, \varphi'_x + t\psi'_x + \xi) \neq 0$ for $t \in]0, 1/C]$, $\xi \in \mathbb{C}^n$, $|\xi| \leq t/C$, and x in a complex neighbourhood of x_0. Also by considering ψ rather as a function on Λ_φ, it is possible to formulate the assumptions above in a canonically invariant way. When (2.1)-(2.3) hold, we shall say that P is micro-non-characteristic at (x_0, ξ_0) along Λ_φ with respect to the hypersurfaces ψ = const. .

Theorem 2.1 (Hanges-Sjöstrand). Suppose that (2.1)-(2.3) hold and that $\psi(x_0) = 0$. Assume further that $u(x)$ is a distribution defined near x_0, such that $(x_0, \xi_0) \notin WF_a(P^* u)$ and $V \cap \{(x, \xi) \in \Lambda_{\varphi, R} ; \psi(x) < 0\} \cap WF_a(u) = \emptyset$, where V is some neighbourhood of (x_0, ξ_0) and $\Lambda_{\varphi, R} = \{(x, \varphi'_x)$; x is real and $\operatorname{Im}\varphi(x) = 0\}$. Then $(x_0, \xi_0) \notin WF_a(u)$.

This result permits to essentially recover the results of Bony[1], Bony-Schapira[2]. In those papers similar results are given , where Λ_φ is replaced by a conic real involutive manifold V . However, with a mild generic condition on the leaves of V , we can construct φ as above .

Consider now the Lie algebra $\mathcal{L}$ generated by the Hamilton fields $H_{\mathrm{Re}\, p}$, $H_{\mathrm{Im}\, p}$. By a theorem of Nagano[17], there is in a neighbourhood of (x_0, ξ_0) a unique connected analytic integral manifold Γ of $\mathcal{L}$ which contains (x_0, ξ_0) . If k is the dimension of Γ , this means in particular that the dimension of the fiber of $\mathcal{L}$ at each point of Γ is constant = k . Hanges[6] introduced the assumption that

(2.4) $$p|_\Gamma = 0 ,$$

and a condition (2.5-) which is slightly weaker than the following one :

(2.5) The projection $\Gamma \to X$ has injective differential.

It is easy to see that (2.4) implies that Γ is isotropic for the symplectic 2-form , so $k \leq n$. Under the assumptions (2.4),(2.5-) and also that $k \geq 1$, Hanges constructed a microlocal solution to $Pu \in a(X)$ such that $WF_a(u) = \Gamma$ near (x_0, ξ_0) . He also conjectured that if u is an arbitrary solution of $Pu \in a(X)$,then either Γ is contained in $WF_a(u)$ or is disjoint from $WF_a(u)$. Later he also showed that the conjecture is correct when k = 1 , i.e. the case when P admits a real bicharacteristic strip through (x_0, ξ_0).

This is a weaker assumption than that p should be real-valued , which was done in the first theorems on propagation of singularities by Hörmander [8] and Sato-Kawai-Kashiwara [20]. The proof is not very difficult , and the reason why this result has been discovered only recently is perhaps that the analogous statement for C^∞-singularities is false without further assumptions .

Using Theorem 2.1 one gets most of Hanges' conjecture :

Theorem 2.2(Hanges-Sjöstrand). Under the assumptions (2.4), (2.5) , if $u \in \mathcal{D}'(X)$, $Pu \in a(X)$, then either Γ is contained in $WF_a(u)$ or disjoint from $WF_a(u)$.

The assumption (2.5) is convenient in the proof but can probably be eliminated . To get Theorem 2.2 from Theorem 2.1 one first constructs a phase φ as in Theorem 2.1 with $\Gamma = \Lambda_{\varphi,R}$. Then ψ will satisfy the assumptions of that theorem whenever $H_{\mathrm{Re}\, p}$ or $H_{\mathrm{Im}\, p}$ - as vector-fields on Γ - are transversal to the hypersurfaces ψ = const. (in Γ). One can than apply a geometric argument of Bony[1].

To end this section we outline the proof of Theorem 2.1. The main technical step is to show that for every μ in a small pointed complex neighbourhood of 0 , there is an analytic symbol $a(x,\mu,\lambda)$ defined for x in a complex μ-independent domain such that

$$(2.6) \qquad e^{-i\lambda\varphi_\mu(x)}\, P\, e^{i\lambda\varphi_\mu} a = 1 + \mathcal{O}_\mu(1)\, e^{-C|\mu|\lambda} \quad ,$$

where $\mathcal{O}_\mu(1)$ denotes a bounded function for each μ, and $C > 0$ is a positive constant (as always in the following) independent of μ. Moreover,

$$(2.7) \qquad \varphi_\mu(x) = \varphi(x) + i\mu\psi(x) \quad .$$

In proving (2.6) we consider the pseudodifferential operator of order 0 :

$$P_\mu = \lambda^{-(\text{order of } P)} e^{-i\lambda\varphi_\mu} P e^{i\lambda\varphi_\mu} \quad .$$

Then if p_μ is the principal symbol, we know that $p_\mu(x,\xi) \neq 0$ for x in a complex μ-independent domain and for $|\xi| \leq C|\mu|$. We then have to invert P_μ, when acting on symbols.

Let Q_μ be the formal pseudodifferential operator with symbol $1/p_\mu$. Then we can define the action of Q_μ on symbols by the integral formula :

$$(2.8) \quad Q_\mu u(x,\lambda) = \left(\frac{\lambda}{2\pi}\right)^n \iint_{\Gamma(x)} e^{i\lambda(x-y)\eta} \, Q_\mu(x,\eta,\lambda) \, u(y,\lambda)\, dy\, d\eta \quad ,$$

where $\Gamma(x)$ is a singular contour parametrized by Re y, Im y of the form $\eta = i\, C|\mu|\, |x-y|^{-1} \overline{(x-y)}$, $|x-y| \leq C$, where C is small but independent of μ. Then (2.8) gives a welldefined integral operator and it is not hard to show that

$$P_\mu \circ Q_\mu = I + \lambda^{-1} A_\mu \quad ,$$

where A_μ is a pseudodifferential operator of the type (2.8) and of order 0. The problem is then to invert an operator of the form $I + \lambda^{-1} A_\mu$ and this can be done by noting that A_μ is uniformly bounded with respect to λ on L^2-spaces with measure $e^{\lambda|\mu|\varphi(x)} L(dx)$, where L is the Lebesgue measure and $\varphi(x)$ is a realvalued Lipschitz function with norm $\leq C/2$, if C is the constant appearing in the definition of $\Gamma(x)$.

Naturally we can still solve (2.6), if we replace ψ by the slightly perturbed function $\tilde{\psi} = \psi(x) + 2\delta(x-x_0)^2 - \delta^3$, where $\delta > 0$ is small.

Let $W = V \times \{\xi \in R^n ; |\xi - \xi_0| \leq C\}$, where V is the ball around x_0 with radius δ. Let

$$I^W = \int_{\alpha \in W} \pi_{\alpha,\lambda}\, d\alpha ,$$

where $\pi_{\alpha,\lambda}$ is a resolution of the identity. Then if $\tilde{\varphi}_\mu$ is defined as in (2.7) with ψ replaced by $\tilde{\psi}$, we have modulo terms which are uniformly bounded in λ:

$$\begin{aligned} (e^{i\lambda\tilde{\varphi}_\mu}, I^W u) &\equiv (P e^{i\lambda\tilde{\varphi}_\mu} a, I^W u) \equiv \\ (e^{i\lambda\tilde{\varphi}_\mu} a, P^* I^W u) &\equiv (e^{i\lambda\tilde{\varphi}_\mu} a, [P^*, I^W] u) \\ \equiv \int_{\partial W} (e^{i\lambda\tilde{\varphi}_\mu} a, \tilde{\pi}_{\alpha,\lambda} u) &. \end{aligned}$$

It is not hard to show (using the assumption on $WF_a(u)$) that the integrand above is bounded, when $|\mu|$ is small and $\mathrm{Re}\,\mu \geq 0$. Thus

$$(2.9) \qquad (e^{i\lambda\tilde{\varphi}_\mu}, I^W u) = \mathcal{O}_\mu(1), \quad \mathrm{Re}\,\mu \geq 0 .$$

This is a strong estimate in the region where $\tilde{\psi} < 0$ and by integrating (2.9) against suitable functions of the form $e^{i\lambda f(\mu)}$, we can build up the Gaussian type factors, used in the definition of the analytic wavefront set in section 1, and show that u has no analytic wavefront set on the part of $\Lambda_{\varphi,R}$, where $|x-x_0|<\delta$, $\tilde{\psi}(x) < 0$. In particular $(x_0, \xi_0) \notin WF_a(u)$.

3. Microhyperbolic operators.

These operators were introduced by Kashiwara-Kawai[11]. Our treatment extends naturally to boundary value problems. Let $P(x,D_x)$ be an operator with analytic coefficients and (x_0, ξ_0) a point, where p vanishes. Let $(y_0, \eta_0) \in T_{(x_0,\xi_0)}(T^*X\setminus 0)$. Then P is <u>microhyperbolic</u> at (x_0, ξ_0) in the direction (y_0, η_0), if there exists a real neighbourhood V_0 of (x_0, ξ_0) and a number $\varepsilon_0 > 0$, such that $p((x,\xi)+it(y_0,\eta_0)) \neq 0$, for $(x,\xi) \in V_0$, $0 < t \leq \varepsilon_0$. Thanks to a local version of Bochner's tube theorem, due to Kashiwara [10], there is a certain stability in the definition of microhyperbolicity. Namely, if P is microhyperbolic as above, then there exist real neighbourhoods V of (x_0, ξ_0) and W of (y_0, η_0), such that $p((x,\xi)+it(y,\eta)) \neq 0$, for $(x,\xi) \in V$, $(y,\eta) \in W$, $0 < t \leq \varepsilon_0$.

We have the following basic result of Kashiwara-Kawai [11].

Theorem 3.1. Let $\tilde{\Psi}(x, \xi)$ be an analytic real-valued function defined in a neighbourhood $\tilde{W}$ of (x_0, ξ_0) with $\tilde{\Psi}(x_0, \xi_0) = 0$ and assume that P is microhyperbolic at (x_0, ξ_0) in the direction $-H_{\tilde{\Psi}} = (-\partial\tilde{\Psi}/\partial\xi, \partial\tilde{\Psi}/\partial x)$. Let u be a distribution defined near x_0, such that Pu is analytic and

$$(3.1) \qquad WF_a(u) \cap \{(x, \xi) \in \tilde{W};\ \tilde{\Psi}(x, \xi) > 0\} = \emptyset .$$

Then $(x_0, \xi_0) \notin WF_a(u)$.

We indicate briefly a proof of this result, that was given in [22]. "Convexify" $\tilde{\Psi}$ by putting

$$(3.2) \quad \Psi(x, \xi) = \tilde{\Psi} + \delta^3 - \delta((x-x_0)^2 + (\xi - \xi_0)^2) ,$$

where $\delta > 0$ is small. Let $W_k \subset T^*X \setminus 0$ be the ball around (x_0, ξ_0) of radius $k\delta$ and put $W = W_2$. Then $\Psi < \tilde{\Psi}$ on ∂W and $\Psi(x_0, \xi_0) > 0$. Hence

$$(3.3) \quad \{(x, \xi) \in \partial W;\ \Psi(x, \xi) \geq 0\} \cap WF_a(u) = \emptyset ,$$

and Theorem 3.1 will follow from

$$(3.4) \qquad WF_a(u) \cap \{(x, \xi) \in W;\ \Psi(x, \xi) > 0\} = \emptyset .$$

Let $\pi_{\alpha,\lambda}$ be a resolution of the identity and consider formally the Fourier integral operator (see [9]) :

$$Q = Q_\mu = \int e^{\mu\lambda\Psi(\alpha)} \pi_{\alpha,\lambda} \, d\alpha \quad .$$

The phase function is of the form $\varphi(x,y,\alpha) - i\mu\Psi(\alpha)$ and one can show that the corresponding canonical transformation is

$$(3.5) \quad \mathcal{H}_\mu = \exp(i\mu H_\Psi) + \mathcal{O}(\mu^2) \quad .$$

Put $P(x,\tilde{D}_x,\lambda) = \lambda^{-(\text{order of } P)} P(x,D_x)$ .Then formally

$$(3.6) \qquad Q_\mu \circ P = P_\mu \circ Q_\mu \quad ,$$

where P_μ is a formal classical pseudodifferential operator whose principal symbol p_μ is given by

$$(3.7) \qquad p_\mu \circ \mathcal{H}_\mu = p \; .$$

It follows from the microhyperbolicity assumption that P_μ is elliptic in a fixed μ-independent <u>real</u> neighbourhood of (x_0, ξ_0) . If we introduce a realization

$$Q_\mu^W = \int_{\alpha \in W} e^{\mu\lambda\Psi(\alpha)} \pi_{\alpha,\lambda} \, d\alpha \; , \; u_{\mu,\lambda} = Q^W u \quad ,$$

and a similar realization $P_\mu^{W_3}$ of P , we can show , using (3.3) , that $P_\mu^{W_3} u_{\mu,\lambda}$ is exponentially decreasing in C_0^∞ as $\lambda \to \infty$, for $\mu > 0$ sufficiently small . Using then that P_μ is elliptic and again (3.3) , we get after some work , that $u_{\mu,\lambda}$ is exponentially decreasing and from that it is not hard to deduce (3.4) .

4. Analytic wavefront sets for boundary value problems.

We now want to study the propagation of singularities for non-characteristic boundary value problems. The problems are local so we consider the manifold

$$M = \{x \in R^n \,;\, |x'| < a \,,\, 0 \leq x^n < a\} = M' \times [0,a[\,,$$

with boundary $\partial M = \{x \in M \,;\, x^n = 0\}$. Let

$$(4.1) \qquad P = D^m_{x^n} + A_1(x,D_{x'})\, D^{m-1}_{x^n} + \ldots + A_m(x,D_{x'}) \qquad ,$$

where A_j are of order j , either scalar or $k_0 \times k_0$ - matrices of scalar operators . By $\mathcal{D}'(\overset{o}{M})$ we denote the space of distributions on $\overset{o}{M}$ that are extendible across ∂M . We recall the wellknown fact , that if $u \in \mathcal{D}'(M)$ and $Pu \in C^\infty(M)$, then $u \in C^\infty([0,a[\,;\, \mathcal{D}'(M'))$. From now on ,we shall always assume that $u \in \mathcal{D}'(M)$ is a solution to

$$(4.2) \qquad P\,u \in a(M) \quad ,$$

where $a(M)$ denotes the space of functions on M that are analytic up to the boundary .

<u>Definition 4.1.</u> We say that $u = 0$ at $(x_0', \xi_0') \in T^*\partial M \setminus 0$ if $(x_0', \xi_0') \notin WF_a(\, D^j_{x^n} u(x',0)\,)$, $0 \leq j \leq m-1$.

Notice that if $u = 0$ at (x_0', ξ') for all ξ' , then the traces of u are analytic at x_0' and it follows from the theorems of Cauchy-Kowalewski and Holmgren that u is analytic up to the boundary at $(x_0',0)$. The following result of Schapira [21] is therefore a microlocal version of Holmgren's uniqueness theorem :

Theorem 4.2. Let u solve (4.2) and assume that $u = 0$ at $(x_0', \xi_0') \in T^*\partial M\setminus 0$. Then for some $\varepsilon > 0$:
$WF_a(u|_{\overset{\circ}{M}}) \cap \{(x, \xi) ; (x'-x_0')^2 + (\xi'-\xi_0')^2 < \varepsilon^2 ,$
$0 < x^n < \varepsilon \} = \emptyset$.

In [23],[22] we gave elementary proofs of this result . The one in [23] is by a simple duality argument , where we apply the Cauchy-Kowaleaski theorem to the equation $P^* v = 0$, prescribing the traces of v on the hypersurface $x^n = \delta$, to be of the form 0 and $e^{i\lambda\varphi(x)}$, where φ is a Gaussian type phase-function . The proof in [22] does not use the Cauchy-Kowalewski theorem , but the L^2 - continuity of pseudodifferential operators instead . We can now define the reduced analytic wavefront set (c.f. Schapira [21]).

Definition 4.3. We put

$$(4.3) \qquad WF_{ba}(u) = WF_a(u|_{\overset{\circ}{M}}) \cup \left(\bigcup_{j=0}^{m-1} WF_a(D^j_{x^n} u(x',0))\right)$$

$$\subset \quad (\det p)^{-1}(0)|_{\overset{\circ}{M}} \cup (T^*\partial M\setminus 0) .$$

On the last set in (4.3) there is a natural topology for which a fundamental system of neighbourhoods of a point $(x_0', \xi_0') \in T^*\partial M\setminus 0$ is given by the sets

$$V \cup \{(x,\xi) \in (\det p)^{-1}(0)\ ;\ (x',\xi') \in V\ ,\ 0 < x^n < \varepsilon\},$$

where $0 < \varepsilon < a$ and V are neighbourhoods of (x_0', ξ_0') in $T^*\partial M \setminus 0$. With this topology $WF_{ba}(u)$ is a closed conic set .Also the projection of $WF_{ba}(u)$ in M is the analytic singular support of u .

There is a slightly more refined definition of analytic wavefront sets . Let $(x_0', \xi_0') \in T^*\partial M \setminus 0$ and let $z_1, \ldots, z_d \in C$ be the distinct roots of the equation $\det p(x_0', 0, \xi_0', z) = 0$. Then one can construct 0:th order pseudodifferential operators $\Pi_j(x, \tilde{D}_x, \lambda)$ with symbols defined for (x, ξ') close to $(x_0', 0, \xi_0')$, such that Π_j is a polynomial in $\tilde{D}_{x^n}$ of degree $m-1$ and such that formally

$$\Pi_j = I + A_j P \quad , \text{ near } (x_0', 0, \xi_0', z_j) \ ,$$

$$\Pi_j = B_j P \quad , \text{ near } (x_0', 0, \xi_0', z_k) \ , \ k \neq j \ .$$

Here A_j , B_j are suitable pseudodifferential operators . If u is a solution to (4.2) , there is a natural way , using the operators Π_j , to define a closed conic set

$$WF_a(u) \subset \Sigma = \{(x', 0, \xi', \xi^n)\ ;\ (x', \xi') \in T^*\partial M \setminus 0\ ;$$
$$\xi^n \in C\ ,\ \det p(x', 0, \xi) = 0\} \cup (\det p)^{-1}(0)|_{\mathring{M}} \ .$$

See [22] for more details . $WF_a(u)$ projects naturally onto $WF_{ba}(u)$.

5. Microhyperbolic boundary value problems.

Let P , M be as in section 4 and let

$$B(x',D_x) = \begin{pmatrix} B_1(x',D_x) \\ \vdots \\ B_{m_+}(x',D_x) \end{pmatrix}$$

be a $m_+ \times k_0$ matrix of operators on M , where B_j is of order m_j and polynomial in D_{x^n} of degree $\leq m-1$.

Let b_j , b , p be the various principal symbols and let γ_0 denote the restriction operator $u \mapsto u|_{x^n=0}$. Recall that (P,B) (or (p,b)) is elliptic at a point $(x'_0, \xi'_0) \in T^*\partial M \setminus 0$ if (5.1)-(5.3) hold :

(5.1) $\det p(x'_0, 0, \xi'_0, \xi^n) \neq 0$, $\xi^n \in R$,

(5.2) $\dim(\operatorname{Ker} p(x'_0, 0, \xi'_0, D_{x^n}) \cap \mathcal{S}(\overline{R_+})) = m_+$,

(5.3) $\gamma_0 b(x'_0, \xi'_0, D_{x^n}) : \operatorname{Ker} p \cap \mathcal{S}(\overline{R_+}) \to C^{m_+}$
is a bijection .

This definition makes sense whenever p , b are polynomials in ξ^n . Let $\mathcal{V} = (\mathcal{V}_{x'}, \mathcal{V}_{\xi}) \in R^{n-1} \times R^n$.

<u>Definition 5.1.</u> (P,B) is microhyperbolic at (x'_0, ξ'_0) in the direction $\mathcal{V}$ if there exist a neighbourhood $V_0 \subset T^*\partial M \setminus 0$ of (x'_0, ξ'_0) and numbers $\varepsilon_0 > 0$, $b_0 > 0$ such that for $(x', \xi') \in V_0$, $-b_0 < x^n < b_0$, $0 < t \leq \varepsilon_0$, $\xi^n \in R$:

(5.4) $\det p((x,\xi)+it\nu) \neq 0$,

(5.5) the problem

$$p(x'+it\nu_{x'},0,\xi'+it\nu_{\xi'},it\nu_{\xi^n}+D_{x^n})u = 0$$

$$\gamma_0\, b(x'+it\nu_{x'},\xi'+it\nu_{\xi'},it\nu_{\xi^n}+D_{x^n})u = 0$$

is elliptic .

Again the local Bochner tube theorem implies a certain stability in the definition , which of course is a microlocal form of the so called weak Lopatinski condition . We have

Theorem 5.2([22]). Assume that $u \in \mathcal{D}'(M)$ satisfies

(5.6) $Pu \in a(M)$, $\gamma_0 Bu \in a(\partial M)$.

Let $(x_0',\xi_0') \in T^*\partial M\setminus 0$ and let $\psi(x,\xi')$ be a real valued analytic function defined in a neighbourhood $V \subset (T^*\partial M\setminus 0)\times [0,a[$ of $(x_0',0,\xi_0')$, vanishing at that point with the property that (P,B) is microhyperbolic at that point in the direction $-H_\psi = (-\partial\psi/\partial\xi',\partial\psi/\partial x)$. Then, if

(5.7) $\{(x,\xi)\in V\times R_{\xi^n}\,;\,\psi(x,\xi')>0\}\cap WF_{ba}(u) = \emptyset$,

we have $(x_0',0,\xi_0') \notin WF_{ba}(u)$.

The proof is a very natural generalization of the one in section 3 . Let $\pi_{\alpha,\lambda}$ be a resolution of the identity over R^{n-1} , defined for α near (x_0',ξ_0') . Then after "convexifying" ψ into a new function $\tilde{\psi}$, we conjugate the problem by using the "tangential Fourier integral operator"

$$Q = \int e^{\mu\lambda\tilde{\varphi}(\alpha,x^n)} \pi_{\alpha,\lambda}\, d\alpha \quad .$$

The new problem is then elliptic and has a left parametrix.

There is also a more refined notion of microhyperbolicity: Let $z_1,\dots,z_d$ be the distinct roots of $\det p(x_0',0,\xi_0',z)=0$, ordered in such a way that $z_1,\dots,z_{d'}$ are real, $z_{d'+1},\dots,z_d$ non real. Let m_j be the multiplicity of z_j and for (x',ξ') close to (x_0',ξ_0'), let $z_{j,k}(x',\xi')$, $1\le k\le m_j$, denote the roots which are close to z_j. At each point $(x_0',0,\xi_0',z_j)$ we fix a real vector $\nu_j = (y_0',\eta_0',\lambda_j)$. Assume first:

(I) For $1\le j\le d'$, p is microhyperbolic at $(x_0',0,\xi_0',z_j)$ in the direction ν_j.

This implies that for (x',ξ') real and sufficiently close to (x_0',ξ_0') and for $s>0$ sufficiently small, $\mathrm{Im}(z_{j,k}((x',\xi')+is(y_0',\eta_0'))) - s\lambda_j \neq 0$. Let K denote the set of such roots with $\mathrm{Im}(z_{j,k})-s\lambda_j > 0$. (K depends on x',ξ',s.) Let E_K be the linear space of exponential solutions associated to K of the equation $p(x'+isy_0',0,\xi'+is\eta_0',D_t)u = 0$. We assume

(II) For (x', ξ') in some small real neighbourhood of (x'_0, ξ'_0) and s in some interval $]0, \varepsilon]$, $\varepsilon > 0$, the problem $(p(x'+isy'_0, 0, \xi'+is\eta'_0, D_t), \gamma_0 b(..))$ is K-elliptic, i.e. $\gamma_0 b(x'+isy'_0, \xi'+is\eta'_0, D_t) : E_K \to C^{m_+}$ is a bijection.

We shall say that (P,B) is microhyperbolic in the multidirection $(\nu_j)_{j=1,..,d}$, if (I) and (II) hold. We recover of course the earlier definition when all the ν_j are equal.

Let $V_j \subset \Sigma$ be some neighbourhoods of $(x'_0, 0, \xi'_0, z_j)$, let $\psi_j(x, \xi')$ be real valued and analytic, defined near $(x'_0, 0, \xi'_0)$ such that

(5.8) $$\psi_j(x'_0, 0, \xi'_0) = 0 ,$$

(5.9) $$\psi_j(x', 0, \xi') = \psi_k(x', 0, \xi') ,$$

Put $\nu_j = - H_{\psi_j}(x'_0, 0, \xi'_0)$.

Theorem 5.3([22]). Assume (5.6),(5.8),(5.9) and that (P,B) is microhyperbolic at $(x'_0, 0, \xi'_0)$ in the multidirection $(\nu_1, \dots, \nu_d)$. If

(5.10) $$V_j \cap \{\psi_j > 0\} \cap WF_a(u) = \emptyset ,$$

then $(x'_0, 0, \xi'_0, z_j) \notin WF_a(u)$, $j=1,..,d$.

The proof is only a slight extension of the proof of Theorem 5.2 . Using the operators Π_j one make a block-decomposition of P .

The first application of these results is the case when (P,B) is elliptic at $(x_0', 0, \xi_0')$. Then we have the following microlocal version of a result of Morrey-Nirenberg , due to Schapira [21]: If (5.6) holds , then $(x_0', 0, \xi_0') \notin WF_{ba}(u)$. In fact , we can apply Theorem 5.2 with $\psi = 0$.

As a second application , we prove a result of Schapira [21] on the transversal reflection of analytic singularities. Assume for simplicity that P is scalar and that p is real valued . Let $\mu_1, \dots, \mu_r, \lambda_1, \dots, \lambda_s, \overline{\lambda}_1, \dots, \overline{\lambda}_s$, $m = r+2s$, be the roots of $p(x_0', 0, \xi_0', z) = 0$, arranged so that μ_j are real , $\operatorname{Im} \lambda_j > 0$, We assume that $\mu_1, \dots, \mu_r$ are distinct . Then at each point $(x_0', 0, \xi_0', \mu_j)$ either H_p or $-H_p$ points into M , transversally to the boundary. For $\delta > 0$ sufficiently small , let $\gamma_j : [0, \delta[\to T^*M \setminus 0$ be the integral curve of $\pm H_p$, starting at $(x_0', 0, \xi_0', \mu_j)$. Let B consist of $r - \rho + s$ operators , where $0 \le \rho \le r$ and assume that

(5.10) (P,B) is K-elliptic at (x_0', ξ_0') , with $K = \{\mu_{\rho+1}, \dots, \mu_r, \lambda_1, \dots, \lambda_s\}$.

<u>Theorem 5.4(Schapira [21]).</u> In the above situation , let u be a solution of (5.6) such that $\gamma_j(]0, \delta]) \cap WF_a(u) = \emptyset$, $1 \le j \le \rho$,Then $(x_0', 0, \xi_0') \notin WF_{ba}(u)$.

Proof. The problem is microhyperbolic in the multidirection given by

$$\nu_j = (0,0,1)\ ,\ 1 \le j \le \rho \ \text{(at the first } \rho \text{ roots } \mu_j),$$

$$\nu_j = (0,0,-1),\ \rho+1 \le j \le r \ \text{(at the remaining } \mu_j),$$

$$\nu_j = (0,0,0),\ r+1 \le j \le m \ \text{(at } \lambda_j, \overline{\lambda}_j).$$

It suffices to apply Theorem 7.4 with $\psi_j = x^n$ for $1 \le j \le \rho$, $= -x^n$ for $\rho+1 \le j \le r$, $= 0$ for $j \ge r+1$.

In the next section, we shall apply microhyperbolicity essentially to the Dirichlet problem for the wave equation. In [22] some other applications were also given: The oblique derivative problem for the Laplacian, the wave equation with more general boundary conditions, and a 4:th order Dirichlet problem with glancing.

6. Application to the wave equation.

We are now interested in the singularities of the solutions to the boundary value problem

$$(6.1)\qquad \Box u \in a(R_t \times X)\ ,\ u|_{R_t \times \partial X} \in a(R_t \times \partial X)\ ,$$

where $\Box = \frac{1}{2}(D_t^2 - D_x^2)$ is the wave operator on $R_t \times R_x^{n-1}$ and $X \subset R^{n-1}$ is a closed set with analytic boundary (having the structure of a manifold with boundary). As mentioned above the Dirichlet boundary condition in (6.1) can be replaced with many others. In particular all our results below are valid with the Neumann condition. (Then it is quite convenient that only the <u>weak</u> Lopatinski condition is needed in section 5.)

Locally (6.1) can be reduced to

(6.2) $\quad Pu \in a(M) \quad , \quad u|_{\partial M} \in a(\partial M)$,

where M is as before and

(6.3) $\quad P = D^2_{x^n} + R(x, D_{x'})$.

Here

(6.4) $\quad R$ is of real principal type ,

and most of our results will now concern the more general problem (6.2), where we always assume (6.3),(6.4) .

Let $r(x, \xi')$ be the principal symbol of R , $r_0(x', \xi') = r(x', 0, \xi')$. Write $T^*\partial M \setminus 0 = \mathcal{H} \cup \mathcal{E} \cup \mathcal{G}$, where $\mathcal{H} : r_0 < 0$, $\mathcal{E} : r_0 > 0$, $\mathcal{G} : r_0 = 0$. Then (6.2) is elliptic in $\mathcal{E}$ so if u solves (6.2) we know that $WF_{ba}(u)$ is a closed conic subset of $\Sigma_b = \mathcal{H} \cup \mathcal{G} \cup p^{-1}(0)|_{\mathring{M}}$.

<u>Definition 6.1([23]).</u> An analytic ray is a continuous curve $\gamma : I \to \Sigma_b$ (where I is some interval) such that for every $t_0 \in I$:

1° If $\gamma(t_0) \in p^{-1}(0)|_{\mathring{M}}$, then γ is differentiable at t_0 with derivative H_p .

2° If $\gamma(t_0) \in \mathcal{H}$ then $\gamma(t) \in p^{-1}(0)|_{\mathring{M}}$ for $|t-t_0| > 0$ sufficiently small .

3° If $\gamma(t_0) \in \mathcal{G}$ and we write (somewhat incorrectly) $\gamma(t) = (x(t), \xi(t))$, then $(x'(t), \xi'(t))$ is differentiable at t_0 with derivative H_{r_0} .

This definition is a slight modification of the one for C^∞-rays in Melrose-Sjöstrand [16]. That definition is obtained simply by replacing $\mathcal{H}$ by $\mathcal{H} \cup \mathcal{G}_+$ in 2° . Here $\mathcal{G}_+ \subset \mathcal{G}$ is the so called diffractive region , defined by $\partial r/\partial x^n < 0$. For the wave equation it corresponds to the points where the obstacle $\complement X$ is strictly convex in the ray direction . Notice that a boundary bicharacteristic of r_0 is always an analytic ray , while it is a C^∞- ray only as long as it avoids $\mathcal{G}_+$. Locally through a point in $\mathcal{G}_+$ there are many analytic rays , while there is essentially only one C^∞-ray .

The main result is now

Theorem 6.2([23]). If u solves (6.2) , then $WF_{ba}(u)$ is a union of maximally extended analytic rays .

Since analytic rays are not always uniquely extendible, many further problems arise , some of which we shall discuss in the last two sections .

In order to prove Theorem 6.2 , we first notice that the propagation away from from $\mathcal{G}$ follows from the results on the propagation in the interior and on transversal reflection, that we have discussed earlier . Let $(x_0', \xi_0') \in \mathcal{G}$ and choose analytic coordinates $t \in R$, $s \in R^{2n-3}$ near (x_0', ξ_0')

with the origin at that point , such that $H_{r_0} = \partial/\partial t$.
For $\varepsilon > 0$ small and $\delta > 0$ small but independent of ε , we put

$$(6.5) \qquad (x, \xi') = \frac{t}{\varepsilon\delta} - \frac{s^2}{\varepsilon^4} - \frac{x^n}{\varepsilon^2} + 1 \text{ .}$$

It is not hard to verify that the problem (6.2) is micro-hyperbolic in the direction $-H_\psi$ at every point where $t \leq 0$, $\psi \geq 0$. Applying Theorem 5.2 , one can then get

Proposition 6.3. Let (x_0', ξ_0') belong to some fixed compact set in $\mathcal{G}$. Then there exist constants $\varepsilon_0 > 0$, $A_0 > 0$ such that if $\varepsilon \in \,]0, \varepsilon_0]$ and

$$WF_{ba}(u) \cap \{(x, \xi) \in \Sigma_b \text{ ; } 0 \leq x^n \leq A_0\, \varepsilon^2 \text{ ,}$$
$$|(x', \xi')-\exp(\varepsilon H_{r_0})(x_0', \xi_0')| \leq A_0 \varepsilon^2\} = \emptyset \text{ ,}$$

then $(x_0', 0, \xi_0') \notin WF_{ba}(u)$, for every solution u of (6.2) .

Applying the same geometric arguments as in Melrose-Sjöstrand [16], we get Theorem 6.2.

7. Further results in the diffractive region.

We still consider the problem (6.2) . Let $\rho_0 = (x_0', \xi_0')$ $\in \mathcal{G}_+$ and for some small $\delta > 0$, let $\alpha : [-\delta, \delta] \to \mathcal{G}_+$, $\gamma : [-\delta, \delta] \to \Sigma_b$ be the unique boundary bicharacteristic respectively C^∞-ray with $\alpha(0) = \gamma(0) = \rho_0$. Let

$$① = \alpha([-\delta, 0[) \quad , \quad ② = \gamma([-\delta, 0[)$$
$$③ = \alpha(]0, \delta]) \quad , \quad ④ = \gamma(]0, \delta])$$

Fig :

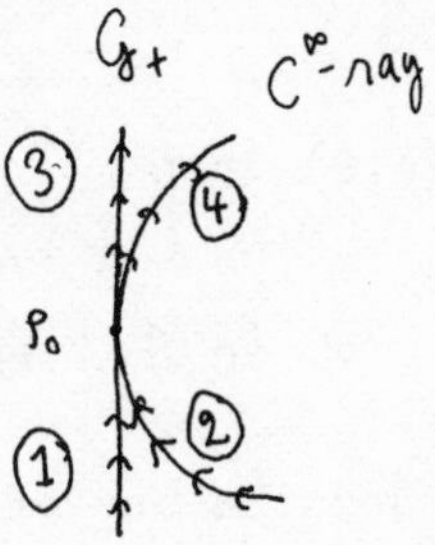

Let u solve (6.2) . Then Theorem 6.2 shows that $(①\cup②)\cap WF_{ba}(u)=\emptyset \Rightarrow \rho_0 \notin WF_{ba}(u)$. On the other hand , it has been proved implicitly by Rauch [18] for the wave equation and by Friedlander-Melrose [5] for a special operator of the form (6.3) , that $② \cap WF_{ba}(u)=\emptyset \not\Rightarrow$ $\rho_0 \notin WF_{ba}(u)$ (contrary to the case of C^∞-singularities). In [24] we showed that

(7.1) $(②\cup④)\cap WF_{ba}(u)=\emptyset \Rightarrow \rho_0 \notin WF_{ba}(u)$.

Independently , K.Kataoka [12] showed a stronger result , namely that no boundary condition is necessary !

Theorem 7.1(Kataoka [12]). Suppose that $Pu \in a(M)$. Then (7.1) holds .

On the other hand we have

Theorem 7.2. There exists a local solution u of (6.2) such that $WF_{ba}(u) = WF_b(u)$ is the cone generated by the C^∞-ray γ . In particular :

$$(①\cup③)\cap WF_{ba}(u)=\emptyset \not\Rightarrow \rho_0 \notin WF_{ba}(u) .$$

In [25] we proved this result for a special operator . We shall outline a construction in the general case below and obtain at the same time a new proof of Theorem 7.1 .

We also have

Theorem 7.3([25]). If u solves (6.2) then

$$(7.2)\quad (② \cup ③) \cap WF_{ba}(u) = \emptyset \Rightarrow \rho_0 \notin WF_{ba}(u) .$$

The proof of Theorem 7.3 combines the use of "weight-factors" as in section 2,3 with the construction of certain "incoming" asymptotic solutions . Since it is quite technical and intended for publication elsewhere we limit ourselves to outline the proof of Theorem 7.1 and 7.2 in the remainder of this section .

We work locally and let $\Psi(x',\theta')$ be a real valued function defined near a point (x_0',θ_0') , such that

$$(7.3)\qquad \frac{\partial\Psi}{\partial x'}(x_0',\theta_0') = \xi_0' \quad ,$$

$$(7.4)\qquad \det \frac{\partial^2\Psi}{\partial x'\partial\theta'} \neq 0 \quad ,$$

$$(7.5)\qquad r_0(x',\frac{\partial\Psi}{\partial x'}) + \theta^1 = 0 .$$

Let $H(x', \xi^n,\theta',t)$ be the solution to the problem

$$(7.6)\qquad \begin{aligned} &\frac{\partial H}{\partial t} + (\xi^n)^2 + r(x',-\frac{\partial H}{\partial \xi^n} , \frac{\partial H}{\partial x'}) = 0 \\ &H\big|_{t=0} = \Psi(x',\theta') \quad . \end{aligned}$$

Then H is well defined for (x',θ') close to (x_0',θ_0') , $\xi^n \in C$ and $|t| \leq C(1+|\xi^n|)^{-1}$.

When all the variables are real and $|\xi^n|$ not too large , we have $\partial^3 H/\partial t^3 < 0$ and $\partial^2 H/\partial t^2 = 0$,for $t = \tilde{t}(x', \xi^n,\theta') = f \cdot \xi^n$, where $f > 0$. Moreover

$\partial H/\partial t = \theta^1$ for $t = \tilde{t}$. We have $\partial H/\partial t = 0$ for $t = t_{\pm} = \tilde{t} \pm \sqrt{\frac{\theta^1}{a}} + \mathcal{O}(|\theta^1|)$, where $a \neq 0$ and $t_{\pm}$ are actually the values of the same multivalued function $\hat{t}(x', \xi^n, \sqrt{\theta^1}, \theta'')$, $\theta' = (\theta^1, \theta'')$. The difference between the two critical values is $\mathcal{O}(|\theta^1|^{3/2})$.

Put $G = H|_{t=\hat{t}}$. Then G is given by

$$(7.7) \qquad \begin{aligned} &(\xi^n)^2 + r(x', -\frac{\partial G}{\partial \xi^n}, \frac{\partial G}{\partial x'}) = 0 \\ &G\Big|_{\xi^n = \sqrt{\theta^1}} = \Psi(x', \theta') \end{aligned}$$

and $G = G\Big|_{\theta^1 = 0} + \mathcal{O}(\theta^1)$. Moreover it is easy to show that $\partial^3 G/\partial(\xi^n)^3 < 0$ when all variables , including $\sqrt{\theta^1}$, are real , and that $(\partial_{\xi^n})^2 G = 0$ on a hypersurface of the form $\xi^n = \tilde{\xi}^n = \mathcal{O}(|\theta^1|)$, which has an x-space projection of the form $x^n + a(x', \sqrt{\theta^1}, \theta')\, \theta^1 = 0$, where $a > 0$ when all the variables are real and $x^n = -\partial_{\xi^n} G$. Let ξ^n_+ , ξ^n_- be the critical points of $\xi^n \mapsto x^n \xi^n + G$. Again , these are different values of a single multivalued function $\hat{\xi}^n(x', (x^n + a\,\theta^1)^{\frac{1}{2}}, (\theta^1)^{\frac{1}{2}}, \theta')$. Let φ be the corresponding multivalued critical value . Then with a convenient choice of branches we have

$$(7.8) \qquad p(x, \varphi'_x) = 0 \quad , \quad \varphi\Big|_{x^n} = \Psi(x', \theta') \quad .$$

As is wellknown from the C^∞-theory of diffraction (Ludwig [13], Taylor [26], Melrose [14]) , we can write

$$(7.9) \qquad \varphi = \varphi_1 \pm \tfrac{2}{3}(\varphi_2)^{3/2} = \varphi_{\pm} \; ,$$

where φ_j are holomorphic in x, $(\theta^1)^{\frac{1}{2}}, \theta''$, and $\varphi_j = \varphi_j|_{\theta^1=0} + \mathcal{O}(|\theta^1|)$ and φ_2 is a positive function times $x^n + a\,\theta^1$. We have

<u>Lemma 7.5.</u> When x', θ' are real and $x^n \geq 0$, we have $\operatorname{Im} \varphi = \mathcal{O}(|\theta^1|^{3/2})$, for the 4 branches of φ.

Solving the appropriate transport equations, we can define a classical analytic symbol $a(x', \xi^n, \theta', t, \lambda)$ such that formally

$$(7.10) \qquad (\tilde{D}_t + P(x', -\tilde{D}_{\xi^n}, \tilde{D}_{x'}, \xi^n, \lambda))(e^{i\lambda H} a) = 0 ,$$

$$(7.11) \qquad a|_{t=0} = 1 .$$

Working near some fixed $\xi^n \in \mathbb{C}$ we can find a contour $\gamma : [0,1] \to \mathbb{C}$, such that $\gamma(0) = 0$, $\operatorname{Im} H|_{t=\gamma(1)} > 0$. Then

$$(7.12) \qquad \hat{u} = \int_\gamma e^{i\lambda H} a \, dt$$

solves formally the problem

$$(7.13) \qquad P(x', -\tilde{D}_{\xi^n}, \tilde{D}_{x'}, \xi^n, \lambda)\, u = e^{i\lambda\psi(x',\theta')} .$$

In a region where $|\xi^n|$ and $1/|\xi^n|$ are bounded there is an obvious choice of γ since $\partial H/\partial t \neq 0$. In such a region we get

$$(7.14) \qquad \hat{u} = e^{i\lambda\psi} b ,$$

where the classical analytic symbol b is uniquely determined by (7.13). Let $r > 0$ be small but fixed and put

$$(7.15) \qquad u(x,\theta',\lambda) = \frac{\lambda}{2\pi} \int_{0_r} e^{i\lambda(x^n \xi^n + \Psi)}\, b\, d\xi^n \ ,$$

where 0_r is the circle around 0 of radius r with the usual orientation. It is not hard to see that modulo holomorphic functions of exponential decrease, we have near $(x_0', 0)$:

$$(7.16) \qquad Pu \equiv 0 \ , \ u\big|_{x^n=0} \equiv 0 \ , \ D_{x^n} u\big|_{x^n=0} \equiv e^{i\lambda\Psi} \ .$$

Changing (7.11) we can also get a similar solution v of

$$(7.17) \qquad Pv \equiv 0 \ , \ v\big|_{x^n=0} \equiv e^{i\lambda\Psi} \ , \ D_{x^n} v\big|_{x^n=0} \equiv 0 \ .$$

Let $A = r_0 e^{i\pi/2}$, $B = r_0 e^{i7\pi/6}$, $C = r_0 e^{i11\pi/6}$. Then, if $|\xi^n| \leq r$, the contours γ_A, γ_B, γ_C, joining 0 to A,B,C have the properties for γ required above. Put

$$\hat{u}_A = \int_{\gamma_A} e^{i\lambda H}\, a\, dt \ ,$$

and define $\hat{u}_B$, $\hat{u}_C$ similarly. Then by deforming the contours so that they pass through the appropriate critical points of H one can investigate in what regions in the ξ^n-plane that any of these three functions coincides modulo an exponentially decreasing error with $b\, e^{i\lambda\Psi}$. We find that

$$\frac{2\pi}{\lambda} u \equiv \int_A^B e^{i\lambda x^n \xi^n} \hat{u}_C \, d\xi^n + \int_B^C \ldots \hat{u}_A \, d\xi^n + \int_C^A \ldots \hat{u}_B \, d\xi^n .$$

We write this as

$$(7.18) \quad \begin{aligned} u &= \frac{\lambda}{2\pi} \int_B^A e^{i\lambda x^n \xi^n} (\hat{u}_A - \hat{u}_C) \, d\xi^n + \\ &\frac{\lambda}{2\pi} \int_A^C e^{i\lambda x^n \xi^n} (\hat{u}_A - \hat{u}_B) \, d\xi^n = u_- + u_+ . \end{aligned}$$

These integrals can again be estimated in terms of the critical values of the functions $\xi^n \mapsto x^n \xi^n + G$, and we get

(I) When $|x^n| \leq C|\theta^1|$, then

$$u_{\pm} = \mathcal{O}(1) \exp(-\lambda \operatorname{Im}(\varphi_1) + C\lambda |\theta^1|^{3/2}) .$$

(II) When $|x^n| \geq C|\theta^1|$ and x^n is in a conic neighbourhood of R_+ , then

$$u_{\pm} = \mathcal{O}(1) \exp(- \operatorname{Im}(\varphi_{\pm}) + C\lambda |\theta^1|^{3/2}) .$$

These estimates are valid for (x',θ') in a small complex domain and it should be noted , that in the region (II) $\varphi_{\pm}$ are well defined functions of θ^1 up to an indetermination which is $\mathcal{O}(|\theta^1|^{3/2})$.

We next choose a classical analytic symbol $f(y',\theta',\lambda)$ such that the pseudodifferential operator

$$u \mapsto \lambda^n \iint e^{i\lambda(\psi(x',\theta') - \psi(y',\theta'))} f(y',\theta',\lambda) u(y') \, dy' d\theta'$$

is formally the identity . For (x_1',θ_1') close to (x_0',θ_0') , we introduce the superposition solution

$$U_{x_1',\theta_1'}(x,\lambda) =$$

(7.19)
$$= \lambda^n \iint u(x,\theta',\lambda) f(y',\theta',\lambda)\ e^{i\lambda(-\varphi(y',\theta')+\psi(y',\theta_1')+i(y'-x_1')^4)}\ dy'\, d\theta' ,$$

where the domain of integration is a rectangle $|y'-x_1'| \le r_1$, $|\theta'-\theta_1'| \le r_2$. It can be showed , using (I),(II) that on the real domain with $x^n \ge 0$:

(7.20) $$U = \mathcal{O}(1)\ \lambda^n \exp(C\lambda|\theta_1^1|^{3/2}) ,$$

and that in a suitable sense U is microlocally concentrated to a neighbourhood of the C^∞-ray γ . Moreover , modulo exponentially decreasing errors ,

(7.21) $$P\, U \equiv 0 ,$$

(7.22) $$U\big|_{x^n=0} \equiv 0 \quad , \quad D_{x^n} U\big|_{x^n=0} \equiv e^{i\lambda(\psi(x',\theta_1')+i(x'-x_1')^4)} .$$

We also have a similar solution V to the Neumann problem .

To prove Theorem 7.2 we now only have to put

$$u(x,\lambda) = \int_1^\infty \lambda^{-(n+2)}\ U_{x_0',\theta_0'}(x,\lambda)\ d\lambda$$

and make the fairly straight forward verification that $WF_a(u)$ is the cone generated by γ .

To prove Theorem 7.1 , we let $U_{x_1',\theta_1'}$, $V_{x_1',\theta_1'}$ solve the Dirichlet and Neumann problems for P^* instead of P . If u satisfies the assumptions of Theorem 7.1 , it is fairly straight forward to show by integration by parts , that

$$\int e^{i\lambda(\psi(x',\theta_1') + i(x'-x_0')^4)}\ \overline{u}(x',0,\lambda)\ dx'$$

and the similar expression for $D_{x^n}u$ are exponentially decreasing, when (x_1', θ_1') is in a small neighbourhood of (x_0', θ_0'). From this it is easy to see that $(x_0', \xi_0') \notin WF_{ba}(u)$.

8. Back to $\Box$.

Let $\Omega \subset R^{n-1}$ be an open convex set with analytic boundary, let X be the complement, let $x_0 \in \overset{o}{X}$ and let $u \in \mathcal{D}'(R_t \times \dot{X})$ be the solution to

$$(8.1) \qquad \Box u = 0 \ , \quad u|_{R_t \times \partial X} = 0 \ ,$$

with initial conditions

$$(8.2) \qquad u|_{t=0} = 0 \ , \quad D_t u|_{t=0} = \delta(x-x_0) \ .$$

Theorem 6.2 shows that $WF_{ba}(u)$ is contained in the union of all analytic rays which pass over $(0,x_0)$. The interesting problem (,completely open when Ω is non-convex ,)is wether we have equality.

Theorem 8.1(Rauch-Sjöstrand[19]). When $n-1 = 2$ and Ω is is convex, then $WF_{ba}(u)$ is the union of all analytic rays that pass over $(0,x_0)$.

The rays which are not obviously in $WF_{ba}(u)$ are the ones that hit ∂X tangentially, then glide along ∂X for some time and possibly leave the boundary later. In the region $0 \leq t \leq l_0$, where $2l_0$ is the length of the shortest curve that winds once around Ω and joins x_0 to itself, these rays have a length-minimizing projection in the x-space and Theorem 8.1 can be proved essentially by applying the Holmgren uniqueness theorem as in Rauch [18].

For larger times, we can replace X by its universal covering space $\tilde{X}$ and the rays can be lifted to rays over $R_t \times \tilde{X}$ which have length-minimizing projections in $\tilde{X}$. Again the argument of Rauch shows that Theorem 8.1 holds with X replaced by $\tilde{X}$. The solution of (8.1),(8.2) is a locally finite sum of values of the corresponding solution on the covering space, and using Theorem 6.2 one can verify the no cancellation of singularities takes place when we sum.

Using the further results of section 7, we can almost eliminate the assumption that $n-1 = 2$.

<u>Theorem 8.2([25]).</u> Let u solve (8.1),(8.2) and assume that Ω is strictly convex. Then $WF_{ba}(u)$ is the union of all analytic rays that pass over $(0,x_0)$.

Again, despite Theorem 7.2, one can show as in the 2-dimensional case that the theorem is valid in a sufficiently short time interval, where all the "non-trivial" rays have length-minimizing x-space projections. The proof for larger time inervals is then an easy application of (7.1) and (7.2).

Since $WF_{ba}(u)$ is closed, the assumption about strict convexity in Theorem 8.2 can be replaced by the weaker assumption that every ray passing over $(0,x_0)$ can be approximated in any finite time interval $[0,T]$ or $[-T,0]$ by such rays which in addition avoid $\mathcal{G} \setminus \mathcal{G}_+$ in that interval.

References.

1. J.M.Bony, Sém. Goulaouic-Schwartz,1975-76,n°17.

2. J.M.Bony,P.Schapira,Ann.Inst.Fourier, 26,1(1976),81-140

3. L.Boutet de Monvel, P.Kree, Ann.Inst.Fourier,17(1967), 295-323.

4. J.Bros,D.Iagolnitzer,Sém.Goulaouic-Schwartz,1975,n°18.

5. F.G.Friedlander,R.B.Melrose,Math.Proc.Camb.Phil.Soc., 81(1977),97-120.

6. N.Hanges,Duke Math.Journal,47(1980),17-25.

7. N.Hanges,"Propagation of analyticity along real bicharacteristics",to appear.

8. L.Hörmander,C.P.A.M.,24(1971),671-704.

9. L.Hörmander,Acta Math.,127(1971),79-183.

10. M.Kashiwara,Sagaku no Ayumi,15(1970),19-72,(in Japanese).

11. M.Kashiwara,T.Kawai,Journal Math.Soc.Japan,27(1975), 359-404.

12. K.Kataoka,"Microlocal theory of boundary value problems II",to appear.

13. D.Ludwig,C.P.A.M. 19(1967),103-138.

14. R.B.Melrose,Duke Math Journal, 42,4(1975),583-635.

15. R.B.Melrose,J.Sjöstrand,C.P.A.M.,31(1978),593-617.

16. R.B.Melrose,J.Sjöstrand,to appear,see also Sém.Goulaouic-Schwartz,1977-78,n°15.

17. Nagano, J.Math.Soc.Japan 18(1966),398-404.

18. J.Rauch,Bull.Soc.R.Sci.Liège,46,5-8(1977),156-161.

19. J.Rauch,J.Sjöstrand,Indiana Univ.Jounal of Math.,To appear.

20. M.Sato,T.Kawai,M.Kashiwara,Springer L.N.Math.,n°287 .

21. P.Schapira,Publ.RIMS,Kyoto Univ.,12 Suppl.(1977),441-453.(See also Sém.Goulaouic-Schwartz,1976-77, n° 9 for further results involving non-microcharacteristic operators.)

22. J.Sjöstrand,"Analytic singularities and micro-hyperbolic boundary value problems", to appear.

23. J.Sjöstrand,Comm.P.D.E.,5(1)(1980),41-94.

24. J.Sjöstrand,Comm.P.D.E.,5(2)(1980),187-207.

25. J.Sjöstrand,"Propagation for second order Dirichlet problems III",to appear.

26. M.Taylor,C.P.A.M. 28(1975),457-478.

DIFFRACTION EFFECTS IN THE SCATTERING OF WAVES

Michael E. Taylor

Rice University

ABSTRACT

This paper describes some recent progress in the use of parametrices for diffractive boundary value problems for studying effects of grazing rays on the behavior of scattered waves.

INTRODUCTION

This paper summarises a number of developments in the study of diffractive boundary value problems over the past five years. In the first two sections, work of Melrose [13], [15] and Taylor [19], [22] on the construction of parametrices for such a grazing ray problem is reviewed. We restrict attention to the Dirichlet problem for the usual scalar wave equation. In sections 3-5 we sketch some joint work of Melrose and Taylor [17] on Fourier integral operators with folding canonical relations, and applications to some problems in scattering theory, and on the corrected Kirchoff approximation. Section 6 describes some results of Farris [3] on the solution operator to the wave equation with diffractive boundary. In the last two sections, some results on some general classes of equations are discussed, with emphasis on cases where a few new techniques are required. Some related problems have been considered by Imai-Shirota [7] and by Kubota [9]. Since these last two sections do not sketch work published elsewhere, they are wordier than the first six.

In this paper we make no mention of non smooth obstacles. For a study of diffraction of waves by cones and polyhedra, see [1], which is summarised in [23].

H. G. Garnir (ed.), Singularities in Boundary Value Problems, 271–316.

We make use of pseudo-differential operators with symbols $p(x,\xi)$ in the classes $S^m_{\rho,\delta}$ of Hörmander [5], i.e., satisfying estimates

$$|D_x^\beta D_\xi^\alpha p(x,\xi)| \le c_{\alpha\beta}(1+|\xi|)^{m-\rho|\alpha|+\delta|\beta|}.$$

A subclass of $S^m_{1,0}$ is S^m, consisting of $p(x,\xi)$ asymptotic to $\sum_{j\ge 0} p_{m-j}(x,\xi)$, with $p_{m-j}(x,\xi)$ homogeneous of degree $m-j$ in ξ, for $|\xi| \ge 1$. Also we say $p(x,\xi,\eta) \in N^m_\rho$ provided

$$|D_x^\beta D_\xi^\alpha D_\eta^j p(x,\xi,\eta)| \le c_{\beta\alpha j}|\xi|^{m-|\alpha|}(|\xi|^\rho+|\eta|)^{-j}.$$

§1. THE GRAZING RAY PARAMETRIX.

In this section we review the construction of a parametrix for the solution of the wave equation on the exterior of a convex domain $K \subset \mathbb{R}^n$, assumed to be smooth, with strictly positive curvature. Such parametrices were constructed in [13] and [19]; see also [15] and [22]. We briefly discuss some refinements.

We look for a parametrix for the solution u to the problem

$$\left(\frac{\partial^2}{\partial t^2} - \Delta\right)u = 0 \quad \text{on } \mathbb{R}^n - K \tag{1.1}$$

$$u\big|_{\mathbb{R}\times\partial K} = f \tag{1.2}$$

$$u = 0 \quad \text{for} \quad t << 0 \tag{1.3}$$

given $f \in \mathcal{E}'(\mathbb{R}\times\partial K)$. The boundary condition (1.2) is the Dirichlet boundary condition. Also of interest is the Neumann boundary condition

$$\frac{\partial}{\partial\nu} u\big|_{\mathbb{R}\times\partial K} = g. \tag{1.4}$$

This, and a large class of other boundary conditions, is amenable to treatment, given the discussion of the Neumann operator which we will provide in section 2.

We will assume $WF(f)$ is contained in a small conic neighborhood of a point in $T^*(\mathbb{R}\times\partial K)$ over which a grazing ray passes, since the non-grazing case is relatively elementary. The parametrix we will construct is of the form

$$u(t,x) = \iint \left[g \frac{A(\xi_1^{-1/3}\rho)}{A(-\xi_1^{-1/3}\eta)} + ih\xi_1^{-1/3} \frac{A'(\xi_1^{-1/3}\rho)}{A(-\xi_1^{-1/3}\eta)} \right] e^{i\theta}\hat{F}(\xi,\eta)\,d\xi d\eta. \tag{1.5}$$

The phase function ρ,θ will solve certain eikonal equations and the amplitudes g,h will satisfy certain transport equations. The function $A(s) = A_\pm(s)$ is a certain Airy function, $A_\pm(s) = Ai(e^{\pm 2/3\pi i}s)$, solving the Airy equation $A''(s) - sA(s) = 0$. $A(s)$ blows up as $s \to +\infty$ and is oscillatory as $s \to -\infty$. In fact, one has

$$A_\pm(s) = \pm \frac{i}{2} e^{\mp 2/3\pi i} F(s)e^{\mp i\chi(s)} \tag{1.6}$$

where $F(s)^{-1} \in S^{1/4}_{1,0}(\mathbb{R})$, $\chi(s) \in S^{3/2}_{1,0}(\mathbb{R})$ have expressions of the form

$$F(s)^2 \sim \frac{1}{\pi\sqrt{-s}} \left[1-a_1(-s)^3 + \ldots\right] \quad \text{as} \quad s \to -\infty$$

$$F(s) \sim \frac{1}{\pi} e^{2/3 s^{3/2}} \quad \text{as} \quad s \to +\infty$$

$$\chi(s) - \frac{\pi}{4} \sim 2/3(-s)^{3/2}\left[1-b_1(-s)^3 + \ldots\right] \quad \text{as} \quad s \to -\infty.$$

χ is real, and $\chi'(s) = -\dfrac{1}{\pi F(s)^2}$. See [18]. It turns out that we can find solutions to the eikonal equations

$$\begin{aligned} &\theta_t^2 - |\nabla_x\theta|^2 + \frac{\rho}{\xi_1}(\rho_t^2 - |\nabla_x\rho|^2) = 0 \\ &\theta_t\rho_t - \nabla_x\theta\cdot\nabla_x\rho = 0 \end{aligned} \tag{1.7}$$

on $\mathbb{R}^n - K$ for $\eta \geq 0$, and to infinite order on ∂K for $\eta \leq 0$, such that

$$\rho\big|_{\mathbb{R}\times\partial K} = -\eta \tag{1.8}$$

and such that

$$\frac{\partial}{\partial\nu}\rho\big|_{\mathbb{R}\times\partial K} < 0. \tag{1.9}$$

(The functions ρ,θ are real valued, smooth, and homogeneous of degree 1 in (ξ,η).) From this, the asymptotic relation (1.6) makes sense out of (1.5) as a Fourier integral operator with singular phase function. The unknown distribution F, with wave front set in a small conic neighborhood of $\{\eta=0\}$, is related to

$u\big|_{\mathbb{R}\times\partial K} = f$ by a Fourier integral operator. Indeed, using (1.8) and the similarly derived fact that one can arrange

$$h\big|_{\mathbb{R}\times\partial K} = 0 \tag{1.10}$$

one gets, with $\theta_0 = \theta\big|_{\mathbb{R}\times\partial K}$,

$$u\big|_{\mathbb{R}\times\partial K} = \iint g e^{i\theta_0}\hat{F}\, d\xi d\eta = JF \tag{1.11}$$

In solving the transport equation for g one can arrange that g be nonvanishing on a small conic set, and the phase function θ_0 can be seen to yield a non-degenerate canonical transformation $\mathcal{J}$, so J is an elliptic Fourier integral operator, and hence is microlocally invertible. Thus, the parametrix to (1.1)-(1.3) is given by (1.5) with

$$F = J^{-1}f. \tag{1.12}$$

We briefly go over the solution to the eikonal equation (1.7), satisfying the condition (1.8), which is more restrictive than the condition $\rho\big|_{\mathbb{R}\times\partial K} = -\eta + O\big(|\xi|(|\eta|/|\xi|)^\infty\big)$ proved and used in [22] (see also [19]). The extra ingredient used to obtain (1.8) is Melrose's result on equivalence of glancing hypersurfaces [14]. Melrose [15] has noted that this result leads to solutions to (1.7) such that $\rho\big|_{\mathbb{R}\times\partial K}$ is independent of (t,x). The argument we sketch here is just a little different from that one.

Let $\Omega = \mathbb{R}\times(\mathbb{R}^n - K)$. The pair of hypersurfaces $J_1 = T^*_{\partial\Omega}(\mathbb{R}^{n+1})$ and $K_1 = \{|\xi|^2 - \tau^2 = 0\}$ in $T^*(\mathbb{R}^{n+1})$ has glancing intersection, in the sense of [14]. Consequently, there is a (microlocally defined) homogeneous symplectic map

$$T^*(\mathbb{R}^{n+1}_+) \xrightarrow{\chi} T^*\Omega \tag{1.13}$$

taking the "canonical pair" of hypersurfaces to J_1, K_1. More precisely, $J_0 = \{x_{n+1} = 0\}$ is taken to J_1 and $K_0 = \{p_0(x,\xi)=0\}$ is taken to K_1 by χ, where

$$p_0(x,\xi) = \xi_{n+1}^2 - x_{n+1}\xi_1^2 + \xi_1\xi_n \tag{1.14}$$

Now, on J_1 and J_0, the symplectic form gives a Hamilton foliation. Let this determine an equivalence relation $\sim$. Then $J_1 \cap K_1/\sim$ has the structure of a symplectic manifold with boundary, and is naturally isomorphic to the closure of the "hyperbolic" region in $T^*(\partial\Omega)$, the region over which real rays pass, and similarly $J_0 \cap K_0$ is naturally isomorphic to the closure of the hyperbolic region in $T^*(\partial\mathbb{R}^{n+1}_+)$. Thus we get a homogeneous symplectic map

$$T^*(\partial\mathbb{R}^{n+1}_+) \xrightarrow{\chi_J} T^*(\partial\Omega), \tag{1.15}$$

defined in the hyperbolic regions, smooth up to the boundary, which consists of the grazing directions. Furthermore, χ_J intertwines the "billiard ball maps" $\delta_0^\pm$ and $\delta^\pm$. Here, the billiard ball maps $\delta^\pm : T^*(\partial\Omega) \to T^*(\partial\Omega)$, defined on the hyperbolic region, continuous up to the boundary, smooth in the interior, are defined at a point (x_0, ξ_0) by taking the two rays that lie over this point, in the variety $K_1 = \{|\xi|^2 - \tau^2 = 0\}$, and following the null bicharacteristics through these points until you pass over $\partial\Omega$ again, projecting such a point onto $T^*(\partial\Omega)$; δ^+ increases the t-coordinate and δ^- decreases it. $\delta_0^\pm$ is defined similarly.

Let $v \in \Lambda'(\partial\Omega)$ be a gradient field corresponding under χ_J to some $(\xi_1, \ldots, \xi_n) = $ const. in $T^*(\mathbb{R}^n)$. Let $S_v \subset T^*\Omega$ be the Hamilton flow-out, where $S_v|_{\partial\Omega}$ is identified with the appropriate point in $K_1 \subset T^*(\Omega)$ lying over $v \in T^*(\partial\Omega)$. Let $S_\xi \subset T^*\mathbb{R}^{n+1}$ be the analogous flow-out in $T^*\mathbb{R}^{n+1}$, so χ takes S_ξ^ξ to S_v.

The functions $\theta(z,\xi)$, $\rho(z,\xi)$, $x \in \Omega$, solving (1.7), with $z = (t,x)$, $\eta = \xi_n$, are obtained as follows. Pick $\Phi \in C^\infty(S_v)$ such that $d\Phi = i^*\alpha$ where α is the canonical 1-form on $T^*\Omega$ and $i = S_v \to T^*\Omega$ is the natural inclusion. Φ is determined up to a term independent of z, so normalize it, e.g., by picking a point $q(\xi) \in S_v$ in some smooth convenient fashion and requiring Φ to vanish there. The convexity hypothesis implies that the natural projection

$$\pi : S_v \to \Omega \tag{1.16}$$

is a simple fold. One has a smooth involution $j : S_v \to S_v$, interchanging points with the same image under π. With respect to this involution, we will break up Φ into even and odd parts. Let $\Psi = \Phi \circ j$. If S_v is regarded as the graph of the field $v^\pm$, over its image $\pi(S_v)$, define

$$\begin{aligned} \theta &= \tfrac{1}{2}(\Phi+\Psi)\circ v^\pm \\ \rho &= \xi_1^{1/3}\,[\tfrac{3}{4}(\Phi-\Psi)\circ v^\pm]^{2/3} \end{aligned} \tag{1.17}$$

It is straightforward to verify that $\phi^\pm = \theta \pm \tfrac{2}{3}\xi_1^{-1/2}\rho^{3/2}$ satisfies the eikonal equation $(\phi_t^\pm)^2 = (\nabla_x \phi^\pm)^2$ on $\pi(S_v)$, and (1.7) follows. The point of the construction (1.17) is that ρ and θ are smooth up to the image under π of the fold set, the "caustic." Consequently they can be continued across in a smooth fashion. If

$\eta = \xi_\eta \geq 0$, S_v projects onto a region containing $\partial\Omega$; if $\eta < 0$, this is no longer the case. Thus ρ,θ are defined on Ω for $\eta \geq 0$ by (1.17). Using a formal power series expansion and the Whitney extension theorem, we can smoothly extend ρ,θ to $\eta < 0$ so that the eikonal equation (1.7) is solved to infinite order at the boundary $\partial\Omega$. This is enough to make distributions defined by (1.5) solve the wave equations, mod C^∞, granted an analogous formal solution to the transport equations.

Now we want to look into the behavior of ρ, and verify (1.8). Note that $\rho = 0$ on the caustic set; in particular, on $\partial\Omega$, $\rho = 0$ at $\eta = 0$. Also, we can see that ρ is independent of x on $\partial\Omega$, by studying the eikonal equations, which gives

$$v = i^*\left(d\theta \pm \sqrt{\frac{\rho}{\xi_1}}\, d\rho\right) \text{ on } \partial\Omega$$

(where $i:\partial\Omega \to \Omega$), since v is invariant under the billiard ball maps. This implies $i^*d\rho = 0$, so $\rho|_{\partial\Omega}$ depends on (ξ,η). To To see that actually $\rho|_{\partial\Omega} = -\eta$, we make use of the fact that ρ, unlike θ, is defined <u>independently</u> of the choice of normalization of Φ. Now define $\Phi_0 \in C^\infty(S_\xi)$ in the same fashion as Φ on $C^\infty(S_v)$. If we normalize Φ_0 to vanish at $q_0(\xi) = \chi_J^{-1}q(\xi)$, where Φ was normalized to vanish on $q(\xi)$, then Φ_0 may give rise to a non-smooth θ_0, but we are only concerned with the value of ρ_0, so we proceed. We see that

$$\Phi_0 = \Phi \circ \chi_J \,.$$

Now we know that $\frac{4}{3}\rho^{3/2}|_{\partial\Omega}$ is the difference in the two values of $\xi_1^{1/2}\Phi$ at two points in S_v lying over a common image point in $\partial\Omega$. To say these points are so related is equivalent to saying that they both lie in $J_1 \cap K_1$ and are equivalent under the relation $\sim$ defined above. Similarly, $\frac{4}{3}\rho_0^{3/2}|_{\partial\mathbb{R}^{n+1}_+}$ is the difference between two values of $\xi_1^{1/2}\Phi_0$ at points lying over a common base point in $\partial\mathbb{R}^{n+1}_+$, which is to say these two points are in $J_0 \cap K_0$ and related by $\sim$. Thus χ_J <u>preserves</u> this pairing, so

$$\rho|_{\partial\Omega} = \rho_0|_{\partial\mathbb{R}^{n+1}_+} \,.$$

However, in the canonical example, one explicitly has

$$\rho_0 = -\xi_n + x_{n+1}\xi_1$$

and in particular $\rho_0 = -\xi_n = -\eta$ on $\partial\mathbb{R}^{n+1}_+$. This establishes (1.8).

In the above construction, that of θ is not canonical. One can arrange that $\theta|_{\partial\Omega}$ generate the canonical transformation χ_J. In general, whatever canonical transformation it generates has in common with χ_J that it conjugates $\delta^{\pm}$ to $\delta_0^{\pm}$.

A parallel but simpler argument produces the amplitudes from certain transport equations, to be solved exactly on Ω for $\eta \geq 0$ and to infinite order for $\eta < 0$, with (1.10) holding.

This sketches the construction of the parametrix (1.5). For more details, and a study of the singularities of (1.5), see chapter X of [22], or the original papers [13], [14], [19]. Of course the basic result on the singularities of (1.5) is that they lie over WF(JF) and propagate forward in time along null bicharacteristics of $\partial^2/\partial t^2 - \Delta$, thus verifying the geometrical optics description in the diffractive case.

§2. THE NEUMANN OPERATOR.

The exact solution to the boundary value problem (1.1)-(1.3) can be written as Kirchoff's integral

$$u(t,x) = \int_{\mathbb{R}\times\partial K} \left[f(s,y) \frac{\partial G}{\partial \nu} (t-s,x-y) - g(s,y)G(t-s,x-y)\right] ds\, dS(y) \tag{2.1}$$

where

$$g = \frac{\partial u}{\partial \nu}\bigg|_{\partial K} = Nf \tag{2.2}$$

defines the Neumann operator.

$G(t,x)$ is the free space fundamental solution to the wave equation on $\mathbb{R} \times \mathbb{R}^n$. For $n = 3$, for example, one has $G = \frac{1}{4\pi} \frac{\delta(|x|-t)}{t}$ for $t > 0$. Evidently it is very useful to analyze the properties of N. When K is convex, so the diffractive hypothesis is satisfied, we can analyze N as a pseudo-differential operator, using the parametrix (1.5), as follows.

Differentiate (1.5) and restrict to $\mathbb{R} \times \partial K$. Use (1.8) and (1.10). Note that (1.8) implies $\nabla\rho$ is normal to ∂K, so, if one takes ρ independent of t, which can be arranged, the second half of (1.7) implies $\theta_\nu|_{\partial K} \equiv 0$. Thus (1.5) gives

$$\left.\frac{\partial u}{\partial \nu}\right|_{\mathbb{R}\times\partial K} = \iint (g\rho_\nu + ih_\nu)\xi_1^{-1/3}\,\frac{A'}{A}(-\xi_1^{-1/3}\eta)\hat{F}(\xi,\eta)e^{i\theta_0}\,d\xi d\eta$$

$$+ \iint g_\nu \hat{F}(\xi,\eta)e^{i\theta_0}d\xi d\eta$$

$$= K_1 QF + K_2 F \tag{2.3}$$

where

$$(QF)\hat{}\,(\xi,\eta) = \xi_1^{-1/3}\,\frac{A'}{A}(-\xi_1^{-1/3}\eta)\hat{F}(\xi,\eta) \tag{2.4}$$

defines $Q \in OPS^0_{1/3,0}$. K_1 and K_2 are Fourier integral operators with the same phase function as J, K_1 elliptic of order 1, K_2 order zero. Egorov's theorem gives $K_1 = JA$, $K_2 = JB$ for certain $A \in OPS^1$ elliptic, $B \in OPS^0$. Comparing with (1.11), we get

$$N = J(AQ+B)J^{-1}. \tag{2.5}$$

Thus N is conjugated to the special form $AQ+B$, by a Fourier integral operator whose associated canonical transformation is the very one χ_J given in (1.15). The fact that this transformation conjugates the billiard ball maps $\delta^\pm$ to standard form has deep connections with the form $\xi_1^{-1/3}\eta$ of the argument of the Airy quotient A'/A in (2.4), as we will see.

Now, we look at the conjugate under J of another Fourier multiplier, $\mathcal{A}i$, defined by

$$(\mathcal{A}iF)\hat{}\,(\xi,\eta) = Ai(-\xi_1^{-1/3}\eta)\hat{F}(\xi,\eta). \tag{2.6}$$

We will see more of this in later sections, as an example of a Fourier integral operator with folding canonical relation. For the moment, just think of it as a Fourier integral operator defined in the conic region $\eta > 0$, via the expansion

$$Ai(s) = F(s)\sin\chi(s) \tag{2.7}$$

where $F(s)$ and $\chi(s)$ are those that appear in (1.6). Thus, in $\{\eta>0\}$, $\mathcal{A}i$ is a sum of two Fourier integral operators, whose canonical transformations are (with $\eta = \xi_n$, $\xi = (\xi_1, \dots, \xi_n)$)

$$\mathfrak{A}^\pm(x,\xi) = \left(x_1 \pm \tfrac{1}{3}\left(\frac{\xi_n}{\xi_1}\right)^{3/2}, x_2, \dots, x_{n-1}, x_n \pm \left(\frac{\xi_n}{\xi_1}\right)^{1/2}, \xi\right)$$

Compare with the "standard" billiard ball map:

$$\delta_0^{\pm}(x,\xi) = \left(x_1 \pm \tfrac{2}{3}\left(\frac{\xi_n}{\xi_1}\right)^{3/2}, x, \ldots, x_{n-1}, x_n \pm 2\left(\frac{\xi_n}{\xi_1}\right)^{1/2}, \xi \right).$$

Clearly $(\mathcal{J}^{\pm})^2 = \delta_0^{\pm}$. This gives the following result. The operator $J(\mathcal{A}i^2)J^{-1}$ is an operator which, when restricted to the "hyperbolic" region, is a sum of three Fourier integral operators, whose three canonical relations are the two billiard ball maps, δ^+ and δ^-, and the identity.

Another geometrical phenomenon, emphasized by Melrose, [16], involving the canonical transformation χ_J versus the argument $\zeta_0 = \xi_1^{-1/3}\eta$, is the following. Define ζ by $\zeta_0 = \zeta\circ\chi_J$. Consider the Hamiltonian vector field $H_{\zeta^{3/2}}$ and consider its time-one flow, $\exp H_{\zeta^{3/2}}$. This is the map δ^+.

As a final remark for this section, note that the Neumann boundary value problem (1.1), (1.3), and (1.4), can be solved using

$$N^{-1} = JQ^{-1}(A+BQ^{-1})^{-1}J^{-1}$$

since $Q^{-1} \in OPS^{1/3}_{1/3,0}$ and $A+BQ^{-1} \in OPS^{1}_{1/3,0}$ is elliptic. A study of the Neumann operator is useful in considering other boundary value problems for the wave equation, including the problem of diffraction of electromagnetic waves by a convex perfect conductor. Details are given in chapter X of [22].

§3. FOURIER INTEGRAL OPERATORS WITH FOLDING CANONICAL RELATIONS.

The operation of convolution by $\delta(x_1 - \tfrac{1}{3}x_n^3)$. $\delta(x_2) \ldots \delta(x_{n-1})$ is a Fourier integral operator with folding canonical relation, i.e., its canonical relation $\Lambda' \subset T^*\mathbb{R}^n \times T^*\mathbb{R}^n$ projects onto each factor as a simple fold. This operation is the same as Fourier multiplication by $\xi_1^{-1/3}\, Ai(-\xi_1^{-1/3}\xi_n)$. Thus the operator $\mathcal{A}i$ defined by (2.6) is a Fourier integral operator with folding canonical relation. So is the operator $\mathcal{A}i'$, defined by

$$(\mathcal{A}i'F)^{\wedge}(\xi\ \eta) = Ai'(-\xi_1^{-1/3}\eta)\hat{F}(\xi,\eta). \tag{3.1}$$

This is (essentially) convolution by the above distribution, multiplied by x_n. One of the aims of this section is to show that any Fourier integral operator $A \in I^m(X_1,X_2;\Lambda')$, for $\Lambda' \subset T^*X_1 \times T^*X_2$ a folding canonical relation (we assume $\dim X_1 = \dim X_2$), can be written in the form

$$A = J(P_1 \mathcal{A}i + P_2 \mathcal{A}i')K \tag{3.2}$$

for some elliptic Fourier integral operators J and K (of order zero) and some $P_1 \in OPS^{m+1/6}$, $P_2 \in OPS^{m-1/6}$. We also want to understand the behavior of A^*A.

Suppose $\Lambda' \subset T^*X_1 \times T^*X_2$ is a folding canonical relation. We first give a condition which guarantees that two elements $A_1, A_2 \in I^m(X_1,X_2;\Lambda')$ generate them all, as a module over OPS^0, at least locally near a point on the image of the fold set $L \subset \Lambda'$, projected onto X_1. First we introduce some geometry. The projection π_j of Λ' to T^*X_j determines an involution, which we denote $\mathcal{J}_j$, such that

$$\mathcal{J}_j(\zeta) = \zeta' \quad \text{if} \quad \pi_j(\zeta) = \pi_j(\zeta'). \tag{3.3}$$

Note that, for any $P \in OPS^0$, PA has principal symbol which is a multiple p of that of A_1, and $\mathcal{J}_2{}^*p = p$ on Λ'. This explains why <u>two</u> operators are needed to generate $I^m(X_1,X_2;\Lambda')$. Indeed, we have the following:

<u>Proposition 3.1</u>. Let $\zeta \in L$ (the fold set in Λ') and suppose $\sigma_{A_1} \neq 0$ at L. Let $\sigma_{A_2} = \beta\sigma_{A_1}$ and suppose $\beta - \mathcal{J}_2^*\beta$ vanishes to precisely first order on L, near ζ. Then, microlocally near $\pi_1\zeta$, for any $A \in I^\nu(X_1,X_2;\Lambda')$, you can write, modulo smoothing operator,

$$A = P_1A_1 + P_2A_2; \ P_j \in OPS^{\nu-m}. \tag{3.4}$$

Here σ_{A_j} denotes the principal symbol of A_j, a section of the Keller-Maslov-Hörmander line bundle over Λ'. β is scalar.

<u>Proof</u>. The hypothesis implies that any homogeneous (scalar) function g on Λ' can (near ζ) be written in the form

$$g = g_1 + g_2\beta$$

where g_1 and g_2 are homogeneous of the appropriate degree and even with respect to $\mathcal{J}_2$; hence $g_j = \pi_2^* p_j$. Letting P_j have principal symbol p_j, if $\sigma_A = g\sigma_{A_1}$, we get (3.4), modulo $I^{\nu-1}(X_1,X_2;\Lambda')$. An inductive argument finishes the proof.

It is easy to see that the operators $\mathcal{A}i$, $\mathcal{A}i'$ satisfy the hypotheses of proposition 3.1, after normalization of their order.

The next thing we want to do is show that, given a folding canonical relation $\Lambda' \subset T^*X_1 \times T^*X_2$ (dim $X_j = n$), there exist homogeneous canonical transformations

$$\chi_j : T^*\mathbb{R}^n \to T^*X_j$$

such that

$$\chi_2^{-1} \circ \Lambda' \circ \chi_1 \subset T^*\mathbb{R}^n \times T^*\mathbb{R}^n$$

is the "standard" folding canonical relation C_0 associated to Ai:

$$C_0(x,\xi) = \left(x_1 \pm 1/3\left(\frac{\xi_n}{\xi_1}\right)^{3/2}, x_2, \ldots, x_{n-1}, x_n \pm \left(\frac{\xi_n}{\xi_1}\right)^{1/2}, \xi\right). \tag{3.5}$$

First we introduce some geometric objects associated with Λ', in addition to the involutions J_j discussed above. We also have "boundary maps"

$$\delta_1^{\pm} = \pi_1 \circ J_2 \circ \pi_1^{-1}, \quad \delta_2^{\pm} = \pi_2 \circ J_1 \circ \pi_2^{-1} \tag{3.6}$$

where $\pm$ depends on the choice of continuous inverse of π_1. The domain and range of $\delta_j^{\pm}$ is the image under π_j of Λ' in T^*X_j. These boundary maps have the same properties as the billiard ball maps discussed in section 1. Indeed, in applications we will see later, $X_1 = \mathbb{R}\times\partial K$ and $\delta_1^{\pm}$ will be the billiard ball map. Furthermore, in the special case when $\Lambda' = C_0$, $\delta_1^{\pm} = \delta_2^{\pm} = \delta_0^{\pm}$, the billiard ball map for the canonical example discussed in section 1. There is a simple formula for $\delta_0^{\pm}$:

$$\delta_0^{\pm}(x,\xi) = \left(x_1 \pm 2/3\left(\frac{\xi_n}{\xi_1}\right)^{3/2}, x_2, \ldots, x_{n-1}, x_n \pm 2\left(\frac{\xi_n}{\xi_1}\right)^{1/2}, \xi\right). \tag{3.7}$$

In section 1 we showed that, if $X_1 = \mathbb{R}\times\partial K$ and $\delta_1^{\pm}$ is the billiard ball map, there is a canonical transformation $\chi_1 = \chi_J$ which conjugates $\delta_1^{\pm}$ to $\delta_0^{\pm}$. This holds generally. In fact, proposition 7.14 of [14] says there exist homogeneous symplectic coordinates (x,ξ) on T^*X_1 with $\xi_n \geq 0$ on $\pi_1(\Lambda')$, such that in these coordinates $\delta_1^{\pm}$ takes the form (3.7). We are now ready to state the main geometrical result.

<u>Proposition 3.2</u>. If $\Lambda' \subset T^*X_1 \times T^*X_2$ is a folding canonical relation, and if $\chi_1 : T^*\mathbb{R}^n \to T^*X_1$ conjugates $\delta_1^{\pm}$ to $\delta_0^{\pm}$, then there exists a canonical transformation $\chi_2 : T^*\mathbb{R}^n \to T^*X_2$ such that

$$\chi_2^{-1} \circ \Lambda' \circ \chi_1 = C_0 \tag{3.8}$$

<u>Proof</u>. Replacing Λ' by $\Lambda' \circ \chi_1$, we can suppose $\delta_1^{\pm} = \delta_0^{\pm}$. We look for χ_2 such that $\chi_2^{-1} \circ \Lambda' = C_0$. First note that there exist natural maps

$$\chi^{\pm} : \Lambda' \to C_0 \tag{3.9}$$

defined as follows. For $p \in \pi_1(\Lambda') = \pi_1(C_0) \subset T^*(\mathbb{R}^n)$, there are two points $q_1(p), q_2(p) \in \Lambda'$ and two points $r_1(p), r_2(p) \in C_0$ mapped to p by π_1, these two points degenerating to one for $p \in \{\xi_n = 0\}$, the image of the fold sets. We can suppose that $q_1(p)$ (resp. $r_1(p)$) belongs to one selected component of the complement of the fold set in Λ' (resp. in C_0), and that $q_2(p)$ (resp. $r_2(p)$) belongs to the other. Then $\chi^{\pm}$ is defined by $\chi^+(q_j(p)) = r_j(p)$, $\chi^-(q_j(p)) = r_{j'}(p)$, where $1' = 2$, $2' = 1$. It is not hard to see $\chi^{\pm}$ are C^∞ and presume the "folded symplectic forms" on Λ' and C_0, which are the pull backs by π_1^* of the symplectic form on $T^*(\mathbb{R}^n)$. Note that, since $\chi^{\pm}$ each conjugate $\delta_1^{\pm}$ to $\delta_0^{\pm}$, these maps take the involution J_2 on Λ' to the analogous involution J_2 on C_0.

We are ready to define χ_2. First we define χ_2^{-1} on the image $\pi_2(\Lambda')$ in T^*X_2, as follows. Let $p \in \pi_2(\Lambda') \subset T^*X_2$. Let $\pi_2(p_1) = \pi_2(p_2) = p$, $p_j \in \Lambda'$, let $\tilde{p}_j = \chi^+(p_j)$. We claim that

$$\pi_2(\tilde{p}_1) = \pi_2(\tilde{p}_2) \in T^*(\mathbb{R}^n). \tag{3.10}$$

Indeed, (3.10) holds if and only if J_2 interchanges $\tilde{p}_1$ and $\tilde{p}_2$. But by the same token J_2 does interchange p_1 and p_2, and since χ^+ conjugates one J_2 to the other, we have (3.10). So set

$$\chi_2^{-1}(p) = \pi_2(\tilde{p}_1) = \pi_2(\tilde{p}_2). \tag{3.11}$$

From the structure of π_2 as a fold, it follows from (3.11) that χ_2^{-1} is C^∞ on the region with boundary $\pi_1(\Lambda')$. So there exists a smooth extension to a neighborhood of the boundary. Pick any one, to define χ_2^{-1}. This completes the proof.

Propositions 3.1 and 3.2 together easily give the following main result.

Theorem 3.3. If $A \in I^m(X_1, X_2, \Lambda')$ with Λ' a folding canonical relation, then there exist elliptic FIOPS J and K, corresponding to the canonical transformations χ_2 and χ_1 of proposition 3.2, such that

$$A = J(P_1 \mathcal{A}i + P_2 \mathcal{A}i')K \tag{3.12}$$

for some $P_1 \in OPS^{m+1/6}$, $P_2 \in OPS^{m-1/6}$. Furthermore one can fix the canonical transformation associated with K (alternatively, with J) to be any one which conjugates the appropriate boundary maps to the standard form $\delta_0^{\pm}$.

One simple corollary of theorem 3.3 gives the sharp order of continuity of these FIOPS on Sobolev spaces.

Corollary 3.4. If $A \in I^m(X_1, X_2, \Lambda')$ as in Theorem 3.3, then

$$A: H^s(X_1) \to H^{s-m-1/6}(X_2) \quad \text{for all } s.$$

Furthermore, $A: H^s(X_1) \to H^{s-m}(X_2)$, if and only if $\sigma_A\big|_L = 0$, where $L \subset \Lambda'$ is the fold set.

Proof. This follows from the representation (3.12) by standard continuity results for the FIOPs J and K, the pseudo-differential operators P_1 and P_2, and the Fourier multipliers $\mathcal{A}i$ and $\mathcal{A}i'$.

Finally, we analyze A^*PA, given $P \in OPS^\mu$. By theorem 3.3 we have

$$A^*PA = K^*(\mathcal{A}iP_1^* + \mathcal{A}i'P_2^*)J^*PJ(P_1\mathcal{A}i + P_2\mathcal{A}i')K$$

with $P_1, P_1^* \in OPS^{m+1/6}$, $P_2, P_2^* \in OPS^{m-1/6}$. By Egorov's theorem, $J^*PJ = \tilde{P} \in OPS^\mu$. By proposition 3.1, all the pseudo-differential operators can be pushed to the left of $\mathcal{A}i$ and $\mathcal{A}i'$, and we get

$$A^*PA = K^*(P_{11}\mathcal{A}i^2 + P_{12}\mathcal{A}i\mathcal{A}i' + P_{22}(\mathcal{A}i')^2)K \tag{3.13}$$

with $P_{11} \in OPS^{\mu+2m+1/3}$, $P_{12} \in OPS^{\mu+2m}$, $P_{22} \in OPS^{\mu+2m-1/3}$. This puts A^*PA in a standard form. Note that $WF(A^*PAu) \subset \mathcal{C}\circ WF(u)$ where the "canonical relation" $\mathcal{C}$ is the union of two Lagrangian manifolds

$$\mathcal{C} = \tilde{\Lambda} \cup \Delta^+ .$$

$\tilde{\Lambda}$ is a folding canonical relation and Δ^+ is a Lagrangian manifold with boundary (a subset of the diagonal) transversally intersecting $\tilde{\Lambda}$, the intersection coinciding with the fold set for $\tilde{\Lambda}$.

§4. THE SCATTERING OPERATOR.

The scattering operator gives information on the behavior at infinity of solutions to the wave equation. It is related to the scattering amplitude $a_s(\theta,\omega,\lambda)$ which gives the large x behavior of the "outgoing" solution to the boundary value problem for $u_s = u_s(\lambda,x,\omega)$:

$$(\Delta+\lambda^2)u_s = 0 \text{ on } \mathbb{R}^3-K, \quad u_s\Big|_{\partial K} = e^{-i\lambda\chi\cdot\omega}, \tag{4.1}$$

namely

$$a_s(\theta,\omega,\lambda) = \lim_{r\to\infty} re^{-i\lambda r} u_s(\lambda, r\theta, \omega). \tag{4.2}$$

The scattering operator is the operator with kernel $\hat{a}_s(\theta,\omega,s-t)$, where

$$\hat{a}_s(\theta,\omega,t) = \int_{-\infty}^{\infty} a_s(\theta,\omega,\lambda)e^{-i\lambda t}\, d\lambda. \tag{4.3}$$

Since the outgoing solution to (4.1) can be written

$$u_s(x) = \int_{\partial K} \left[u_s(y)\,\frac{\partial}{\partial\nu}\,G_\lambda(x-y) - \frac{\partial u_s}{\partial\nu}\,G_\lambda(x-y)\right] dS(y) \tag{4.4}$$

where $G_\lambda(x) = e^{i\lambda|x|}/|x|$, applying (4.2) gives

$$a_s(\theta,\omega,\lambda) = \int_{\partial K} e^{-i\lambda\theta\cdot y}\left[i\lambda(\nu\cdot\theta)u_s(\lambda,y,\omega) + \frac{\partial}{\partial\nu}\,u_s(\lambda,y,\omega)\right] dS(y). \tag{4.5}$$

From (4.3) we get a formula for the kernel of the scattering operator:

$$\hat{a}_s(\theta,\omega,t) = \int_{\partial K} \left(\frac{\partial}{\partial\nu} - (\nu\cdot\theta)\,\frac{\partial}{\partial t}\right) w(t+y\cdot\theta,y,\omega)\,dS(y) \tag{4.6}$$

where w solves the boundary value problem

$$\left(\frac{\partial^2}{\partial t^2} - \Delta\right) w = 0, \quad w\Big|_{\mathbb{R}\times\partial K} = \delta(t-y\cdot\omega), \quad w = 0 \quad \text{for} \quad t \ll 0.$$

We can write (4.6) as

$$S = T(N+P)U \tag{4.7}$$

where N is the Neumann operator of section 2, $P = -(\nu\cdot\theta)\frac{\partial}{\partial t} \in OPS^1$, and the operators T and U are defined as follows. $T: \mathcal{E}'(\mathbb{R}\times\partial K) \to \mathcal{E}'(\mathbb{R}\times S^2)$ is given by

$$TF(t,\theta) = \int_{\partial K} F(t+y\cdot\theta,y)\,dS(y) \tag{4.8}$$

and $U: \mathcal{E}'(\mathbb{R}\times S^2) \to \mathcal{E}'(\mathbb{R}\times\partial K)$ is given by

$$Uf(t,y) = \int_{S^2} f(t-y\cdot\omega,\omega)\,d\omega. \tag{4.9}$$

We first point out some basic geometric properties of the operators T and U, so that the results of section 3 will be seen to apply. We assume $K \subset \mathbb{R}^3$ is convex, with smooth boundary of strictly positive curvature.

Proposition 4.1. T and U are Fourier integral operators of order -1 with folding canonical relations, which are inverse to each other. The boundary maps $\delta^{\pm}$ on $T^*(\mathbb{R}\times\partial K)$ are the billiard ball maps.

Proof. Direct consequence of formula (4.8) and (4.9).

Consequently we have

$$U = J(P_1 Ai + P_2 Ai')K \quad \text{and} \quad T = K^{-1}(\tilde{P}_1 Ai + \tilde{P}_2 Ai')J^{-1}$$

for certain $P_1, \tilde{P}_1 \in OPS^{-1+1/6}$, $P_2, \tilde{P}_2 \in OPS^{-1-1/6}$. Here J is the same elliptic FIOP which puts the Neumann operator in standard form; see (2.5). Thus (4.7) yields

$$S = K^{-1}(\tilde{P}_1 Ai + \tilde{P}_2 Ai')(AQ+B+J^{-1}PJ)(P_1 Ai + P_2 Ai')K \tag{4.11}$$

Collapsing terms gives

$$S = K^{-1}(P_{00} Ai^2 + P_{01} Ai Ai' + P_{11}(Ai')^2)K \tag{4.12}$$

where $P_{00} \in OPN_{1/3}^{-1+1/3}$ has the form

$$P_{00} \sim P_{00}^{c} + \sum_{\alpha\geq 0} P_{00\alpha} Q^{(\alpha)} \tag{4.13}$$

with $P_{00}^{c}, P_{00\alpha} \in OPS_{1}^{-1+1/3}$, $Q^{(\alpha)}$ having symbol $D_{\xi,\eta}^{\alpha}\sigma_Q$, and similarly $P_{01} \in OPN_{1/3}^{-1}$, $P_{11} \in OPN_{1/3}^{-1-1/3}$, with asymptotic expansions similar to (4.13).

We next investigate the fact that the wave front relation of S is smaller than that of general operators whose form is given by the right side of (4.12). Indeed, in the shadow region one has $(N+(\nu\cdot\omega)\partial/\partial t)\delta(t-y\cdot\omega) \in C^{\infty}$. Meanwhile, Green's theorem implies

$$\int_{\partial K} \nu\cdot(\omega+\theta)\delta'(t-x\cdot(\omega-\theta))dS(x) = 0 ,$$

so

$$S\delta_{(0,\omega)} = T(N + (\nu\cdot\omega)\frac{\partial}{\partial t})\delta(t-y\cdot\omega).$$

Thus $S\delta_{(0,\omega)}$ is singular at $t = \min_{y\in\partial K} y\cdot(\theta-\omega)$, but the singularity at $t = \max_{y\in\partial K} y\cdot(\theta-\omega)$, which would occur for most distributions of the form (4.12) is absent.

In fact, we can get an alternate formula for $S = T(N+P)U$ whose form more closely reflects this restriction on the wave front relation of S, as follows. We have

$$S\delta_{(0,\omega)} = T(N+(\nu\cdot\omega)\frac{\partial}{\partial t})U\delta_{(0,\omega)} \tag{4.14}$$

and, for certain $P_1^{\#} \in OPS^{-1+1/6}$, $P_2^{\#} \in OPS^{-1-1/6}$, we have

$$A^{-1}J^{-1}(N+(\nu\cdot\omega)\frac{\partial}{\partial t})UK^{-1} = Q(P_1 \mathcal{A}i + P_2 \mathcal{A}i') + P_1^{\#}\mathcal{A}i + P_2^{\#}\mathcal{A}i'. \tag{4.15}$$

Rewrite the right side of (4.15) as

$$Q(\mathcal{A}i\tilde{P}_1 + \mathcal{A}i'\tilde{P}_2) + \mathcal{A}i\tilde{P}_3 + \mathcal{A}i'\tilde{P}_4. \tag{4.16}$$

Using the Wronskian relation

$$Ai' = \frac{\alpha}{A_-} + \frac{A_-'}{A_-} Ai, \quad \alpha \neq 0 ,$$

write (4.16) as

$$\mathcal{A}i(Q^2\tilde{\tilde{P}}_2 + Q(\tilde{P}_1 + \tilde{\tilde{P}}_4) + P_3) + \alpha A_-^{-1}(Q\tilde{P}_2 + \tilde{P}_4). \tag{4.17}$$

We now use the following result.

<u>Lemma 4.2</u>. Suppose $A,B,C \in OPS^0$ and

$$T = AQ^2 + BQ + C \in OPS^{-\infty} \text{ on } \{\eta > 0\}. \tag{4.18}$$

Then the terms in the asymptotic expansion of the symbols of $A,B,$ and C all vanish to infinite order at $\eta = 0$ and

$$(AQ^2 + BQ + C)\mathcal{A}i \in OPS^{-\infty}. \tag{4.19}$$

<u>Proof</u>. Note that the symbols of Q and Q^2 have, respectively, the asymptotic expansions

$$q \sim \xi_1^{-1/3}(\beta_0(\xi_1^{-1/3}\eta)^{1/2} + \beta_1(\xi_1^{-1/3}\eta)^{-1} + \beta_2(\xi_1^{-1/3}\eta)^{-5/2} + \dots)$$

$$q^2 \sim \xi_1^{-2/3}(\gamma_0\xi_1^{-1/3}\eta + \gamma_1(\xi_1^{-1/3}\eta)^{-1/2} + \gamma_2(\xi_1^{-1/3}\eta)^{-2} + \dots)$$

Let the symbol of A be asymptotic to ΣA_j, etc. Then the part homogeneous of degree $-j$ in the expansion of T, in $\eta > 0$, is

$$T_j = A_j + \beta_0\xi_1^{-1/2}\eta^{1/2}B_j + \gamma_0\xi_1^{-1}\eta C_j$$

$$+ \beta_1 \eta^{-1} B_{j-1} + \gamma_1 \xi_1^{-1/2} \eta^{-1/2} C_{j-1}$$

$$+ \ldots$$

$$+ \beta_j \zeta_1^{-1/2+j/2} \eta^{1/2-3/2 j} B_0 + \gamma_j \xi_1^{-1+j/2} \eta^{1-3/2 j} C_0 = 0$$

Separating terms into integer or non-integer powers of η, we get a pair of equations holding to infinite order at $\eta = 0$, for each j. For each k, we get $2j$ equations in the $3k$ unknowns $\eta^\ell A_\ell, \eta^\ell B_\ell, \eta^\ell C_\ell$, mod $O(\eta^k)$ $(0 \le \ell \le k-1)$ and if j is picked so $j > k$, one has uniqueness: $\eta^\ell A_\ell = \eta^\ell B_\ell = \eta^\ell C_\ell = 0 \bmod O(\eta^k)$, $0 \le \ell < k$. Taking k arbitrarily large gives A_j, B_j, C_j all vanishing to infinite order at $\eta = 0$. From this fact, (4.19) is a simple consequence.

To see how the lemma applies to (4.17), note that the first term must have wave front relation contained in that of A_-^{-1}, and since Ai is a sum of two elliptic FIOPs in the open cone $\eta > 0$, this implies that $Q^2\tilde{\tilde{P}}_2 + Q(\tilde{P}_1 + \tilde{\tilde{P}}_4) + P_3$ belongs to $OPS^{-\infty}$ on $\eta > 0$. Taking adjoints, we can apply the lemma, and taking adjoints back implies

$$Ai(Q^2\tilde{\tilde{P}}_2 + Q(\tilde{P}_1 + \tilde{\tilde{P}}_4) + P_3) \in OPS^{-\infty} . \tag{4.20}$$

We also have all the terms in the asymptotic expansion of $\tilde{\tilde{P}}_2$, and hence of $\tilde{P}_2$, vanishing to infinite order at $\eta = 0$, which gives

$$\alpha(Q\tilde{P}_2 + \tilde{P}_4) = P_5 \in OPS^{-1-1/6} .$$

Thus, (4.15) gives

$$(N+(\nu\cdot\omega)\frac{\partial}{\partial t})U = J\ A\ A_-^{-1}\ P_5\ K \quad . \tag{4.21}$$

Consequently, using the representation

$$T = K^{-1}(\tilde{P}_1 Ai + \tilde{P}_2 Ai')J^{-1}$$

we get

$$S = K^{-1}(\tilde{P}_1 Ai + \tilde{P}_2 Ai')A\ A_-^{-1}\ P_5\ K$$

or

$$S = K^{-1}(P_1^b Ai + P_2^b Ai')A_-^{-1}\ P_5 K. \tag{4.22}$$

We remark that P_5 is elliptic. [This follows from the ellipticity of $\tilde{P}_4$, or of $\tilde{\tilde{P}}_4$, which in turn follows from the ellipticity of $\tilde{P}_1$, hence of P_1.] Thus one could replace P_5K by K in (4.22) and effectively absorb the P_5 term. We summarize as follows.

Theorem 4.3. The scattering operator has the form

$$S = K^{-1}(P_1^{\#}Ai + P_2^{\#}Ai')A_-^{-1}\ K$$

where K is an elliptic FIOP of order 0, $P_1^{\#} \in OPS^{-1}$, $P_2^{\#} \in OPS^{-1-1/3}$.

For further details, and results on the scattering amplitude, we refer to [17].

§5. THE CORRECTED KIRCHOFF APPROXIMATION.

As one can see from section 4, it is desirable to have a good hold on the normal derivative $\partial u_s/\partial\nu$ of the solution to the boundary value problem

$$(\Delta+\lambda^2)u_s = 0 \quad \text{on} \quad \mathbb{R}^3-K, \quad u_s\Big|_{\partial K} = e^{-i\lambda x\cdot\omega} \tag{5.1}$$

satisfying the outgoing condition

$$u_s = O(|x|^{-1}),\ \frac{\partial u_s}{\partial r} - i\lambda u_s = o(|x|^{-1}) \quad \text{as} \quad |x| \to \infty. \tag{5.2}$$

This is a very classical problem, and what has been used for many years in calculations in non-rigorous scattering theory is the Kirchoff approximation:

$$\frac{\partial u_s}{\partial\nu}\Bigg|_{\partial K} \approx i\lambda|\nu\cdot\omega|e^{-i\lambda x\cdot\omega}\ . \tag{5.3}$$

This approximation was given by G. Kirchoff in [8] in an effort to cast light on the Fresnel theory of diffraction, and was motivated by the idea that the scattered field, for large λ, is approximately obtained, at a point $x\in\partial K$, where $\nu\cdot\omega > 0$ (the "illuminated region") by replacing ∂K by its tangent plane at x and solving the wave equation exactly, and at a point $x\in\partial K$ where $\nu\cdot\omega < 0$ (the "shadow region") by the consideration that the total field should be essentially zero.

Rigorous affirmation of (5.3), for K smooth and of positive curvature, was first given in [22], where it was shown that, with

$$\frac{\partial u_s}{\partial\nu}\Bigg|_{\partial K} = K(x,\lambda,\omega)e^{-i\lambda x\cdot\omega} \tag{5.4}$$

there is the estimate

$$\left| K(x,\lambda,\omega) - i\lambda|\nu\cdot\omega| \right| \le C\, \lambda^{3/4+\varepsilon} \quad . \tag{5.5}$$

This estimate used certain L^p inequalities. An analogous estimate for the validity of the Kirchoff approximation for the Neumann boundary condition was given by Yingst [24].

Here we sketch work in [17], obtaining a complete asymptotic expansion for the coefficient K in (5.4), giving a corrected form of the Kirchoff approximation. A trivial by product is a sharpening of the estimate (5.5) to

$$\left| K(x,\lambda,\omega) - i\lambda|\nu\cdot\omega| \right| \le C\, \lambda^{2/3}. \tag{5.6}$$

But the most interesting aspect of the result is of course the analysis of the nature of the transition of the normal derivative of the scattered wave across the shadow boundary.

We apply our study of the Neumann operator. First, a simple calculation gives the formula:

$$N\left(e^{-i\lambda(x\cdot\omega-t)}\right) = e^{i\lambda t} \left.\frac{\partial u_s}{\partial\nu}\right|_{\partial K} . \tag{5.7}$$

So, with $\psi = x\cdot\omega - t\big|_{\mathbb{R}\times\partial K}$, we have

$$\frac{\partial u_s}{\partial\nu} = e^{-i\lambda t}\, N\left(e^{-i\lambda\psi}\right) \tag{5.8}$$

We are led to applying the Neumann operator N to the oscillatory term $e^{-i\lambda\psi}$. Because the Neumann operator is not a pseudo-differential operator of classical type, the main technical problem is to figure out how to do this. The first key result, about the geometrical relation between the phase function ψ and the operator J, proved by Melrose [16], is:

<u>Lemma 5.1</u>. The Neumann operator can be written in the form (2.5) with J so chosen that

$$J^{-1}\left(e^{-i\lambda\psi}\right) = a(y,x,\lambda)e^{-i\lambda\tilde{\psi}} \tag{5.9}$$

where $a \in S^0$ and

$$\tilde{\psi}(x,y) = x_1 + y^3/3 \tag{5.10}$$

Here x are the variables to which ξ are dual, with η dual to y.

The next step is to examine

$$\frac{A'}{A}\left(e^{-i\lambda\tilde{\psi}}\right).$$

A calculation gives

$$\frac{A'}{A}\left(e^{-i\lambda\tilde{\psi}}\right) = e^{-i\lambda x_1}\ \Phi(\lambda^{1/3}y) \tag{5.11}$$

where Φ is given by an integral of Fock type:

$$\Phi(\tau) = \int \frac{A'}{A}(s)\ Ai(s)\ e^{-i\tau s}ds = \frac{A'}{A}(D_\tau)(e^{i\tau^3/3}). \tag{5.12}$$

The asymptotic expansion of $\Phi(\tau)$ is given as follows. Set

$$r(s) = \frac{A'}{A}(s). \tag{5.13}$$

Lemma 5.2. $\Phi(\tau) = e^{i\tau^3/3}\ \Psi(\tau)$ with $\Psi \in S^1(\mathbb{R})$; indeed

$$\Psi(\tau) \sim \sum_{k\geq 0} \frac{a^{(3k)}(0)}{(3k)!}\ r^{(3k)}(\tau^2) \tag{5.14}$$

where $a(\sigma) = e^{-i\sigma^3/3}$.

Proof. We can rewrite the far right side of (5.12) as

$$e^{i\tau^3/3}\ r(D_s)\left(e^{-is^3/3}e^{i\phi(s,\tau^2)}\right)\Big|_{s=0}$$

with $\phi(s,\lambda) = \lambda s - \lambda^{1/2}s^2 \in S^1_{1,0}$ and show that, mod $O(\lambda^{-\infty})$, all the contribution comes from a small neighborhood of $s = 0$. The fundamental asymptotic expansion lemma applies to the pseudo-differential operator $r(D_s)$ acting on its oscillatory argument, to give (5.14).

Note in particular that

$$\Psi(\tau) = \frac{A'}{A}(\tau^2) + \rho(\tau);\quad \rho(\tau) \in S^{-5}(\mathbb{R}). \tag{5.15}$$

Consequently, we have so far that

$$\frac{A'}{A}\left(e^{-i\lambda\tilde{\psi}}\right) = \Psi(\lambda^{1/3}y)e^{-i\lambda\tilde{\psi}}. \tag{5.16}$$

In a similar fashion, one establishes that

$$(AQ+B)(J^{-1}e^{-i\lambda\psi}) = \tilde{b}(x,y,\lambda)e^{-i\lambda\tilde{\psi}} \tag{5.17}$$

with

$$\tilde{b} \sim \sum_{j,k,\ell\geq 0} \sigma_{jk\ell}(x,y,\lambda)\, \Psi^{(\ell)}_{jk}(\lambda^{1/3}y) + \sum_{j\geq 0} \tau_j(x,y,\lambda) \tag{5.18}$$

where $\sigma_{jk\ell} \in S^{2/3-j/3-k-2\ell/3}$ and $\tau_j \in S^{-j}$. Here Ψ_{jk} is defined by

$$\Psi_{jk}(\tau) = e^{-i\tau^{3}/3}\, r_{jk}(D_\tau)(e^{i\tau^{3}/3})$$

$$\sim \sum_{\ell\geq 0} \frac{a^{(3\ell)}(0)}{(3\ell)!}\, r^{(3\ell)}_{jk}(\tau^2) \in S^{1-2j}(\mathbb{R}) \tag{5.19}$$

where

$$r_{jk}(\lambda) = \lambda^k r^{(j+k)}(\lambda). \tag{5.20}$$

It remains to apply the Fourier integral operator J to the right side of (5.17). Note that $\tilde{b} \in S^1_{2/3,1/3}$. Since $2/3 > 1/2$, classical methods apply to the asymptotic expansion of $J(\tilde{b}e^{-i\lambda\tilde{\psi}})$, and one achieves the desired result:

Theorem 5.3. The corrected Kirchoff formula is

$$\left.\frac{\partial u_s}{\partial \nu}\right|_{\partial K} = K(x,\lambda,\omega)e^{-i\lambda x\cdot\omega} \tag{5.21}$$

where $K \in S^1_{2/3,1/3}$ has the expansion

$$K(x,\lambda,\omega) \sim \sum_{j,k,\ell\geq 0} \kappa_{jk\ell}(x,\lambda,\omega)\Psi^{(\ell)}_{jk}(\lambda^{1/3}Z) + K^c(x,\lambda,\omega) \tag{5.22}$$

with $K^c \in S^0$ and $\kappa_{jk\ell} \in S^{2/3-j/3-k-2\ell/3+a(\ell)}$ where $a(0) = a(1) = 0$ and $a(\ell) = 1$ for $\ell \geq 2$. Furthermore,

$$Z \text{ vanishes to first order on } \{\nu\cdot\omega=0\}. \tag{5.23}$$

We record all the terms of order greater than zero. We have

$$K = -\frac{\nu\cdot\omega}{Z}\lambda^{2/3}\left[\frac{A'}{A}(\lambda^{2/3}Z^2) + \rho(\lambda^{1/3}Z)\right] \tag{5.24}$$
$$+ \kappa_{100}\Psi_{10}(\lambda^{1/3}Z) + \kappa_{002}\Psi^{(2)}_{00}(\lambda^{1/3}Z) + S^0_{2/3,1/3}$$

where ρ is given by (5.15), Ψ_{10} and Ψ_{00} by (5.19), and κ_{100}, $\kappa_{002} \in S^{1/3}$. Note that, from (5.24), it is apparent that $\lambda^{2/3}$ is the best possible power of λ that could go on the right side in (5.6).

§6. A REPRESENTATION FOR THE WAVE EVOLUTION OPERATOR.

This section discusses some results of Farris [2], [3], to which we refer for further details. We want to look at the structure of the solution operator $e^{iT\sqrt{-\Delta}}$ at a fixed time T. Δ is defined on $\mathbb{R}^n - K$ with Dirichlet boundary condition on ∂K, or more generally Δ could be defined on any complete Riemannian manifold M with compact diffractive boundary. Let $\mathcal{O}$ be bounded away from ∂M, and suppose T is picked so that, if we consider all the geodesics issuing from $\overline{\mathcal{O}}$, reflecting off ∂M by the usual rules of geometrical optics, the set U of endpoints of distance T from their origins, avoids ∂M. It follows that, for $u \in \mathcal{E}'(\mathcal{O})$, $e^{iT\sqrt{-\Delta}} u$ is C^∞ near ∂M. The goal here is to show that this operator is of the form (mod C^∞)

$$e^{iT\sqrt{-\Delta}} u = K_2 \frac{A_+}{A_-} K_1 u, \quad u \in \mathcal{E}'(\mathcal{O}) \tag{6.1}$$

where K_1, K_2 are elliptic FIOPs (depending on T) and the operator A_+/A_- is Fourier multiplication:

$$(A_+/A_- F)^\wedge(\xi,\eta) = \frac{A_+}{A_-}(-\xi_1^{-1/3}\eta)\hat{F}(\xi,\eta). \tag{6.2}$$

To start, let Δ_0 be the free space Laplacian, on $\mathbb{R}^n$, or more generally on some complete boundaryless manifold $\tilde{M}$ containing M. Let $F = e^{iT\sqrt{-\Delta_0}}$ and let $R: \mathcal{E}'(\mathcal{O}) \to \mathcal{D}'(\mathbb{R}\times\partial M)$ be given by

$$Ru = e^{it\sqrt{-\Delta_0}} u\Big|_{\mathbb{R}\times\partial M}. \tag{6.3}$$

Also let $E: \mathcal{E}'(\mathbb{R}\times\partial M) \to \mathcal{D}'(M)$ be defined as follows: Ef is the value at $t = T$ of the outgoing solution w to the wave equation on $\mathbb{R}\times M$, ($w = 0$ for $t \ll 0$) with boundary condition $w\big|_{\mathbb{R}\times\partial M} = f$. Then ER is well defined, and

$$e^{iT\sqrt{-\Delta}} = F - ER. \tag{6.4}$$

The map E is gotten by taking (1.5) (with $F = J^{-1}f$) and evaluating at $t = T$. Note that, if WF(F) is in a small conic neighborhood of $\eta = 0$, which we may assume without loss of generality, away from ∂M one can replace $A(\xi_1^{-1/3}\rho)$ and $A'(\xi_1^{-1/3}\rho)$ by their asymptotic expansions, by (1.9), and write

$$E = LA_-^{-1}J^{-1} \tag{6.5}$$

where (using a cutoff $\phi(\xi_1^{-1}\eta)$)

$$LF = \int [gA_-(\xi_1^{-1/3}\rho) + ih\xi_1^{-1/3}A_-'(\xi_1^{-1/3}\rho)]e^{i\theta}\hat{F}\, d\xi d\eta \tag{6.6}$$

the integral being evaluated at $t = T$ and restricted to $x \in U$. L is a classical FIOP, and if we restrict our attention to near the boundary ∂M, L is elliptic.

The map R is a Fourier integral operator with folding canonical relation, and the boundary maps $\delta^{\pm}$ on $T^*(\mathbb{R}\times\partial M)$ are easily seen to coincide with the billiard ball maps. Thus, by section 3, one can write

$$J^{-1}RK = \tilde{P}_1\mathcal{A}i + \tilde{P}_2\mathcal{A}i' \tag{6.7}$$

for some pseudo differential operators $\tilde{P}_j$ and some elliptic Fourier integral operators J and K, and in fact, J can be taken to be the operator (1.11), which enters into the formula for the Neumann operator (2.5). Combining (6.5) and (6.7) gives

$$ER = LA_-^{-1}(\tilde{P}_1\mathcal{A}i + \tilde{P}_2\mathcal{A}i')K^{-1}. \tag{6.8}$$

In fact, we claim K can be taken to be the operator

$$KF = \int [g\overline{A}_-(\xi_1^{-1/3}\rho) + ih\overline{A}_-'(\xi_1^{-1/3}\rho)]e^{i\theta}\hat{F}\, d\xi d\eta, \tag{6.9}$$

the integral being evaluated at $t = 0$. Note that $\overline{A}_- = A_+$ and

$$E^- = KA_+^{-1}J^{-1} \tag{6.10}$$

where E^-f is the value at $t = 0$ of the <u>incoming</u> solution $\tilde{w}$ to the wave equation on $\mathbb{R} \times M$ ($\tilde{w} = 0$ for $t \gg 0$) with boundary condition $\tilde{w}\big|_{\partial M} = f$. A study of the geometry of these operators shows that $J^{-1}RK$ is a FIOP whose folding canonical relation is the standard model C_0. One use of this explicit representation of K is that one can prove the following.

<u>Lemma 6.1</u>. $L^{-1}FK$ and its inverse $K^{-1}F^{-1}L$ are (elliptic) pseudo-differential operators on $\{\eta \geq 0\}$.

Proof. It suffices to prove it for $\{\eta > 0\}$. Use the representation

$$FK = FE^{-}JA_{+}, \quad L = EJA_{-} \quad \text{on} \quad \{\eta > 0\}.$$

Each of these is an elliptic FIOP in this region, and to see that they move wave front sets in the same fashion, it suffices to note that

$$J \frac{A_{-}}{A_{+}} J^{-1} \tag{6.11}$$

has, in $J\{\eta > 0\}$, the canonical transformation equal to the billiard ball map δ_{+}. This is established by the same argument treating $J\,Ai^2 J^{-1}$ in section 2.

Given this lemma, we have, in addition to (6.3),

$$J^{-1}R(F^{-1}L) = P_1 Ai + P_2 Ai'$$

and hence, as a convenient modification of (6.8), we have

$$ER = LA_{-}^{-1}(P_1 Ai + P_2 Ai')L^{-1}F. \tag{6.12}$$

Returning to (6.4), we see that, acting on $\mathcal{E}'(\mathcal{O})$,

$$e^{iT\sqrt{-\Delta}} = L[1-A_{-}^{-1}(P_1 Ai + P_2 Ai')]L^{-1}F. \tag{6.13}$$

In order to simplify the term in brackets, let us note that, by virtue of the known propagation of singularities for the operator $e^{iT\sqrt{-\Delta}}$, the term in brackets must move wave front sets the same way A_{+}/A_{-} does. Using

$$Ai = -\bar{\omega}A_{-} - \omega A_{+}, \quad \omega = e^{2/3\pi i}$$

one gets

$$1-A_{-}^{-1}(P_1 Ai + P_2 Ai') = \frac{A_{+}}{A_{-}}\,[-\omega^2 - A_{+}^{-1}((P_1+\omega)Ai + P_2 Ai')] \tag{6.14}$$

and using the Wroskian relation

$$A_{-}'Ai - A_{-}Ai' = \alpha, \quad \alpha \neq 0$$

we rewrite the term in brackets on the right side of (6.14) as

$$-\omega^2 - \alpha A_{+}^{-1}P_2 A_{-}^{-1} - A_{+}^{-1}(P_1+\omega - P_2\frac{A_{-}'}{A_{-}})Ai. \tag{6.15}$$

Now this operation, which in the region $\{\eta > 0\}$ is a classical FIOP, is supposed to preserve wave front sets. In particular, the last term in (6.15) is supposed to preserve wave front sets. This implies that

$$A_+^{-1}(P_1+\omega - P_2\frac{A'_-}{A_-})\ A_- \in OPS^{-\infty} \quad \text{on} \quad \{\eta > 0\}.$$

and hence

$$P_1+\omega - P_2\frac{A'_-}{A_-} \in OPS^{-\infty} \quad \text{on} \quad \{\eta > 0\} \tag{6.16}$$

Since P_1 and P_2 are classical pseudo-differential operators, (6.16) is a very stringent condition. Indeed, one has the following, which is in fact a special case of lemma 4.2.

Lemma 6.2. Let $A, B \in OPS^m$. Suppose

$$A + BQ \in OPS^{-\infty} \quad \text{on} \quad \{\eta > 0\}. \tag{6.17}$$

Then all the terms in the asymptotic expansions of the symbols of A and B must vanish to infinite order at $\eta = o$, and

$$(A+BQ)A_\pm^{-1}, \quad (A+BQ)A\acute{\iota} \in OPS^{-\infty}. \tag{6.18}$$

Proof. Replacing $A'/A\ (-\xi_1^{-1/3}\eta)$ by its asymptotic expansion gives an infinite set of identities, from (6.17), a priori satisfied for $\eta > 0$ but, by continuity, satisfied for $\eta \geq 0$. For the principal symbols one gets

$$a_0 + b_0\sqrt{\eta/\xi_1} = 0$$

which implies a_0 and b_0 vanish to infinite order at $\eta = 0$. Such vanishing of higher order terms follows inductively, and from this, (6.18) is an elementary consequence.

Applying the lemma to (6.16), we conclude that the expression (6.15) is equal to

$$-\omega^2 - \alpha A_+^{-1}P_2A_-^{-1} \quad \text{mod } OPS^{-\infty} \tag{6.19}$$

and that

$$P_2 \text{ vanishes to infinite order at } \eta = 0. \tag{6.20}$$

From (6.20) one can adapt a proof of Egorov's theorem to get

$$\alpha A_+^{-1}P_2A_-^{-1} \;=\; P_3 \in OPS^0. \tag{6.21}$$

Putting together (6.13) - (6.15) with the evaluation (6.19) and (6.21), we have the main result:

Theorem 3.3. Acting on $\mathcal{E}'(\mathcal{O})$, mod $OPS^{-\infty}$, one has

$$e^{iT\sqrt{-\Delta}} = L \frac{A_+}{A_-} (-\omega^2 - P_3)\ L^{-1} F. \tag{6.22}$$

This representation allows one to analyze

$$e^{-iT\sqrt{-\Delta}}\ P\ e^{iT\sqrt{-\Delta}}$$

given $P \in OPS^m(M)$, with symbol supported in U, as a pseudo-differential operator, of non-classical type, with symbol essentially supported in $\mathcal{O}$. This new operator can be regarded as having a "principal symbol" which is continuous, but not smooth, as the cosphere bundle of M, and then Egorov's theorem holds; the two principal symbols are related by the (non smooth) canonical transformation associated with $e^{iT\sqrt{-\Delta}}$. For details, see [3].

§7. FIRST ORDER SYSTEMS OF DIFFERENTIAL EQUATIONS.

We consider here a $k \times k$ first order system of differential equations, of the form

$$G(x, D_x)\ u = 0 \tag{7.1}$$

in the half space $\Omega = R^n_+ = \{x_n > 0\}$, with boundary condition at $x_n = 0$,

$$\beta(x', D_{x'}) u(0, x') = f \tag{7.2}$$

where $x = (x', x_n)$. Here $\beta \in OPS^0$ is a $k \times k'$ matrix. The typical case to keep in mind is a boundary value problem for a hyperbolic system, with the variable x_1 representing time. We suppose $p(x,\xi) = \det G_1(x,\xi)$ has simple characteristics, G_1 denoting the principal symbol of G, and more precisely we suppose

$$p(x,\xi) = 0 \Rightarrow \frac{\partial}{\partial \xi_1}\ p(x,\xi) \neq 0 \quad (\xi \neq 0). \tag{7.3}$$

We also assume the boundary is noncharacteristic for G.

Our goal is to construct a microlocal parametrix for solutions to (7.1), (7.2), satisfying the "outgoing" condition that u is zero for $x_1 << 0$, given $f \in \mathcal{E}'(R^{n-1})$, under the diffractive hypothesis on the boundary with respect to $p(x,\xi)$, which we now give.

We shall suppose that over a point $(x_0', \xi_0') \in T^*(R^{n-1}) = T^*(\partial R^n_+)$ there pass j grazing rays and ℓ non-grazing rays.

There are thus $j+\ell$ points $\zeta_\nu \in T^*_{(x_0',0)}(R^n)$ for which G is characteristic, such that if $\pi: T^*_{(x_0',0)}(R^n) \to T^*_{x_0'}(R^{n-1})$ is the natural projection, then $\pi(\zeta_\nu) = \xi_0'$, $\nu = 1, \ldots, j+\ell$. Through these points pass null bicharacteristic strips γ_ν, the first j grazing and the rest non-grazing. Each grazing ray is assumed to satisfy the following convexity hypothesis:

γ_ν stays inside R^n_+ (so $x_n \geq 0$) and makes only second order contact with ∂R^n_+. (7.4)

The equations for the bicharacteristic strips of p are $\dot{x}_j = \partial/\partial\xi_j\, p$, $-\dot{\xi}_j = \partial/\partial x_j\, p$. In particular $\dot{x}_n = \partial/\partial\xi_n\, p(x,\xi)$, so if a grazing ray passes over $(x_0',0,\zeta_\nu) = (x,\xi)$, we have

$$\frac{\partial}{\partial\xi_n} p(x,\xi) = 0. \tag{7.5}$$

Assume (x,ξ) is characteristic; $p(x,\xi) = 0$. The convexity assumption (7.4) implies $\{p,\{p,x_n\}\} > 0$ at this point. We make a further hypothesis to the effect that the hypersurfaces $\mathrm{Char}\ p = \{p=0\}$ and $T^*_{\partial\Omega}(\Omega)$ have "glancing intersection" in the terminology of [14], namely $\{x_n,\{x_n,p\}\} > 0$ at this point, so

$$\frac{\partial^2}{\partial\xi_n^2} p(x,\xi) \neq 0 \text{ at } (x,\xi) = (x_0',0,\zeta_\nu). \tag{7.6}$$

We remark that, in the special case when $p(x,\xi)$ is a second order polynomial in ξ, (7.6) is automatically satisfied, provided $\partial\Omega$ is non-characteristic. We will deduce from (7.6) a certain "transversality" condition, which was a hypothesis in [20].

The "characteristic variety" is the subset of $T^*(\partial\Omega)\backslash 0$ over which grazing null bicharacteristics pass, so $p(x,\xi) = 0$ and $\frac{\partial}{\partial\xi_n} p(x,\xi) = 0$. Since we assume $\frac{\partial^2}{\partial\xi_n^2} p(x,\xi) \neq 0$, we can locally define roots $\xi_n = a_\nu(x,\xi')$ of $\partial/\partial\xi_n p(x,\xi',a_\nu) = 0$. Then the characteristic variety Σ_ν in $T^*(\partial\Omega)\backslash 0$ is defined by

$$A_\nu(x,\xi') = p(x,\xi',a_\nu(x,\xi')) = 0 \qquad (x=(x',0)). \tag{7.7}$$

Note that $\frac{\partial}{\partial\xi_1} A_\nu(x,\xi') = \frac{\partial}{\partial\xi_1} p(x,\xi',a_\nu(x,\xi')) + \frac{\partial}{\partial\xi_n} p(x,\xi',a_\nu(x,\xi')) \frac{\partial a_\nu}{\partial\xi_1}$, but the latter term vanishes. Now hypothesis (7.3) implies

$$A_\nu(x,\xi') = 0 \Rightarrow \frac{\partial}{\partial\xi_1} A_\nu(x,\xi') \neq 0. \tag{7.8}$$

Thus the characteristic variety in $T^*(\partial\Omega)\backslash 0$ is a union of smooth conic hypersurfaces Σ_ν, and $\Sigma_\nu \cap T^*_{x'}(\partial\Omega)$ is a hypersurface in each fiber. We are assuming that j of these sets intersect at $(x_0', \xi_0') \in T^*(\partial\Omega)$. We make no further assumption about how the Σ_ν intersect.

Let $S_{\alpha,\nu}$ be any smoothly varying one-parameter family of hypersurfaces of R^n, with $S_{0,\nu} = \partial\Omega = R^{n-1}$, α belonging to an interval centered at 0. Suppose the surfaces "move" at non-zero speed with respect to α, i.e., $S_{\alpha,\nu}$ is defined by $f_\nu(\alpha,x)=0$, and on $S_{\alpha,\nu}$, $\frac{\partial}{\partial\alpha} f_\nu(\alpha,x) \neq 0$ and $\nabla_x f_\nu(\alpha,x) \neq 0$. We want to partition the null bicharacteristic strips in Char p near one of the $\gamma_\nu (1 \le \nu \le j)$ up into smooth families $F_{\alpha,\nu}$ of rays which graze $S_{\alpha,\nu}$. Indeed, it follows from (7.8) that, if we define $F_{\alpha,\nu}$ to consist of the rays near (x,ζ_ν) in $\{p=0\}$ grazing $S_{\alpha,\nu}$, then for α small this one parameter family smoothly foliates a conic neighborhood of (x,ζ_ν) in $\{p=0\}$.

We now construct solutions mod C^∞ to (7.1) of the form (for $x_n \ge 0$)

$$u_\nu = \iint \left[\frac{A(\xi_1^{-1/3}\rho_\nu)}{A(-\xi_1^{-1/3}\eta)} g_\nu + ih_\nu \xi_1^{-1/3} \frac{A'(\xi_1^{-1/3}\rho_\nu)}{A(-\xi_1^{-1/3}\eta)}\right] e^{i\theta_\nu} \hat{F}_\nu(\xi,\eta)\, d\xi\, d\eta \quad (7.9)$$

$1 \le \nu \le j$. The phase functions ρ_ν, θ_ν (real valued) and the amplitudes g_ν, h_ν, taking values in C^K, will be constructed as solutions to certain eikonal equations and transport equations; put $g_\nu \sim \sum_{j\ge 0} g_\nu^{(j)}$, with $g_\nu^{(j)}$ homogeneous of degree -j in (ξ,η), and similarly $h_\nu \sim \sum_{j\ge 0} h_\nu^{(j)}$. Applying $G(x,D_x)$ to (7.9) and setting highest order terms zero yields

$$\begin{aligned} G_1(x,\nabla_x\theta_\nu)g_\nu^{(0)} + G_1(x,\tfrac{\rho}{\xi_1}\nabla_x\rho_\nu)h_\nu^{(0)} &= 0 \\ G_1(x,\nabla_x\rho_\nu)g_\nu^{(0)} + G_1(x,\nabla_x\theta_\nu)h_\nu^{(0)} &= 0. \end{aligned} \quad (7.10)$$

If we set

$$\phi_\nu^\pm = \theta_\nu \pm \tfrac{2}{3}\, \xi_1^{-1/2}\, \rho_\nu^{3/2}$$

since $G_1(x,\xi)$ is linear in ξ, (7.10) implies

$$G_1(x, \nabla_x\phi_\nu^\pm)(g_\nu^{(0)} \pm \sqrt{\rho/\xi_1}\, h_\nu^{(0)}) = 0. \quad (7.11)$$

In particular, we get the characteristic equation

$$p(x,\nabla\phi_\nu^\pm) = 0. \quad (7.12)$$

With $\alpha = \eta/\xi_1$, let $S_{\alpha,\nu}$ be a one parameter family of hypersurfaces in R^n as before, $S_{0,\nu} = \partial\Omega$; the $S_{\alpha,\nu}$ may depend also on $\omega = \xi_1^{-1}\xi$ as a parameter, $S_{\alpha,\nu} = S_{\alpha,\omega,\nu}$. Suppose $\theta_{\nu\alpha}(x) = \theta_\nu(x,\xi,\eta)$ has the property that, for all $x \in S_{\alpha,\nu}$, $(x,\nabla\theta_{\nu\alpha}(x)) \in$

$T^*(R^n)$ is a point in $\{p=0\}$, near $(x',0,\zeta_\nu)$, through which passes a ray grazing $S_{\alpha,\nu}$ at x. This condition gives rise to the "surface eikonal equation" on $S_{\alpha,\nu}$, and many such solutions can be constructed; see [22]. Given such $\theta_{\nu\alpha}\big|_{S_{\alpha,\nu}}$, let $\Lambda_{\nu\alpha}$ be the Hamilton flow-out of the graph of $\nabla_x\theta_{\nu\alpha}\big|_{S_{\alpha,\nu}}$; $\Lambda_{\nu\alpha} \subset \{p=0\} \subset T^*(R^n)$ and under $\pi:T^*(R^n) \to R^n$, $\Lambda_{\nu\alpha}$ projects onto the "illuminated side" of $S_{\alpha,\nu}$ as a simple fold. As in the proof of proposition 3.1 in chapter 10 of [22], there is a smooth Φ on $\Lambda_{\nu\alpha}$ such that $\Phi = \theta_{\nu\alpha}$ on the fold, and such that $d\Phi$ is the pull-back to $\Lambda_{\nu\alpha}$ of the canonical 1-form on $T^*(R^n)$, and hence taking even and odd parts with respect to the natural involution j on $\Lambda_{\nu\alpha}$ given by the fold, so if $\Psi = \Phi\circ j$,

$$\theta_{\nu\alpha} = \tfrac{1}{2}(\Phi+\Psi)\circ\pi^{-1}$$

$$\rho_{\nu\alpha} = (\tfrac{3}{4})^{2/3}\,[(\Phi-\Psi)^2\circ\pi^{-1}]^{1/3}$$

define smooth functions on the closure of the illuminated side of $S_{\alpha,\nu}$, such that (7.12) holds. As in the treatment of [22], one can extend these phase functions to the shadow side of $S_{\alpha,\nu}$ such that the characteristic equation is solved to infinite order on $\partial\Omega$, and by choosing $S_{\alpha,\omega,\nu}$ appropriately one can arrange that

$$\rho_\nu(x,\xi,\eta)\Big|_{\partial\Omega} = -\eta. \tag{7.13}$$

We also remark that, for α small, x near $\partial\Omega$,

$$-\frac{\partial}{\partial x_n}\rho_\nu \geq C|\xi| > 0. \tag{7.14}$$

For the next step in solving (7.10), we must construct the amplitudes $g_\nu^{(0)}$ and $h_\nu^{(0)}$. Note that, by the hypothesis that G has simple characteristics, i.e., by (3), $\ker G_1(x,\xi)$ is a smoothly varying family of one-dimensional vector spaces on $\{p=0\}$. Since $\Lambda_{\nu\alpha}$ is imbedded smoothly in $\{p=0\}$, there is a smooth C^k-valued function V_ν on $\Lambda_{\nu\alpha}$, $V_\nu(x,\xi) \in \ker G_1(x,\xi)$. Break up V_ν into its even and odd parts with respect to the involution j on $\Lambda_{\nu\alpha}$, to write (with $\pi^{-1}:R^n \to \Lambda_{\nu\alpha}$, double valued, domain $\pi(\Lambda_{\nu\alpha})$).

$$V_\nu\circ\pi^{-1} = \tilde{g}_\nu \pm \sqrt{\rho/\xi_1}\,\tilde{h}_\nu = R_\pm \tag{7.15}$$

for $\tilde{g}_\nu$, $\tilde{h}_\nu$ smooth on the closure of the illuminated side of $S_{\alpha,\nu}$; convenient smooth extensions across $S_{\alpha,\nu}$ are then chosen. To satisfy (7.11), we must find scalar functions $\sigma_\pm$ such that

$$g_\nu^{(0)} \pm \sqrt{\rho/\xi_1}\,h_\nu^{(0)} = \sigma_\pm(\tilde{g}_\nu \pm \sqrt{\rho/\xi_1}\,\tilde{h}_\nu). \tag{7.16}$$

We will look for $\sigma_\pm$ in the form

$$\sigma_{\pm} = \sigma_0 \pm \sqrt{\rho/\xi_1}\sigma_1 \tag{7.17}$$

so (7.16) becomes

$$g_\nu^{(0)} \pm \sqrt{\rho/\xi_1}\, h_\nu^{(0)} = \sigma_0 \tilde{g}_\nu + \frac{\rho}{\xi_1}\sigma_1 \tilde{h}_\nu \pm \sqrt{\rho/\xi_1}\,(\sigma_1 \tilde{g}_\nu + \sigma_0 \tilde{h}_\nu). \tag{7.18}$$

To obtain the transport equations for σ_0, σ_1, we use the $j=1$ equation from among the continuations of (7.10) to higher transport equations:

$$\begin{aligned} G_1(x,\nabla\theta_\nu)g_\nu^{(j)} + G_1(x,\frac{\rho_\nu}{\xi_1}\nabla\rho_\nu)h_\nu^{(j)} &= -G(g_\nu^{(j-1)}) \\ G_1(x,\nabla\rho_\nu)g_\nu^{(j)} + G_1(x,\nabla\theta_\nu)\, h_\nu^{(j)} &= -G(h_\nu^{(j-1)}). \end{aligned} \tag{7.19}$$

In fact, let $L_{\pm} = \tilde{\tilde{g}}_\nu \pm \sqrt{\rho/\xi_1}\,\tilde{\tilde{h}}_\nu$ be constructed as (7.15), in $\ker G_1^*$. With

$$E_j^{\pm} = g_\nu^{(j)} \pm \sqrt{\rho_\nu/\xi_1}\;\; h_\nu^{(j)} \tag{7.20}$$

one gets from (7.19)

$$G_1(x,\nabla\phi^{\pm})E_j^{\pm} = -G(g_\nu^{(j-1)}) \mp \sqrt{\rho/\xi_1}\, G(h_\nu^{(j-1)}) \tag{7.21}$$

the right side being defined as 0 for $j=0$ (so (7.11) is obtained). Equation (7.21) determines $E_j^{\pm}$ up to an additive multiple of $R_{\pm}$:

$$E_j^{\pm} = k_j^{\pm} + \sigma_{\pm}^{(j)}R_{\pm} \tag{7.22}$$

where one sees that $k_j^{\pm}$ can be written in the form

$$k_j^{\pm} = k_j^{(0)} \pm \sqrt{\rho/\xi_1}\, k_j^{(1)} \tag{7.23}$$

with $k_j^{(m)}$ smooth. The transport equations for $\sigma_{\pm} = \sigma_{\pm}^{(0)}$ and the other $\sigma_{\pm}^{(j)} = \sigma_0^{(j)} \pm \sqrt{\rho/\xi_1}\,\sigma_1^{(j)}$ are obtained by taking the inner product of (7.21) with $L_{\pm}$.

$$L_{\pm}\cdot[G(g_\nu^{(j-1)}) \pm \sqrt{\rho/\xi_1}\, G(h_\nu^{(j-1)})] = 0. \tag{7.24}$$

Replacing $j-1$ by j in (7.24) and rewriting in terms of $E_j^{\pm}$ yields

$$L_{\pm}\cdot[G(E_j^{\pm}) + \frac{\xi_1}{4\rho}G_1(x,\nabla\rho_\nu)(E_j^{+}-E_j^{-})] = 0. \tag{7.25}$$

Using the fact that $L_{\pm}\cdot G_1(x,\nabla\rho_\nu)R_{\mp} = \tfrac{1}{2}\sqrt{\xi_1/\rho}\; L_{\pm}\cdot G_1(\nabla\phi_\nu^{+}-\nabla\phi_\nu^{-})R_{\mp} = 0$, we see that $L_{\pm}\cdot G_1(x,\nabla\rho_\nu)(E_j^{+}-E_j^{-})$

$$\begin{aligned} &= 2\sqrt{\rho/\xi_1}\; L_{\pm}\cdot G_1(x,\nabla\rho_\nu)k_j^{(1)} \pm L_{\pm}\cdot G_1(x,\nabla\rho_\nu)\sigma_{\pm}^{(j)}(R_{\pm}-R_{\mp}) \\ &= 2\sqrt{\rho/\xi_1}\; L_{\pm}\cdot G_1(x,\nabla\rho_\nu)k_j^{(1)} + 2\sqrt{\rho/\xi_1}\;\sigma_{\pm}^{(j)}L_{\pm}\cdot G_1(x,\nabla\rho_\nu)\tilde{h}_\nu. \end{aligned}$$

Consequently (7.25) can be rewritten as

$$\mp \sqrt{\rho/\xi_1}\, L_\pm \cdot G(\sigma_\pm^{(j)} R_\pm) \mp \sqrt{\rho/\xi_1}\, L_\pm \cdot G(k_j) \qquad (7.26)$$

$$+ \tfrac{1}{2} L_\pm \cdot G_1(x, \nabla\rho_\nu) k_j^{(1)} + \tfrac{1}{2}\, \sigma_\pm^{(j)} L_\pm \cdot G_1(x, \nabla\rho_\nu) \tilde{h}_\nu = 0.$$

Now $X_\pm \sigma_\pm = \mp \sqrt{\rho/\xi_1}\, L_\pm \cdot G(\sigma_\pm R_\pm)$ defines a pair of vector fields $X_\pm$ on the illuminated side of $S_{\alpha,\nu}$ which lift to a smooth vector field X on $\Lambda_{\nu\alpha}$, and X is transverse to the fold. Also, $\pm \sqrt{\rho/\xi_1}\, L_\pm \cdot G(k_j^\pm)$, $L_\pm \cdot G_1(x, \nabla\rho_\nu) k_j^{(1)}$, and $L_\pm \cdot G_1(x, \nabla\rho_\nu) \tilde{h}_\nu$ all lift to smooth functions on $\Lambda_{\nu\alpha}$, so the transport equation (7.26) becomes, for $\sigma^{(j)}$ on $\Lambda_{\nu\alpha}$,

$$(X + \psi_j)\sigma^{(j)} = \lambda_j$$

with $\psi_j, \lambda_j \in C^\infty(\Lambda_{\nu\alpha})$. One can solve for smooth $\sigma^{(j)}$, with $\sigma^{(j)}$ prescribed on the fold, and taking even and odd parts as usual yields $\sigma_\pm^{(j)} = \sigma_0^{(j)} \pm \sqrt{\rho/\xi_1}\, \sigma_1^{(j)}$. In particular, $\sigma_0(x,\xi,\eta)$ can be prescribed to be nonzero for $x \in \partial\Omega$, $\eta = 0$. With $\sigma_\pm$ determined, and hence the left side of (7.16) determined, it follows that (7.11) is satisfied, and due to the smoothness of the various quantities and $\frac{\partial}{\partial x_n} \rho_\nu \neq 0$ at $x \in \partial\Omega$, $\eta = 0$, it follows that (7.10) must be satisfied. Similarly one shows the $\sigma_\pm^{(j)}$, $j \geq 1$, lead to solutions of (7.19).

We remark that we can certainly arrange that the term $\tilde{g}_\nu(x,\xi,\eta)$ be nonzero on $\partial\Omega$ at $\eta = 0$ and hence that $g_\nu^{(0)}$ be nonzero there. Depending on G_1, it may or may not happen that $\tilde{h}_\nu$ is linearly independent of $\tilde{g}_\nu$ at that point, and hence, by (7.18), it may or may not be the case that $h_\nu^{(0)}$ is linearly independent of $g_\nu^{(0)}$ at such a point. Whether or not we have such linear independence affects the study of boundary value problems of "Neumann type" which we shall consider at the end of this section.

We now have the amplitude and phase functions for (7.9). The F_ν are scalar valued distributions to be determined so that the boundary condition (7.2) holds, mod C^∞. First, there are two more types of solutions of (7.1) to write down. The first comes from the ℓ families of non-grazing rays hitting $\partial\Omega$ near (x_0', ξ_0'). These give rise to solutions to (7.1) of the form

$$u_\nu = \int a_\nu(x,\xi)\, e^{i\phi_\nu(x,\xi)}\, \hat{F}_\nu(\xi)\, d\xi\,, \quad k+1 \leq \nu \leq k+\ell \qquad (7.27)$$

determined by the usual methods of geometrical optics. The amplitudes $a_\nu(x,\xi)$ take values in C^k and the F_ν are scalar valued distributions. The final term is the "elliptic term", of the form

$$u_- = \int b_-(x,\xi)\, e^{ix'\cdot\xi}\, \hat{F}_-(\xi)\, d\xi \qquad (7.28)$$

where $b_-(x,\xi)$ is an amplitude of Poisson type, i.e.,

$$x_n^{\ell} D_{x_n}^{j} b_-(x,\xi) \text{ bounded in } S_{1,0}^{j-\ell}(R^{n-1}) \text{ for } 0 \le x_n \le 1, \qquad (7.29)$$

and $\hat{F}_-$ takes values in C^μ, μ the dimension of the sum of the generalized eigenspaces of $H(x,\xi')$ with negative real part, for (x,ξ') near $(x_0',0,\xi_0')$, where $G_1(x,\xi) = \xi_n G_n + G(x,\xi')_k$ and $H(x,\xi') = \frac{1}{i} G_n^{-1} G(x,\xi')$. $b_-(x,\xi)$ takes values in $L(C^\mu,C^k)$.

Our goal is to construct a solution mod C^∞ to (7.1), (7.2) which is smooth along half of each grazing ray hitting $\partial\Omega$ (near WF f) and is smooth along certain specified non-grazing rays $\gamma_{j+\lambda+1}, \dots, \gamma_{j+\ell}$. The typical case, when G is hyperbolic with respect to the "time" variable x_1, is that u is demanded to be smooth along those rays going into Ω from $\partial\Omega$ in the <u>negative</u> x_1 direction. In particular, this choice fixes the choice $A=A_+$ or A_- in (7.9); $A_\pm(s) = Ai(e^{\pm 2\pi i/3}s)$. We want to construct such u as a superposition of solutions of the form (7.9) $(1 \le \nu \le j)$, (7.27) $(j+1 \le \nu \le j+\lambda)$ and (7.28). We get a Fourier integral equation for $F = (F_\nu(1 \le \nu \le j+\lambda), F_-)$ as follows.

Restricting (7.9) to $\partial\Omega = \{x_n=0\}$, we get

$$u_\nu\Big|_{\partial\Omega} = \iint \left[g_\nu + i\, h_\nu\, \xi_1^{-1/3} \frac{A'}{A}(-\xi_1^{-1/3}\eta)\right] e^{i\theta_\nu}\, \hat{F}_\nu(\xi,\eta)\, d\xi d\eta \qquad (7.31)$$

$$= J_\nu^{(0)} F_\nu + J_\nu^{(1)} Q F_\nu$$

$$= J_\nu F_\nu$$

where $J_\nu^{(0)}$ and $J_\nu^{(1)}$ are Fourier integral operators and $(QF_\nu)\hat{} = \xi_1^{-1/3}\, A'/A\, (-\xi_1^{-1/3}\eta)\hat{F}_\nu$. Define a Fourier integral operator A_ν by

$$A_\nu f = \iint \hat{f}(\xi,\eta)\, e^{i\theta_\nu}\, d\xi d\eta. \qquad (7.32)$$

Thus A_ν is elliptic. Denote by A_ν^{-1} a microlocal inverse. From (7.27) we can write $u_\nu\big|_{\partial\Omega} = J_\nu F_\nu$, i.e.,

$$u_\nu\Big|_{\partial\Omega} = \int a_\nu(x',0,\xi)\, e^{ix'\cdot\xi} \hat{F}_\nu(\xi)\, d\xi, \quad j+1 \le \nu \le j+\lambda \qquad (7.33)$$

provided we prescribe as initial data for the phase function that $\phi_\nu(x',0,\xi) = x'\cdot\xi$, which is reasonable. Thus $J_\nu \in OPS_{1,0}^{0}$, $j+1 \le \nu \le j+\lambda$. Finally,

$$u_-\Big|_{\partial\Omega} = \int b_-(x',0,\xi)\, e^{ix'\cdot\xi}\hat{F}_-(\xi)\, d\xi = J_- F_- \qquad (7.34)$$

and $J_- \in OPS^0$. Thus the boundary condition (7.2) is

$$\beta\left(\sum_{\nu=1}^{j+\lambda} J_\nu F_\nu + J_- F_-\right) = f \quad \text{mod } C^\infty. \tag{7.35}$$

To pass from (7.35) to a "pseudo-differential" equation, it is convenient to replace F by $\tilde{F} = (\tilde{F}_\nu(1 \le \nu \le j+\lambda), F_-)$ with

$$\tilde{F}_\nu = A_\nu F, \quad 1 \le \nu \le j; \quad \tilde{F}_\nu = F_\nu, \quad j+1 \le \nu \le j+\lambda.$$

Now $J_\nu F_\nu = J_\nu A_\nu^{-1}\tilde{F}_\nu = J_\nu^{(0)} A_\nu^{-1} F_\nu + J_\nu^{(1)} A_\nu^{-1} A_\nu Q A_\nu^{-1} F_\nu = A_\nu^{(0)} + A_\nu^{(1)} A_\nu Q A_\nu^{-1} F_\nu$ where $A_\nu^{(0)}, A_\nu^{(1)} \in OPS^0$. Thus the left side of (7.35) is

$$\beta\left[\sum_{\nu=1}^{j} (A_\nu^{(0)}\tilde{F}_\nu + A_\nu^{(1)} A_\nu Q A_\nu^{-1}\tilde{F}_\nu) + \sum_{\nu=j+1}^{j+\lambda} J_\nu \tilde{F}_\nu + J_- F_-\right]$$

$$= \beta\left(\sum_{\nu=1}^{j} A_\nu^{(0)}\tilde{F}_\nu + \sum_{\nu=j+1}^{j+\lambda} J_\nu \tilde{F}_\nu + J_- F_-\right) + \beta \sum_{\nu=1}^{j} A_\nu^{(1)} A_\nu Q A_\nu^{-1}\tilde{F}_\nu \tag{7.36}$$

$$= \beta_0 \tilde{F} + \sum_{\nu=1}^{j} P_\nu A_\nu Q A_\nu^{-1} \pi_\nu \tilde{F}$$

where $\beta_0 \in OPS^0$ is given by

$$\beta_0(x', D_{x'}) = \beta\left(\sum_{\nu=1}^{j} A_\nu^{(0)} \pi_\nu + \sum_{\nu=j+1}^{j+\lambda} J_\nu \pi_\nu + J_- \pi_-\right) \tag{7.37}$$

and $\pi_\nu \tilde{F}$, $\pi_- \tilde{F}$ are the projections of $\tilde{F}$ to $\tilde{F}_\nu$ or F_-, respectively. $P_\nu = \beta A^{(1)} \in OPS^0$. Our construction yields the following result.

<u>Theorem 7.1</u>. Provided the operator

$$\tilde{\beta}_0 = \beta_0 + \sum_{\nu=1}^{j} P_\nu A_\nu Q A_\nu^{-1} \pi_\nu \tag{7.38}$$

is microlocally invertible near (x_0', ξ_0'), $WF(f)$ contained in a small conic neighborhood of this point, we can find F solving (7.35) and hence (7.9), (7.27) and (7.28) give a parametrix for (7.1), (7.2).

Note that $\tilde{\beta}_0 = \beta_0 + \sum_{\nu=1}^{j} P_\nu A_\nu Q A_\nu^{-1} \pi_\nu$ is a microlocal operator; $WF(\tilde{\beta}_0 \tilde{F}) \subset WF(\tilde{F})$. In fact, it is possible to show that $\tilde{\beta}_0 \in OPS^0_{1/3, 2/3}$, but this fact is not convenient for our present purposes.

The criterion for checking microlocal invertibility of $\tilde{\beta}_0$ in the case of "regular" boundary value problems is the following.

Theorem 7.2. Suppose β_0 is elliptic in a conic neighborhood of (x_0',ξ_0'). Then $\tilde{\beta}_0$ is microlocally invertible near (x_0',ξ_0').

The treatment of Kreiss-type boundary conditions in [20] essentially involved establishing this result. However, an error was made at one point in the argument of [20], as was advertised in [21], p. 605. Namely, it was asserted that in (7.9) one could arrange that $h_\nu = O((|\eta|/|\xi|)^\infty)$ on $\partial\Omega$. If this were true, the remainder terms $P_\nu A_\nu Q A_\nu^{-1}\pi_\nu$ would belong to $OPS^0_{1,0}$ and have small symbol on a small conic neighborhood of (x_0',ξ_0'), so theorem 7.2 would be obvious. However, such an assertion about the behavior of h_ν, based on a false analogy with the second order scalar case, is not correct, as is apparent from (7.18). One can adjust the scalar quantity σ_1, but no matter how this is adjusted, if $\tilde{h}_\nu$ is linearly independent of $\tilde{g}_\nu$ at $\eta = 0$, on $\partial\Omega$, then $h_\nu^{(0)}$ must be linearly independent of $g_\nu^{(0)}$ there, provided $\sigma_0 \neq 0$. We proceed to the proof of theorem (7.2).

Proof. We have

$$\tilde{\beta}_0 = \beta_0 + \sum_{\nu=1}^{j} K_\nu Q L_\nu \tag{7.39}$$

where $\beta_0 \in OPS^0$ is elliptic, K_ν and L_ν are Fourier integral operators of order zero whose canonical transformations are inverses of each other, and $Q \in OPN^0_{1/3}$ has been defined before. The image of (x_0',ξ_0') under the canonical transformation associated with L_ν is contained in $\{\eta = 0\}$.

To invert $\tilde{\beta}_0$, we split Q into 2 pieces, as follows. Pick a, $1/2 < a < 2/3$, pick $\psi(s) \in C_0^\infty(\mathbb{R})$ with $\psi(s) = 1$ for $|s| \leq 1$, and let

$$\phi(\xi,\eta) = \psi(\xi_1^{-a}\,\eta) \in S^0_{a,0}.$$

Let $R = \phi(D)Q$ and $S = (1-\phi(D))Q$. Straightforward calculations show

$$R \in OPN_{1/3}^{-(1-a)/2} \subset OPS_{1/3,0}^{-(1-a)/2} \quad \text{and} \quad S \in OPN^0_a \subset OPS_{a,0}.$$

Details on such calculations are given in [12] and in chapter 11 of [22]. Note that S has small symbol on a small conic neighborhood of $\{\eta = 0\}$. In fact,

$$|\sigma_S(\xi,\eta)| \leq C(|\eta|/|\xi|)^{1/2} + C|\xi|^{-1/3}.$$

It follows from Egorov's theorem that

$$T = \sum_{\nu=1}^{j} K_\nu S L_\nu \in OPS^0_{a,1-a}$$

and furthermore, on a small conic neighborhood of (x_0',ξ_0'),

$$|\sigma_T(x',\xi')| \le C\rho(x',\xi')^{1/2} + C|\xi'|^{-(2a-1)}$$

where $\rho(x',\xi')$ denotes the distance from $(x',\xi'/|\xi'|)$ to $(x_0',\xi_0'/|\xi_0'|)$ in $S^*(\partial\Omega)$. Consequently $\beta_1 = \beta_0 + \sum_{\nu=1}^{j} K_\nu SL_\nu \in OPS^0_{a,1-a}$ is elliptic in a small conic neighborhood of (x_0',ξ_0'). Denoting by β_1^{-1} a microlocal parametrix for β_1, we are reduced to proving microlocal invertibility near (x_0',ξ_0') of

$$\begin{aligned}\beta_1^{-1}\tilde{\beta}_0 &= I + \beta_1^{-1} \sum_{\nu=1}^{j} K_\nu RL_\nu \\ &= I + \tilde{R}.\end{aligned} \tag{7.40}$$

Since L_ν and K_ν have zero order on all Sobolev spaces, $\beta_1^{-1} \in OPS^0_{a,1-a}$, and $R \in OPS^{-(1-a)/2}_{1/3,0}$, it follows that

$$\tilde{R}: H^s \to H^{s+(1-a)/2}.$$

Hence $\tilde{R}^\ell: H^s \to H^{s+(1-a)\ell/2}$, so one can define a microlocal operator

$$U \sim I - \tilde{R} + \tilde{R}^2 - \ldots$$

such that $U(I+R) = (I+R)U = I \bmod OPS^{-\infty}$. Consequently $U\beta_1^{-1}$ is the desired microlocal inverse of $\tilde{\beta}_0$, and the theorem is proved.

We note that, in the special case when $j=1$ (so only one grazing ray passes over $(x_0',\xi_0') \in T^*(\partial\Omega)$), if we set L_1 of (7.39) equal to H_1A_1, with $H_1 \in OPS^0$, A_1 the elliptic FIOP of (7.32), and set $K_1 = A_1^{-1}H_2$, $H_2 \in OPS^0$, then

$$\tilde{\beta}_0 = A_1^{-1}(A_1\beta_0A_1^{-1} + H_2QH_1)A_1. \tag{7.41}$$

By Egorov's theorem, $A_1\beta_0A_1^{-1} \in OPS^0$ is elliptic, and we see that $H_2QH_1 \in OPS^0_{1/3,0}$ has small symbol in a small conic neighborhood of $\{\eta = 0\}$, $A_1\beta_0A_1^{-1} + H_2QH_1 \in OPS^0_{1/3,0}$ is elliptic, and thus, near (x_0',ξ_0'),

$$\tilde{\beta}_0^{-1} = A_1^{-1}(A_1\beta_0A_1^{-1} + H_2QH_1)^{-1}A_1. \tag{7.42}$$

Such a construction was used in [20], section 1, to treat the transmission problem, involving a wave equation whose sound speed jumps across a hypersurface, which is typical of systems where two families of grazing rays do not intersect.

The condition that β_0 be elliptic implies that the dimension counts match up:

$$k' = j + \lambda + \mu. \tag{7.43}$$

This is a natural condition in view of well-posedness considerations.

At (x_0', ξ_0'), the principal symbol of β_0 is the product of the principal symbol b of β and the map $\Delta(x',\xi'): C^{j+\lambda+\mu} \to C^k$ defined as follows. Let e_ν denote the standard basis of $C^{j+\lambda+\mu}$, and imbed C^μ into $C^{j+\lambda+\mu}$ as the last factor. Then $\Delta(x',\xi')$ takes e_ν to $g_\nu^{(0)}$ for $0 \le \nu \le j$, it takes e_ν to a_ν for $j+1 \le \nu \le j+\lambda$ (see (7.27)) and it coincides with the map b_- on C. The content of theorems 7.1 and 7.2 is that, if the composite map $b\Delta$:

$$C^{k'} \xrightarrow{\Delta} C^k \xrightarrow{b} C^{k'} \tag{7.44}$$

is an isomorphism, then (7.9), (7.27), (7.28) give a microlocal parametrix for (7.1), (7.2).

We want to generalize the above conditions, to include boundary value problem of Neumann type, such as the problem of diffraction of electromagnetic waves treated in [20] and in chapter 10 of [22]. For the purposes of this generalization, it is convenient to rewrite $\tilde{\beta}_0$, rather than in the form (7.38), as

$$\tilde{\beta}_0 = \beta\left(\sum_{\nu=1}^{m} A_\nu^{(1)} A_\nu Q A_\nu^{-1} \pi_\nu + \sum_{\nu=m+1}^{j} A_\nu^{(0)} \pi_\nu + \sum_{\nu=j+1}^{j+\lambda} J_\nu \pi_\nu + J_- \pi_- \right)$$
$$+ \beta\left(\sum_{\nu=1}^{m} A_\nu^{(0)} \pi_\nu + \sum_{\nu=m+1}^{j} A_\nu^{(1)} A_\nu Q A_\nu^{-1} \pi_\nu \right). \tag{7.45}$$

Replace $\tilde{F} = (\tilde{F}_\nu, F_-)$ by $G = (A_\nu Q A_\nu^{-1} \tilde{F}_\nu (1 \le \nu \le m), \tilde{F}_\nu (m+1 \le \nu \le j+\lambda), F_-) = (G_\nu, G_-)$. Thus $\tilde{\beta}_0 \tilde{F} = YG$ with

$$Y = \beta\left(\sum_{\nu=1}^{m} A_\nu^{(1)}\pi_\nu + \sum_{\nu=m+1}^{j} A_\nu^{(0)}\pi_\nu + \sum_{\nu=j+1}^{j+\lambda} J_\nu\pi_\nu + J_-\pi_-\right)$$

$$+ \beta\left(\sum_{\nu=1}^{m} A_\nu^{(0)}A_\nu Q^{-1}A_\nu^{-1}\pi_\nu + \sum_{\nu=m+1}^{j} A_\nu^{(1)}A_\nu QA_\nu^{-1}\pi_\nu\right)$$

$$= X_m + \sum_{\nu=1}^{m} \beta A_\nu^{(0)} Q^{-1}A_\nu^{-1}\pi_\nu + \sum_{\nu=m+1}^{j} K_\nu QL_\nu . \qquad \mathbf{(7.46)}$$

For our Neumann-like modification of the regularity condition of theorem 7.3, which was $b\Delta$ invertible, suppose instead that the following hold.

$b\Delta$ annihilates $e_1, \ldots, e_m$ (i.e., b annihilates $g_1^{(0)}, \ldots, g_m^{(0)}$) on $\{\eta = 0\} = \{\text{grazing sets}\}$; (7.47)

$b\tilde{\Delta}$ is invertible, where $\tilde{\Delta}$ takes e_ν to $h_\nu^{(0)}$ for $1 \le \nu \le m$, and coincides with Δ in the complementary subspace. (7.48)

Theorem 7.3. Under the hypotheses (7.47), (7.48), Y is microlocally invertible, and hence the above construction yields a parametrix for the boundary value problem.

Proof. Hypothesis (7.48) implies X_m is elliptic. Hypothesis (7.47) implies the remainder terms in (7.46) are small, so as in the proof of theorem 7.2, the ellipticity of X_m implies the microlocal invertibility of Y.

Some other results for symmetric hyperbolic systems with dissipative boundary conditions are given in Kubota [9].

§8. A TRANSMISSION PROBLEM.

In this last section we analyze a diffractive boundary value problem that does not fall under the category of problems analyzed in section 7, although the problem here arises from mathematical physics. The method we use to resolve it has some points in common with the methods of Imai-Shirota [7] in theirstudy of some diffractive boundary value problems.

Suppose a manifold Ω is divided into two retions Ω_1 and Ω_2, separated by a smooth hypersurface Γ. Suppose u solves the second order wave equations

$$(\partial^2/\partial t^2 - A_1)u = 0 \quad \text{on} \quad \Omega_1 \tag{8.1}$$

$$(\partial^2/\partial t^2 - A_2)u = 0 \quad \text{on} \quad \Omega_2 \tag{8.2}$$

where A_j are given negative symmetric second order elliptic operators on Ω_j, and that, on the surface Γ, u satisfies the transmission conditions

$$u_1 - u_2 = 0, \frac{\partial}{\partial \nu_1} u_1 + a(x) \frac{\partial}{\partial \nu_2} u_2 = 0 \tag{8.3}$$

where u_1 is the limiting value of u from Ω_1 and the other quantities are similarly defined. $\partial/\partial\nu_j$ is the conormal vector to Γ, pointing into Ω_j, determined by the metric $a^{(j)}_{k\ell}$, where $-\sum_{k,\ell\geq 1} a^{(j)}_{k\ell}\xi_k\xi_\ell$ is the principal symbol of A_j; thus if Γ is given by $w_j = 0$, w_j vanishes simply, $w_j > 0$ in Ω_j, then $\partial/\partial\nu_j$ is proportional to $\sum_{k,\ell\geq 1} a^{(j)}_{k\ell}\, \partial w_j/\partial x_k \ \partial/\partial x_\ell$. The factor $a(x)$ in (8.2) is assumed smooth and positive.

Such a transmission problem, in the diffractive case, was treated in [20] under the hypothesis that over each point $\zeta_0 \in T^*(R \times \Gamma)$ at most one grazing ray can pass, which would be the case if the sound speeds for the two equations had a nonzero jumpeverywhere across Γ. Such a case also satisfies the Kreiss-Sakamoto condition, and so the analysis of section 7 applies. Here we study the case where the grazing rays from Ω_1 and Ω_2 can match up, on some closed conic subset of $T^*(S)$, $S = R \times \Gamma$.

We will suppose Γ is diffractive from either side, Ω_1 or Ω_2, so null bicharacteristics of $\partial^2/\partial t^2 - A_j$ in Ω_j, which hit Γ tangentially, stay in $\overline{\Omega}_j$ and have precisely second order contact with Γ. Picking $\zeta_0 \in T^*(S)$, we will construct a parametrix for (8.1), with boundary condition

$$u_1 - u_2 = f, \quad \frac{\partial}{\partial \nu_1} u_1 + a(x) \frac{\partial}{\partial \nu_2} u_2 = g \quad \text{on} \quad S, \tag{8.3}$$

where $f,g \in \mathcal{E}'(R\times\Gamma)$, WF f and WF g contained in a small conic neighborhood of ζ_0, and satisfying the "outgoing condition"

$$u = 0 \quad \text{for} \quad t << 0. \tag{8.4}$$

We suppose that, over $\zeta_0 \in T^*(S)$, two grazing rays pass, one from Ω_1 and one from Ω_2, since the other possibilities have been taken care of, as mentioned above.

The parametrices will be of the form

$$u = \iint \left[g_j \frac{A(\xi_1^{-1/3}\rho)}{A(-\xi_1^{1/3}\eta)} + ih_j \xi_1^{-1/3} \frac{A'(\xi_1^{-1/3}\rho)}{A(-\xi_1^{-1/3}\eta)} \right] e^{i\theta_j} \hat{F}_j(\xi,\eta)d\xi d\eta \tag{8.5}$$

in Ω_j. The phase functions ρ_j, θ_j satisfy the usual eikonal equations

$$\sum_{k,\ell\geq 0} a_{k\ell}^{(j)} \left(\frac{\partial\theta_j}{\partial x_k} \frac{\partial\theta_j}{\partial x_\ell} + \frac{\rho_i}{\xi_1} \frac{\partial\rho_i}{\partial x_k} \frac{\partial\rho_j}{\partial x_\ell} \right) = 0$$

$$\sum_{k,\ell\geq 0} a_{k\ell}^{(j)} \frac{\partial\rho_j}{\partial x_k} \frac{\partial\theta_j}{\partial x_\ell} = 0. \tag{8.6}$$

Here, $x_0 = t$. $a_{k\ell}^{(j)}$ is the quadratic form given by the principal symbol of $\partial^2/\partial t^2 - A_j$. The amplitudes g_j, h_j have asymptotic expansions obtained from solving the usual transport equations. We remark that one can arrange $g_j \neq 0$ and $h_j \equiv 0$ on Γ. We also arrange, as usual, that

$$\rho_j = -\eta \quad \text{on} \quad \Gamma \tag{8.7}$$

and consequently we can arrange that

$$\frac{\partial\theta_j}{\partial\nu_j} \equiv 0 \quad \text{on} \quad \Gamma. \tag{8.8}$$

Next, we compute $\partial/\partial\nu_j \; u_j$; using the above properties of ρ, θ, g, h, we get

$$\begin{aligned} \frac{\partial}{\partial\nu_j} u_j &= K_j F_j \\ &= \iint \left(g_j \frac{\partial\rho_j}{\partial\nu_j} + i \frac{\partial h_j}{\partial\nu_j} \right) \xi_1^{-1/3} \frac{A'}{A}(-\xi_1^{-1/3}\eta) e^{i\theta_j} \hat{F}_j(\xi,\eta)d\xi d\eta \\ &\quad + \iint \frac{\partial g_j}{\partial\nu_j} e^{i\theta_j} \hat{F}_j \, d\xi d\eta \\ &= K_j^{(0)} QF_j + K_j^{(1)} F_j. \end{aligned} \tag{8.9}$$

Q is defined, as usual, by $(QF)^\wedge = \xi_1^{-1/3} \frac{A'}{A}(-\xi_1^{-1/3}\eta)\hat{F}$, so

$Q \in OPN^0_{1/3} \subset OPS^0_{1/3,0}$. The operators $K_j^{(0)}$ and $K_j^{(1)}$ are Fourier integral operators associated to the same canonical transformation $\mathcal{J}_j$ as J_j, so

$$A_j = J_j^{-1}K_j^{(0)} \in OPS^1, \quad B_j = J_j^{-1}K_j^{(1)} \in OPS^0. \tag{8.10}$$

Thus

$$K_j = J_j(A_jQ + B_j). \tag{8.11}$$

Noting that $\frac{\partial}{\partial \nu_j}\rho_j > 0$ on Γ at $\eta = 0$, we see that the operator $A_j \in OPS^1$ is elliptic, with positive principal symbol.

Our task is to produce F_j such that the boundary conditions (8.3) are satisfied. To do this, we produce some Fourier integral equations for F_1, F_2 in terms of f and g, namely

$$\begin{aligned} J_1F_1 - J_2F_2 &= f \\ K_1F_1 + K_2F_2 &= g. \end{aligned} \tag{8.12}$$

Here we have absorbed the factor $a(x)$ into K_2. Rewrite the second equation as

$$J_1(A_1Q+B_1)F_1 + J_2(A_2Q+B_2)F_2 = g. \tag{8.13}$$

We obtain

$$F_1 - J_1^{-1}J_2F_2 = J_1^{-1}f = \tilde{f}$$

$$F_1 + (A_1Q+B_1)^{-1}\, J_1^{-1}J_2(A_2Q+B_2)F_2 = \tilde{g}$$

or, upon eliminating F_1,

$$\left[(A_1Q+B_1)^{-1}J_1^{-1}\, J_2(A_2Q+B_2) + J_1^{-1}J_2\right]F_2 = \tilde{g}-\tilde{f} = \tilde{h} \tag{8.14}$$

or, with $J = J_1^{-1}J_2$,

$$\left[J(A_2Q+B_2) + (A_1Q+B_1)J\right]F_2 = (A_1Q+B_1)\tilde{h} = h \tag{8.15}$$

or

$$\left[(A_2Q+B_2) + J^{-1}(A_1Q+B_1)J\right]F_2 = J^{-1}h = w. \tag{8.16}$$

We rewrite (8.16) as

$$\mathcal{P}F_2 = 2. \tag{8.17}$$

Our goal is to show a parametrix $\mathcal{P}^{-1}$ exists and is microlocal. We will first establish some energy estimates. Suppose the operators in (8.16) are extended to the torus $\mathbf{T}^n$, in which we have imbedded a small open set in $R \times \Gamma$, and in which a conic subset of $T^*(\mathbf{T}^n)$ is identified with a conic neighborhood of $J_2^{-1}(\zeta_0)$. There is no loss of generality in supposing J is unitary, and we can suppose the relevant properties of A_j and B_j established above hold globally. In particular, the principal symbol of A_j is positive. We use the following fact about the Airy quotient, which has also been used by Imai-Shirota [7].

Lemma 8.1. For $\lambda \in R$, $\frac{A'}{A}(\lambda)$ takes values in the open second quadrant of the complex plane if $A = A_+$ and in the open third quadrant if $A = A_-$.

Choose ω equal to $e^{-3\pi i/4}$ if $A = A_+$, $\omega = e^{3\pi i/4}$ if $A = A_-$, so $\omega \frac{A'}{A}(\lambda)$ takes values in an open proper subcone of the right half plane $\{\mathrm{Re}\ z > 0\}$. Now we intend to establish the following energy estimate.

Lemma 8.2. For $F_2 \in C^\infty(\mathbf{T}^n)$,

$$\mathrm{Re}(\omega\mathcal{P}F_2,F_2) \geq C_1\|F_2\|^2_{H^{1/3}} - C_2\|F_2\|^2_{H^\sigma} \quad \sigma \ll 0. \tag{8.18}$$

Proof. Since $B_j \in OPS^0$ and J is unitary, it suffices to get

$$\mathrm{Re}(\omega A_jQF,F) \geq C_1\|F\|^2_{H^{1/3}} - C_2\|F\|^2_{H^\sigma}, \quad j = 1,2. \tag{8.19}$$

Now the left side of (8.19) equals

$$\tfrac{1}{2}((\omega A_jQ+\bar{\omega}Q^*A_j^*)F,F) = \tfrac{1}{2}(T_jF,F). \tag{8.20}$$

Note that $A_j^* = A_j \bmod OPS^0$ and $Q^*A_j = A_jQ^* \bmod OPN^{1/3}_{1/3}$, so

$$T_j = A_j\mathcal{Q} + S_j \tag{8.21}$$

with $S_j \in OPN^{1/3}_{1/3}$ and $\mathcal{Q} = \omega Q+\bar{\omega}Q^* \in OPN^0_{1/3}$. Note that $\mathcal{Q}^{1/2} \in OPN^0_{1/3}$ has symbol $\alpha = \xi_1^{-1/6}[2\mathrm{Re}\ \omega\ A'/A\ (-\xi_1^{-1/3}\eta)]^{1/2}$. Hence $\nabla_{\xi,\eta}\alpha \in N^{-1/2}_{1/3}$, which implies

$$[OPN^1_{1/3}, \mathcal{Q}^{1/2}] \subset OPN^{1/2}_{1/3} .$$

Thus

$$(T_j F,F) = \mathrm{Re}\,(A_j \mathcal{Q}^{1/2} F, \mathcal{Q}^{1/2} F) + \mathrm{Re}\,(\tilde{S}_j F,F) \tag{8.22}$$

with $\tilde{S}_j = S_j + [A_j, \mathcal{Q}^{1/2}] \in OPN^{1/2}_{1/3}$. Gårding's inequality gives

$$\begin{aligned}(T_j F,F) &\geq C\,\|\mathcal{Q}^{1/2}F\|^2_{H^{1/2}} - C'\,\|F\|^2_{H^{1/4}} \\ &\geq C_1\,\|F\|^2_{H^{1/3}} - C_2\,\|F\|^2_{H^\sigma}\end{aligned}$$

which proves (8.19).

An equivalent restatement of (8.18) is

$$\mathrm{Re}(\omega\Lambda^{-1/3}\mathcal{P}\Lambda^{-1/3}F,F) \geq C_1\,\|F\|^2_{L^2} - C_2\,\|F\|^2_{H^\sigma} . \tag{8.23}$$

Thus the operator $\Lambda^{-1/3}\mathcal{P}\Lambda^{-1/3} \in OPN^{1/3}_{1/3}$, considered as an unbounded operator on $L^2(\mathbf{T}^n)$, with domain $\mathcal{D} = \{u \in L^2(\mathbf{T}^n): \Lambda^{-1/3}\mathcal{P}\Lambda^{-1/3} u \in L^2(\mathbf{T}^n)\}$, has closed range and finite dimensional kernel. The same is true of the adjoint of this operator. Thus $\Lambda^{-1/3}\mathcal{P}\Lambda^{-1/3}$ is Fredholm. Furthermore, $\lambda + \omega\Lambda^{-1/3}\mathcal{P}\Lambda^{-1/3}$ is Fredholm for each $\lambda \in R^+$; if λ is large enough this operator is invertible, by (8.23). Hence the Fredholm index of $\Lambda^{-1/3}\mathcal{P}\Lambda^{-1/3}$ is zero.

Lemma 8.3. $\mathrm{Ker}\,\mathcal{P} \subset C^\infty(\mathbf{T}^n)$.

Proof. First note that the derivation of (8.18) works with $\mathcal{P}$ replaced by $\Lambda^s\mathcal{P}\Lambda^{-s}$; then replacing F_2 by $F_2 = \Lambda^s G$ yields

$$\mathrm{Re}(\omega\,\Lambda^s\,\mathcal{P}\,G,\ \Lambda^s\,G) \geq C_1\,\|G\|^2_{H^{s+1/3}} - C_2\,\|G\|^2_{H^\sigma} . \tag{8.24}$$

Suppose $u \in \mathrm{Ker}\,\mathcal{P} \subset L^2(\mathbf{T}^n)$. Apply (8.24) to $G = J_\varepsilon u$, where J_ε is a Friedrichs mollifier.

$$\begin{aligned}\|J_\varepsilon u\|^2_{H^{s+1/3}} &\leq C_1\,|(\Lambda^s\,\mathcal{P}\,J_\varepsilon u,\ \Lambda^s\,J_\varepsilon u)| + C_2\,\|u\|^2_{L^2} \\ &\leq C_1\,\|[J_\varepsilon,\mathcal{P}]u\|_{H^{s-1/6}}\,\|J_\varepsilon u\|_{H^{s+1/6}} + C\,\|u\|^2_{L^2}.\end{aligned} \tag{8.25}$$

Now $[J_\varepsilon,\mathcal{P}]$ is a bounded family of operators of order 1/3, for $\varepsilon \in (0,1]$, so the estimate (8.25) shows that

$$u \in \mathrm{Ker}\,\mathcal{P},\quad u \in H^{s+1/6} \ \text{ implies } \ u \in H^{s+1/3} . \tag{8.26}$$

Starting with $s = -1/6$ and iterating (8.26), we prove the lemma.

Thus, the operator $\Lambda^{-1/3} \, P \, \Lambda^{-1/3}$, shown to be Fredholm of index zero, has its kernel in $C^\infty(\mathbb{T}^n)$. Similarly the kernel of its adjoint belongs to $C^\infty(\mathbb{T}^n)$. It follows that, altering P by a smoothing operator, we can suppose $\Lambda^{-1/3} \, P \, \Lambda^{-1/3}$ is invertible on $L^2(\mathbb{T}^n)$, and indeed on $H^s(\mathbb{T}^n)$ for any s. Thus

$$P^{-1}: H^s(\mathbb{T}^n) \to H^{s+2/3}(\mathbb{T}^n). \tag{8.27}$$

We have obtained a global inverse for the operator P. It is now necessary to show P^{-1} is microlocal, i.e.,

$$WF(P^{-1}u) \subset WF(u). \tag{8.28}$$

To do this, we introduce some useful classes of operators. We say

$$A \in O(m) \quad \text{if} \quad A: H^s(\mathbb{T}^n) \to H^{s-m}(\mathbb{T}^n) \quad \text{for all } s. \tag{8.29}$$

Let $P_j \in OPS^0$ on $\mathbb{T}^n$ be arbitrary. Let $0 < \rho \le 1$. We say

$$A \in O(m,\rho) \quad \text{if} \quad A \in O(m) \quad \text{and} \quad AdP_1 \ldots AdP_k \cdot A \in O(m-k\rho) \tag{8.30}$$

where $AdP \cdot A = [P,A] = PA-AP$. (8.30) is to hold for all k. Finally, we consider a subclass of $O(m,\rho)$ whose <u>first order</u> commutators have lower order.

$$A \in O(m,\rho,\tilde{\rho}) \quad \text{if} \quad A \in O(m) \quad \text{and} \quad AdP_1 \cdot A \in O(m-\tilde{\rho},\rho). \tag{8.31}$$

Using the commutation properties of A_j, B_j, and Q, plus Egorov's theorem, we immediately obtain the following.

<u>Lemma 8.4</u>. $P \in O(1,1/3,2/3)$.

Our next result is on the behavior of P^{-1}.

<u>Proposition 8.5</u>. $P^{-1} \in O(-2/3,1/3)$.

<u>Proof</u>. If $P_j \in OPS^0$, then $AdP_1 \ldots AdP_k(P^{-1})$ is a sum of terms of the form: products of j terms like $AdA_{1\ell} \ldots AdA_{\mu_\ell \ell} \cdot P$ and of $j+1$ terms P^{-1}. $A_{i\ell}$ are rearrangements of the P_j, and $\mu_1 + \ldots + \mu_j = k$. This follows by an inductive argument, using the derivation property of AdP_ν on a product of operators. Each factor P^{-1} has order $-2/3$, by (8.27). Meanwhile, by lemma 8.4, each factor $AdA_{1\ell} \ldots AdA_{\mu_\ell \ell} \cdot P$ (note $\mu_\ell \ge 1$) has order $2/3 - \mu_\ell/3$. Thus each term described as a product above has order

$$-\frac{2}{3}(j+1) + \sum_{(j \text{ terms})} \left(\frac{2}{3} - \frac{1}{3}\mu_\ell\right)$$

$$= -\frac{2}{3} - \frac{1}{3}\sum \mu_\ell = -\frac{2}{3} - \frac{k}{3}.$$

This completes the proof.

The microlocal property of $\mathcal{P}^{-1}$ follows from the next assertion.

Proposition 8.6. If $T \in O(m,\rho)$, $\rho > 0$, then T is microlocal, i.e.,

$$WF(Tu) \subset WF(u).$$

Proof. Let $\psi_1(x,D)$, $\psi_2(x,D)$ be two elements of OPS^0 whose symbols are essentially supported on disjoint closed conic sets. It suffices to show that $\psi_1(x,D)T\psi_2(x,d) \in O(-\infty)$. Pick $\tilde{\psi}_2$ supported on a slightly larger cone than ψ_2, such that $\tilde{\psi}_2\psi_2 = \psi_2 \bmod O(-\infty)$. Then, mod $O(-\infty)$,

$$\begin{aligned}\psi_1 T\psi_2 &= \psi_1 T\tilde{\psi}_2\psi_2 \\ &= \psi_1\tilde{\psi}_2 T\psi_2 + \psi_1[T,\tilde{\psi}_2]\psi_2 \\ &= \psi_1[T,\tilde{\psi}_2]\psi_2 \mod O(-\infty).\end{aligned}$$

Note that $[T,\tilde{\psi}_2] \in O(m-\rho,\rho)$. Iterating this argument gives

$$\psi_1 T\psi_2 = \psi_1 T_j\psi_2 \mod O(-\infty)$$

with $T_j \in O(m-j\rho,\rho)$. Taking $j \to \infty$ finishes the proof.

REFERENCES

1. Cheeger, J. and Taylor, M., "Diffraction of waves by conical singularities," to appear.

2. Farris, M., "Egorov's theorem for a diffractive boundary value problem," Thesis, Rice University, 1980.

3. __________, "Egorov's theorem on a manifold with diffractive boundary," to appear.

4. Golubitsky, M. and Guillemin, V., Stable Mappings and Their Singularities, Springer-Verlag, New York, 1973.

5. Hörmander, L., "Pseudo-differential operators and hypoelliptic equations," Proc. Symp. Pure Math. X., Amer. Math. Soc., Providence, R.I. (1967), pp. 138-183.

6. ________, "Fourier integral generators, I," Acta Math. 127 (1971), pp. 79-183.

7. Imai, M. and Shirota, T., "On a parametrix for the hyperbolic mixed problem with diffractive lateral boundary," Hokkaido Math. J. 7 (1978), pp. 339-352.

8. Kirchoff, G., Vorlesungen über math. Physik 2 (Optik), Leipzig, 1891.

9. Kubota, K., "A Microlocal parametrix for an exterior mixed problem for symmetric hyperbolic systems," Preprint.

10. Ludwig, D., "Uniform asymptotic expansions at a caustic," Comm. Pure Appl. Math. 19 (1966), pp. 215-250.

11. ________, "Uniform asymptotic expansions of the field scattered by a convex object at high frequencies," Comm. Pure Appl. Math. 19 (1967), pp. 103-138.

12. Majda, A. and Taylor, M., "The asymptotic behavior of the diffraction peak in classical scattering," Comm. Pure Appl. Math. 30 (1977), pp. 639-669.

13. Melrose, R., "Microlocal parametrices for diffractive boundary value problems," Duke Math. J. 42 (1975), pp. 605-635.

14. ________, "Equivalence of glancing hypersurfaces," Invent. Math. 37 (1976), pp. 165-191.

15. ________, "Parametrices at diffractive points," Unpubl. manuscript, 1975.

16. ________, "Forward scattering by a convex obstacle," Comm. Pure Appl. Math. 33 (1980), to appear.

17. Melrose, R. and Taylor, M., "The corrected Kirchoff approximation and the uniform analysis of the amplitude in classical scattering," to appear.

18. Miller, J., The Airy Integral, Cambridge University Press, 1946.

19. Taylor, M., "Grazing rays and reflection of singularities of solutions to wave equations," Comm. Pure Appl. Math. 29 (1976), pp. 1-38.

20. ________, "Grazing rays and reflection of singularities of solutions to wave equations, Part II (systems)," Comm. Pure Appl. Math. 29 (1976), pp. 463-481.

21. ________, "Propagation, reflection, and diffraction of singularities of solutions to wave equations," Bull. Amer. Math. Soc. 84 (1978), pp. 589-611.

22. ________, Pseudo Differential Operators, Princeton University Press, 1981.

23. ________, "Diffraction of waves by cones and polyhedra," Nijmegen Conference on Asymptotic Methods in Analysis, June 1980, North Holland, to appear.

24. Yingst, D., "The Kirchoff approximation for the Neumann problem," Indiana J. Math., to appear.

SINGULARITIES OF ELEMENTARY SOLUTIONS OF HYPERBOLIC EQUATIONS WITH CONSTANT COEFFICIENTS

Mikio Tsuji

Department of Mathematics, Kyoto Sangyo University
Kita-ku, Kyoto 603, Japan.

§1. Introduction.

Let $P(D_x)$ be a hyperbolic operator with constant coefficients with respect to $\theta \in R^n-\{0\} \equiv \dot{R}^n$ where $x \in R^n$ and $D_x = \partial/i\partial_x$, i.e. its characteristic polynomial $P(\xi)$ satisfies i) $P_m(\theta) \neq 0$ where P_m is the principal part of P, and ii) $P(\xi-i\gamma\theta) \neq 0$ for all $\xi \in R^n$ and $\gamma > \gamma_0 > 0$ where γ_0 is constant. Then

$$E(x) = (2\pi)^{-n} \int_{R^n-i\gamma\theta} e^{i<x,\zeta>} P(\zeta)^{-1} d\zeta$$

defines the unique elementary solution of $P(D_x)$ whose support is contained in $\{x \in R^n; <x,\theta> \geq 0\}$. Let $u(x)$ be a distribution in R^n. Su means the support of u, and a singular support, written by SSu, is the smallest and closed subset such that $u \in C^\infty(R^n-SSu)$. We denote an elementary solution of $P(D_x)$ by $E(P)$. Our problem is to obtain the complete description of SSE(P) for any hyperbolic operator $P(D_x)$. Though this problem has a long history, we start from the work of Atiyah-Bott-Gårding [2] and [3].

We develop $s^mP(s^{-1}\xi+\zeta)$ in ascending power of s,

$$s^mP(s^{-1}\xi+\zeta) = s^pP_\xi(\zeta) + O(s^{p+1}),$$

H. G. Garnir (ed.), Singularities in Boundary Value Problems, 317–326.

where $P_\xi(\zeta) \not\equiv 0$. The number $p=m_\xi(P)$ is the multiplicity of ξ relative to P and the polynomial $P_\xi(\zeta)$ is called the localization of P at ξ. Then one can see that $P_\xi(\zeta)$ is also hyperbolic with respect to θ. Atiyah-Bott-Gårding [2] proved the following

Theorem 1.

$$\lim_{s\to\infty} s^{m-p} e^{-is\langle x,\xi\rangle} E(P) = E(P_\xi)$$

in the space of distributions, and

$$U_{\xi\neq 0}\, SE(P_\xi) \subset SSE(P) \subset U_{\xi\neq 0}\, c.h.SE(P_\xi)$$

where c.h.A means the convex hull of a set A.

In [2], they gave the conjecture such that the former inclusion is not proper, i.e.

$$SSE(P) = U_{\xi\neq 0}\, SE(P_\xi) \quad . \tag{1}$$

But, in general, this is not true. In Section 2, we shall give the counter examples to it. To begin with, let's try to generalize their localization theorem as in [5].

We develop $s^{m-p}P(s\xi+\zeta)^{-1}$ in the formal power series of 1/s,

$$\frac{s^{m-p}}{P(s\xi+\zeta)} = \frac{1}{P_\xi(\zeta)} + \Sigma_{j=1}^{\infty} \frac{Q_j(\zeta)}{P_\xi(\zeta)^{j+1}} \left(\frac{1}{s}\right)^j$$

and we define

$$E_0(P_\xi) = E(P_\xi) \quad \text{and}$$

$$E_j(P_\xi) = (2\pi)^{-n} \int_{R^n - i\gamma\theta} e^{i\langle x,\zeta\rangle} Q_j(\zeta) P_\xi(\zeta)^{-j-1}\, d\zeta.$$

Then we have the following

Theorem 2. For any integer $N > 0$, we have

$$\lim_{s\to\infty} s^{N+1}\left[\, s^{m-p} e^{-is\langle x,\xi\rangle} E(P) - \Sigma_{j=0}^{N} E_j(P_\xi)\left(\frac{1}{s}\right)^j \,\right] = E_{N+1}(P_\xi)$$

in the space of distributions, and

$$\cup_{\xi\neq 0}\cup_{j=0}^{\infty} SE_j(P_\xi) \subset SSE(P) .$$

The aim of this talk is to prove that, for any hyperbolic operator with double characteristics, it follows

$$SSE(P) = \cup_{\xi\neq 0}\cup_{j=0}^{\infty} SE_j(P_\xi) . \quad (2)$$

Moreover, as it is tired to calculate $SE_j(P_\xi)$ for large j, we shall give the more concrete representation of SSE(P) by using $SE(P_\xi)$ only.

§2. Examples.

Example 1. (Non-homogeneous case)

Let $P(\xi) = \xi_1(\xi_1^2 - \xi''^2) + \xi_n(a^2\xi_1{}^2 - \xi''^2)$ where $\xi' = (\xi'', \xi_n)$ and $\xi'' = (\xi_2,\dots,\xi_{n-1})$. We suppose "a>1" in order that $P(\xi)$ is irreducible. Obviously P is hyperbolic with respect to $\theta = (1,0,\dots,0)$. The double characteristic point of P is $\xi^0 = (0,..,0,1)$ only on $\{ |\xi| = 1 \}$, and

$$P_{\xi^0}(\xi) = a^2\xi_1^2 - \Sigma_{j=2}^{n-1} \xi_j^2 .$$

i) When n is even, then $SE(P_{\xi^0}) = c.h.SE(P_{\xi^0}) = \{x_1 \geq a|x''|, x_n = 0\}$ where $x'' = (x_2,\dots,x_{n-1})$. Therefore Theorem 1 shows that (1) is true.

ii) When n is odd and $n\geq 5$, then we can prove that (1) is true, and $SSE(P)\cap\{x_1 = 1\}$ is the continuous curve in Figure 1. Moreover, E(P) is not zero in the interior of its curve. Therefore, the domain whose boundary is SSE(P) is not convex, though such domains are, in general, convex.

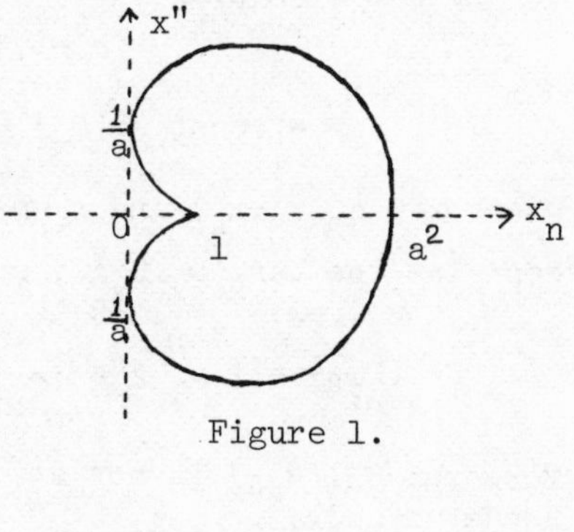

Figure 1.

iii) At last, we give the counter example to the conjecture (1). Let $Q(\xi) = P(\xi) + c$ $(c\neq 0)$, and assume n=5. Then, as $Q_{\xi^0}(\xi) = P_{\xi^0}(\xi)$, we have

$$E_0(Q_{\xi^0}) = E(Q_{\xi^0}) = \text{const. } \delta(x_1{}^2 - a^2|x''|^2)\,\delta(\mathbf{x}_5) ,$$

$$E_1(Q_{\xi^0}) = (2\pi)^{-5} \int_{R^5 - i\gamma\theta} e^{i\langle x,\zeta\rangle} Q(\zeta)\, P_{\xi^0}(\zeta)^{-2}\, d\zeta$$

$$= \text{const.}(\ P(D_x) + c\ )\ H(x_1{}^2 - a^2|x''|^2)\ \delta(x_5)\ ,$$

where $H(r)=1$ for $r > 0$ and $=0$ for $r < 0$. As $c \neq 0$, we see

$$SE_1(Q_{\xi^0}) = \{\ x_1 \geq a|x''|,\ x_5=0\} = \text{c.h.}SE(Q_{\xi^0}) \supsetneqq SE(Q_{\xi^0}).$$

Hence Theorem 1 and 2 give the following relation:

$$SSE(Q) = \cup_{\xi\neq 0}\cup_{j=0}^{\infty} SE_j(Q_\xi) \supsetneqq \cup_{\xi\neq 0} SE(Q_\xi)\ ,$$

which means that (1) is not true.

Example 2. (Homogeneous case)

Let $P(\xi) = \xi_1(\xi_1{}^2 - \Sigma_{j=2}^{5}\ \xi_j{}^2) + \xi_6(a^2\xi_1{}^2 - \Sigma_{j=2}^{4}\xi_j{}^2)$ where $a > 1$. Then $P(\xi)$ is hyperbolic with respect to $\theta = (1,0,\ldots,0)$. In this case, the double characteristic point of P is also $\xi^0 = (0,\ldots,0,1)$ only, and

$$P_{\xi^0}(\xi) = a^2\ \xi_1{}^2 - \ \Sigma_{j=2}^{4}\ \xi_j{}^2\ .$$

Then, as we have

$$\frac{1}{P(s\xi^0+\zeta)} = \Sigma_{j=0}^{\infty}\frac{(-P(\zeta))^j}{P_{\xi^0}(\zeta)^{j+1}}\left(\frac{1}{s}\right)^{j+1}\ ,$$

the third term of our localization theorem is given by

$$E_2(P_{\xi^0}) = -\left(\frac{1}{2\pi}\right)^6 \int_{R^6 - i\gamma\theta} P(\zeta)^2 P_{\xi^0}(\zeta)^{-3}\ e^{i\langle x,\zeta\rangle}\ d\zeta$$

$$= \text{const.}\ P(D_x)^2\ (x_1{}^2 - a^2|x''|^2)_+\ \delta(x_5)\ \delta(x_6)\ ,$$

where $x''=(x_2,x_3,x_4)$ and $r_+=\max\{r,0\}$. Notice that $P(D_x)$ contains the term $D_{x_1}D_{x_5}{}^2$, so we have

$$SE_2(P_{\xi^0}) = \{\ x_1 \geq a\ |x''|,\ x_5 = x_6 = 0\} \supsetneqq SE(P_{\xi^0})\ .$$

Moreover $SE_2(P_{\xi^0})$ is not attained by another $SE(P_\xi)$ for $\xi\neq\xi^0$. Hence we see

$$SSE(P) \supsetneqq \cup_{\xi\neq 0}\ SE(P_\xi)\ .$$

But, in this case, we can prove that $SSE(P)$ is equal to the closure of the right hand in the above relation of inclusion.

Remark. By the letter of K. G. Andersson of April 9, 1980 to the author, we saw that he also constructed the examples as ours. As the example which corresponds to the example 2, he gave the following operator,

$$P(\xi) = \frac{\partial^2}{\partial\xi_1^2}[\ (\ \xi_1^2 - \Sigma_{j=2}^{4}\ \xi_j^2)\ \{(\xi_1 + \xi_5)^2 - \xi_5^2 - \xi_6^2\}^2\ ]\quad .$$

Then $P(D_x)$ does not satisfy the conjecture (1) and SSE(P) is equal to the closure of the right hand of (1).

§3. Singularities of elementary solutions.

Going on from the examples in §2, we try to describe the singular supports of elementary solutions of general hyperbolic operators. We suppose

The multiplicity of $P(\xi)$ is at most double. (H)

At first, we classify the characteristic points of $P(\xi)$. Let $P_m(\xi)$ be the principal part of $P(\xi)$. We define

$$C_1 = \{\xi \in R^n\ ;\ P_m(\xi) = 0, \text{ and grad } P_m(\xi) \neq 0\ \}\quad ,$$

$$C_2 = \{\xi \in R^n\ ;\ P_m(\xi) = 0 \text{ and grad } P_m(\xi) = 0\ \}\ .$$

Consider the localization of P at $\xi^0 \in C_2$. Then we can assume without restriction that $\xi^0=(1,0,\ldots,0)$, $\theta=(1,1,0,..,0)$ and

$$P_{\xi^0}(\xi) = \xi_2^{\ 2} - \Sigma_{j=3}^{\ell}\ \xi_j^{\ 2} + c\quad .$$

The crucial point is the localization of P at ξ^0 such that the convex hull of $SE(P_{\xi^0})$ is not equal to $SE(P_{\xi^0})$. Such a localization $P_{\xi^0}(\xi)$ must satisfy the following condition:

"$c = 0$, $\ell \geq 5$ and ℓ is odd". (*)

Moreover, the next condition plays the important role:

$$P(\xi) \equiv \frac{\partial P}{\partial \xi_j}(\xi) \equiv 0 \quad \text{on } \{\xi_2 = \ldots = \xi_\ell = 0\} \text{for any } j=2,..,\ell. \tag{**}$$

For instance, the example 2 does not satisfy (**). By the use of (*) and (**), we divide the set C_2 into two subsets:

$$S_0 = \{\xi \in C_2\ ;\ P_\xi \text{ satisfies (*), but (**) is violated}\}\ ,$$

and

$$S_1 = \{\xi \in C_2\ ;\quad \xi \notin S_0\ \}\quad .$$

Then we have the following main

Theorem 3. Suppose the condition (H), then

$$SSE(P) = \cup_{\xi\neq 0}\cup_{j=0}^{\infty} SE_j(P_\xi)$$

$$= (\cup_{\xi\varepsilon C_1\cup S_1} SE(P_\xi)) \cup (\cup_{\xi\varepsilon S_0} c.h.SE(P_\xi)). \qquad (3)$$

Before going to the proof, let's write a lemma on the set defined in the right hand of (3).

Lemma 1. Assume the condition (H), then it follows,

1) If $\xi \varepsilon S_0$, then $c.h.SE(P_\xi) \neq SE(P_\xi)$.

2)
$$\cup_{j=0}^{\infty} SE_j(P_\xi) = \begin{cases} SE(P_\xi) & \text{for } \xi \varepsilon C_1\cup S_1 \\ c.h.SE(P_\xi) & \text{for } \xi \varepsilon S_0 \end{cases}$$

3) The set $(\cup_{\xi\varepsilon C_1\cup S_1} SE(P_\xi)) \cup (\cup_{\xi\varepsilon S_0} c.h.SE(P_\xi))$ is closed.

Proof of Theorem 3. By the Theorem 1 and 2, we have

$$\underset{\xi\neq 0}{\cup} (\overset{\infty}{\underset{j=0}{\cup}} SE_j(P_\xi)) \times \{\xi\} \subset WF(E(P)) \subset \underset{\xi\neq 0}{\cup} c.h.SE(P_\xi) \times \{\xi\} .$$

Therefore, when we write $S^* = \{\xi \varepsilon S_1;\ c.h.SE(P_\xi) \neq SE(P_\xi)\}$, our aim is to prove that $(x,\xi) \notin WF(E(P))$ for any $\xi \varepsilon S^*$ and $x \varepsilon c.h.SE(P_\xi) - SE(P_\xi)$.

Let's pick up any point $\xi^0 \varepsilon S^*$. We can assume without restriction that $\xi^0=(1,0,\dots,0)$, $\theta =(1,1,0,..,0)$ and

$$P_{\xi^0}(\xi) = \xi_2^2 - \Sigma_{j=3}^{\ell} \xi_j^2 .$$

Here we put $x'=(x_2,\dots,x_n) \varepsilon R^{n-1}$, $x'_{(1)}=(x_2,\dots,x_\ell)$ and $x'_{(2)}=(x_{\ell+1},\dots,x_n)$. Then it follows

$$c.h.SE(P_{\xi^0}) - SE(P_{\xi^0}) = \{x_1=0,\ x_2>(\Sigma_{j=3}^{\ell} x_j^2)^{1/2},\ x'_{(2)}=0\}. \qquad (4)$$

Let x^0 be in $c.h.SE(P_{\xi^0})-SE(P_{\xi^0})$. By 3) of Lemma 1, we can take a neighborhood U of x^0 such that the closure of U does not intersect with the right side hand of (3). To prove $(x^0,\xi^0) \notin WF(E(P))$, we would like to show that there exists a conical neighborhood Γ of ξ^0 such that, for any $\phi \varepsilon C_0^\infty(U)$, $\eta\varepsilon\Gamma$ and any N>0,

$$|\widehat{\phi(x) E(P)} (\eta)| < C_N(1 + |\eta|)^{-N} \qquad (5)$$

where C_N is a constant independent of $\eta\varepsilon\Gamma$, and $\hat{\phi}(\eta)$ means the Fourier transform of $\phi(x)$. To obtain the estimate (5), we divide our proof into three steps.

The first step. We have

$$P(\zeta) = \zeta_1^{m-2}(\ P_{\xi^0}(\zeta'_{(1)}) - \zeta_1^{-1}R(\zeta)\ )$$

for $\zeta\varepsilon C^n$, where $\zeta'_{(1)}=(\zeta_2,..,\zeta_\ell)$ and $\zeta'=(\zeta_2,...,\zeta_n)$ and

$$R(\zeta) = R_1(\zeta') + \zeta_1^{-1}R_2(\zeta') + \dots + \zeta_1^{-m+3}R_{m-2}(\zeta').$$

When $P(\zeta)\neq 0$ and $P_{\xi^0}(\zeta'_{(1)})\neq 0$, we can expand $P(\zeta)^{-1}$ in a finite geometric series:

$$\frac{1}{P(\zeta)} = \sum_{j=0}^{N-1} R(\zeta)^j\zeta_1^{k-m-j}P_{\xi^0}^{-j-1} + R(\zeta)^N\zeta_1^{-N}P(\zeta)^{-1}P_{\xi^0}^{-N}$$

$$= A_N(\zeta) + B_N(\zeta)\ .$$

If we write

$$F(x) = (2\pi)^{-n}\int_{R^n-i\gamma\theta} A_N(\zeta)\ e^{i\langle x,\zeta\rangle}\ d\zeta\ ,$$

then we have

$$F(x) = H(x_1)\sum_{j=0}^{N-1}\Big(\sum_{k=m+j-2}^{(m-2)(j+1)} x_1^{k-1}\ Q_{j,k}(D_{x'})\Big)\ E(P_{\xi^0}^{j+1})$$

where $Q_{j,k}(D_{x'})$ is a differential operator of order $(-m+k+2j+2)$. As ξ^0 belongs to S*, the condition (**) is satisfied. The necessary and sufficient condition of (**) is that any $R_j(\zeta')$ is written as

$$R_j(\zeta') = \Sigma_{|\alpha|=2}\ r_{j,\alpha}(\zeta')(\zeta'_{(1)})^\alpha,$$

where $\alpha=(\alpha_2,..,\alpha_\ell)$ is multi-indeces. Hence we have

$$Q_{j,k}(\zeta') = \Sigma_{|\alpha|=2j}\ q_{j,k,\alpha}(\zeta')\ (\zeta'_{(1)})^\alpha$$

for any $k\ \varepsilon\ \{m+j-2,...,(m-2)(j+1)\}$. Using this property of $Q_{j,k}$, we can calculate the support of $F(x)$:

$$SF = \{x_1 \geq 0,\ x_2 = (\Sigma_{j=3}^{\ell} x_j^2)^{1/2}\ ,\ x'_{(2)}=0\ \}.$$

On the other hand, (4) means that, when $x \in U$, $x_2 > (\Sigma_{j=3}^{\ell} x_j^2)^{1/2}$. Hence, as $S\phi$ is contained in U, we see $< A_N(\zeta), \hat{\phi}(\eta-\zeta) >_\zeta = 0$ where $<f,g>_\zeta$ is the integral of f.g with respect to ζ, i.e.,

$$\widehat{E(P)_\phi}\ (\eta) = < B_N(\zeta), \hat{\phi}(\eta-\zeta) >_\zeta . \tag{6}$$

The second step. To estimate the right side hand of (6), we shall change the chain of integration as in Atiyah-Bott-Gårding [2], [3] and in Andersson [1]. But, in this case, we can not use their method directly. We perform its procedure three times. This is the essential point of our proof.

For any hyperbolic homogeneous polynomial $P(\xi)$, we denote by $\Gamma(P,\theta)$ the component of $R^n-\{\xi;\ P(\xi)=0\}$ containing θ which is an open convex cone.

At first, we choose a C^∞-vector field $\xi'_{(1)} \to w(\xi'_{(1)}) \in \Gamma((P_{\xi^0})_{\xi'_{(1)}}, -\theta)$, homogeneous of degree zero such that, for any $x \in U$ and $s \in [0,1]$,

$$< x'_{(1)}, w > \geq \varepsilon > 0 ,$$

$$P_{\xi^0}(\ \xi'_{(1)} + i(sw(\xi'_{(1)}) - (1-s)\gamma\theta\ )) \neq 0 .$$

It is important that we can take $w(\xi'_{(1)})$ as $P_{\xi^0}(w) < 0$ for any $\xi'_{(1)} \in \dot{R}^{\ell-1}$.

We now introduce many functions which are used to construct the various vector fields.

Let $\Phi_i \in C^\infty(R^n)$ be homogeneous of degree 0 such that $S\Phi_i \subset \Gamma_{i-1}$ and $\Phi_i = 1$ on Γ_i where $\Gamma_i = \{\xi \in R^n;\ \xi_1 > (M+i)|\xi'|\ \}$ for i=1,2,3, and $\xi' = (\xi_2,..,\xi_n)$. Moreover we put

$$\chi(\xi'_{(1)}) = \begin{cases} 1 & \text{in } \{\ |\xi'_{(1)}| \leq 1/2\ \} \\ 0 & \text{in } \{\ |\xi'_{(1)}| \geq 1\ \} \end{cases}$$

$$\rho_i(\xi') = \begin{cases} 1 & \text{in } \{\ |\xi'| \geq r_i+1\ \} \\ 0 & \text{in } \{\ |\xi'| \leq r_i\ \} \end{cases} \qquad \text{for i=1 and 2}$$

where $r_1=1$ and $r_2=\max\ \{3,\ 2|P_{\xi^0}(w)|^{-1}\} - 1$. At last, we define

$\psi_1(\xi) = \chi(\xi'_{(1)})\ \Phi_1(\xi)$ and $\psi_2(\xi) = \chi\ \Phi_1\ \{\ \rho_2(1-\Phi_3) + (1-\rho_2)\}$.

Next, we define vector fields v_i (ξ) $(i=1,2,3)$ as follows:

$$v_0 = -\gamma\theta \ , \quad v_1 = -\gamma|\xi'_{(1)}|^{\psi_1}\theta \ ,$$

$$v_2 = |\xi'_{(1)}|^{\psi_1} \{ \rho_1\Phi_2|\xi'| \hat{w}(\xi'_{(1)}) - (1 - \rho_1\Phi_2)\gamma\theta \} \ ,$$

$$v_3 = |\xi'_{(1)}|^{\psi_2} \{ \rho_1\Phi_2|\xi'| \hat{w}(\xi'_{(1)}) - (1 - \rho_1\Phi_2)\gamma\theta \}$$

where $\hat{w}(\xi'_{(1)})=(0,w(\xi'_{(1)}),0) \in R^n$. If M is sufficiently large where it is used in the definition of Γ_i, then it holds for $\xi'_{(1)}\neq 0$

$$P_{\xi^0}(\xi'_{(1)} + i(sv_j+(1-s)v_{j+1})'_{(1)}) \neq 0 \tag{7}$$

$$P(\xi + i(sv_j+(1-s)v_{j+1})) \neq 0 \ . \tag{8}$$

It is evident that $\langle x,v_j\rangle$ is bounded below for any j. The proof of (7) and (8) depends principally on the following inequality: If we put $\zeta'_{(1)}=\xi'_{(1)} + i(s|\xi'|w-(1-s)\gamma\theta)$, then there exists a constant $c>0$ such that

$$| P_{\xi^0}(\zeta'_{(1)}) | \geq c|\zeta'_{(1)}|^2$$

for all $\xi'\in R^{n-1}$, $s\in[0,1]$ and $\xi'_{(1)} \in \{ \xi'_{(1)}\in R^{\ell-1}; |\xi'_{(1)}|=1 \}$.

Let V_j be the chain given by $\zeta=\xi+iv_j(\xi)$ $(j=1,2,3)$. When we replace the chain of integration V_j by V_{j+1} in (6), we see by (7), (8) and Stokes' theorem that it does not change the value of (6), that is to say, we have

$$\widehat{E(P)\ \phi}(\eta) = \lim_{\varepsilon\to 0} \int_{V_3\cap\{|\xi'_{(1)}|>\varepsilon\}} B_N(\zeta)\ \hat{\phi}(\eta-\zeta)\ d\zeta \ , \tag{9}$$

where $\hat{\phi}$ is the Fourier transform of ϕ. From now on, we suppose that the integral over V_3 is taken in the sense of (9), even if we don't write the symbol of limit.

<u>The third step.</u> Since $\langle x,v_3\rangle$ is bounded below, it is well known that, for any $k>0$,

$$| \hat{\phi}(\eta-\zeta) | \leq C_k(1 + |\eta-\zeta|)^{-k} \ . \tag{10}$$

Let Γ be a conical neighborhood of ξ^0 with $\overline{\Gamma}-\{0\} \Subset \Gamma_3$, and V_{Γ_3} be the restriction of V_3 on Γ_3. Then

$$\widehat{E(P)\phi}\,(\eta) = \int_{V_3} B_N(\zeta)\,\hat{\phi}(\eta-\zeta)\,d\zeta$$

$$= \int_{V_3-V_{\Gamma_3}} (P(\zeta)^{-1} - A_N(\zeta))\,\hat{\phi}(\eta-\zeta)\,d\zeta + \int_{V_{\Gamma_3}} B_N(\zeta)\,\hat{\phi}(\eta-\zeta)\,d\zeta .$$

The estimate of the first term is easily obtained by the use of (10), because there exists a constant $\delta>0$ such that $|\zeta-\eta| \geq \delta(|\zeta|+|\eta|)$ for $\zeta\varepsilon V_3-V_{\Gamma_3}$ and $\eta\varepsilon\Gamma$.

To estimate the second term, we have on V_{Γ_3}

$$|B_N(\zeta)| \leq C_N|\zeta'|^N|\zeta_1|^{-N} .$$

When $|\zeta-\eta| \geq \delta\,(|\zeta|+|\eta|)$, its estimate is trivial. If $|\zeta-\eta| < \delta(|\zeta|+|\eta|)$, then there exists a constant $C>0$ such that $|\eta| < C|\zeta_1|$. On the other hand, since we have $|\zeta-\eta| \geq C'|\zeta'|$ for large $\zeta\varepsilon V_{\Gamma_3}$ where C' is a constant >0, we can obtain the estimate (5) for the second term. This completes the proof.

Final Remark. $E(P)(x)$ is also analytic outside SSE(P), which is proved by a little modification of the above proof.

REFERENCES

[1] K. G. Andersson, Localization and wave fronts. Séminaire Goulaouic-Schwartz 1972-73, Expose 25.

[2] M. F. Atiyah, R. Bott and L. Gårding, Lacunas for hyperbolic differential operators with constant coefficients I, Acta Math., 124(1970), pp. 109-189.

[3] ______________ , II, Acta Math., 131(1973), pp.145-206.

[4] L. Hörmander, On the singularities of solutions of partial differential equations, Comm. Pure Appl. Math., 23(1970), pp. 361-410.

[5] M. Tsuji, Propagation of the singularities for hyperbolic equations with constant coefficients, Jap. J. Math. (new series), 2(1976), pp. 361-410.

THE MIXED PROBLEM FOR HYPERBOLIC SYSTEMS

Seiichiro Wakabayashi

Institute of Mathematics, the University of Tsukuba
Ibaraki, Japan

1. INTRODUCTION

Lax [27] and Mizohata [34] proved that for the non-characteristic Cauchy problem to be C^∞ well-posed it is necessary that the characteristic roots are real. In the mixed problem Kajitani [23] obtained the results corresponding to those in the Cauchy problem under some restrictive assumptions. In §3 we shall relax his assumptions (see, also, [52]). We note that well-posedness of the mixed problems has been investigated by many authors ([1], [2], [19], [20], [22], [26], [32], [33], [38]). We shall consider the mixed problem for hyperbolic systems with constant coefficients in a quarter-space in §§4-6. Hersh [13], [14], [15] studied the mixed problem for hyperbolic systems with constant coefficients. He gave the necessary and sufficient condition for the mixed problem to be C^∞ well-posed. However, his proof seems to be incomplete (see [25]). Sakamoto [37] justified his results for single higher order hyperbolic equations (see, also, [40], [41], [42]). In §4 we shall consider C^∞ well-posedness of the mixed problem. Duff [11] studied the location and structures of singularities of the fundamental solutions of the mixed problems for single higher order hyperbolic equations, using the stationary phase method. Inner estimates of the location of singularities of the fundamental solutions of the hyperbolic mixed problems were given in [29], [46], [47], [48], using the localization method developed by Atiyah, Bott and Gårding [5] and Hörmander [17]. Outer estimates were given in [49]. In [51] the author studied the wave front sets of the fundamental solutions of single higher order hyperbolic mixed

H. G. Garnir (ed.), Singularities in Boundary Value Problems, 327–370.

problems under the only assumption that the mixed problems are C^∞ well-posed. In §6 we shall apply the methods in [51] to the mixed problem for hyperbolic systems and investigate the wave front sets of the Poisson kernels. Of course, we can investigate the wave front sets of the fundamental solutions in a similar way. We must note that Garnir studied the mixed problems for hyperbolic systems (see [28]). We also note that singularities of solutions of the hyperbolic mixed problems with variable coefficients have been studied by many authors ([4], [9], [10], [12], [30], [31], [36], [43], [45], [50]).

2. PRELIMINARIES

Let $\mathbb{R}^n$ denote the n-dimensional Euclidean space and write $x'=(x_1,\cdots,x_{n-1})$ for the coordinate $x=(x_1,\cdots,x_n)$ in $\mathbb{R}^n$ and $\xi'=(\xi_1,\cdots,\xi_{n-1})$ for the dual coordinate $\xi=(\xi_1,\cdots,\xi_n)$. We shall also denote by $\mathbb{R}^n_+$ the half-space $\{x=(x', x_n)\in\mathbb{R}^n;\ x_n>0\}$. For differentiation we shall use the symbols $D\equiv(D_1,\cdots,D_n) =-i(\partial/\partial x_1,\cdots,\partial/\partial x_n)$, $D'=-i(\partial/\partial x_1,\cdots,\partial/\partial x_{n-1})$. Let $L(x, D) \equiv(L_{ij}(x, D))$ be a partial differential operator with $N\times N$ matrix coefficients whose entries are infinitely differentiable in x and let $B(x', D)\equiv(B_{ij}(x', D))$ be a partial differential operator with $\ell\times N$ matrix coefficients whose entries are infinitely differentiable in x'. For an open neighborhood Ω of $x=0$ in $\overline{R^n_+}$ we write

$$\Omega_T^\pm = \{x\in\Omega;\ \pm(x_1-T)\geqq 0\},\quad T\in\mathbb{R}.$$

We consider the mixed problem for $\{L, B\}$:

$$\left.\begin{aligned} &L(x, D)u(x) = f(x) \quad \text{in } \Omega, \\ &B(x', D)u(x)|_{x_n=0} = g(x') \quad \text{in } \hat{\Omega}, \\ &\operatorname{supp} u(x) \subset \Omega_T^+, \end{aligned}\right\} \tag{2. 1}$$

where $u = {}^t(u_1,\cdots,u_N)$, $f = {}^t(f_1,\cdots,f_N)$, $g = {}^t(g_1,\cdots,g_\ell)$ and $\hat{\Omega} = \{x'\in\mathbb{R}^{n-1};\ (x', 0)\in\Omega\}$ and T is fixed in $\mathbb{R}$.

Definition 2.1. The mixed problem (2. 1) is said to be C^∞ well-posed in Ω_T^+ if the following conditions hold:

(E) For any $f \in C^\infty(\Omega)$ and $g \in C^\infty(\hat{\Omega})$ with $\text{supp}\, f \subset \Omega_T^+$ and $\text{supp}\, g \subset \hat{\Omega}_T^+$ there exists u in $C^\infty(\Omega)$ which satisfies (2. 1).

(U) $u=0$ in Ω_t^- for each $t>T$, provided that $u \in C^\infty(\Omega)$, $\text{supp}\, u \subset \Omega_T^+$, $Lu=0$ in Ω_t^- and $Bu|_{x_n=0}=0$ in $\hat{\Omega}_t^-$.

Moreover the mixed problem (2. 1) is said to be C^∞ well-posed if it is C^∞ well-posed in $\Omega_T^+ \equiv \{x \in \overline{\mathbb{R}_+^n};\ x_1 \geqq T\}$ for any T in $\mathbb{R}$.

Definition 2.2. Let Γ be a proper convex open cone in $\mathbb{R}^n$ with its vertex at the origin such that $\overline{\Gamma} \subset \{x \in \mathbb{R}^n;\ x_1 > 0\} \cup \{0\}$. We say that the mixed problem (2. 1) has the finite propagation property with the cone Γ if the following statement holds for every x^0 in Ω_T^+ with $x_1^0 > T$ and $\Gamma(x^0) \cap \Omega_T^+ \subset\subset \Omega_T^+$: If $u \in C^\infty(\Omega)$, $\text{supp}\, u \subset \Omega_T^+$, $Lu=0$ in $\Gamma(x^0) \cap \Omega_T^+$ and $Bu|_{x_n=0} = 0$ in $\hat{\Gamma}(x^0) \cap \hat{\Omega}_T^+$, then $u=0$ in $\Gamma(x^0) \cap \Omega$. Here the notation $K \subset\subset \Omega$ means that $\overline{K} \subset \Omega$ and $\overline{K}$ is compact and we write

$$\Gamma(x^0) = \{x^0\} - \Gamma, \quad \hat{\Gamma}(x^0) = \{x' \in \mathbb{R}^{n-1};\ (x', 0) \in \Gamma(x^0)\}.$$

Lemma 2.1 ([52]). Assume that the mixed problem (2. 1) is C^∞ well-posed in Ω_T^+. Then for any compact set K in Ω_T^+ and any non-negative integer p there exist a positive constant $C_{p,K}$ and a non-negative integer q such that

$$|u|_{p,K_t^-} \leqq C_{p,K}\{|Lu|_{q,K_t^-} + |Bu|_{x_n=0}|_{q,\hat{K}_t^-}\}$$

for every $u \in C^\infty(\Omega)$ with $\text{supp}\, u \subset K$ and every $t \in \mathbb{R}$, where

$$|f|_{p,K} = \sup_{x \in K,\ |\alpha| \leqq p} |D^\alpha f(x)|,$$

$$|g|_{p,\hat{K}} = \sup_{x' \in \hat{K},\ |\alpha'| \leqq p} |D'^{\alpha'} g(x')|,$$

$$\alpha = (\alpha_1, \cdots, \alpha_n) = (\alpha', \alpha_n).$$

Lemma 2.2 ([52]). Assume that the mixed problem (2. 1) is C^∞ well-posed in Ω_T^+ and has the finite propagation property with Γ. Then for any compact set K in Ω_T^+ and any non-negative integer p there exist a positive constant $C_{p,K}$ and a non-negative integer q such that

$$|u|_{p,\Gamma(x^0)\cap\Omega} \leqq C_{p,K}(x_1^0-T)^{-q}\{|Lu|_{q,\Gamma(x^0)\cap\Omega} + |Bu|_{x_n=0}|_{q,\hat{\Gamma}(x^0)\cap\hat{\Omega}}\}$$

for every $u\in C^\infty(\Omega)$ with $\operatorname{supp} u\subset K$ and every $x^0=(x^{0\prime}, 0)\in\Omega_T^+$ with $x_1^0>T$ and $\Gamma(x^0)\cap\Omega_T^+\subset\subset\Omega_T^+$.

Let us define the Lopatinski determinant $\Delta(x', \xi')$ of the system $\{L, B\}$ for each fixed $x'\in\hat{\Omega}_T^+$. Put

$$P(x, \xi) = \det L(x, \xi).$$

In this section we assume that

(i) $P(x', 0, \xi)$ is a hyperbolic polynomial with respect to $\vartheta = (1,0,\cdots,0)\in\mathbb{R}^n$ for each fixed $x'\in\hat{\Omega}_T^+$, i.e., there is a constant γ_0 such that

$$P(x', 0, \xi) \neq 0 \quad \text{for} \quad \operatorname{Im}\xi_1<-\gamma_0 \quad \text{and} \quad \xi'' = (\xi_2,\cdots,\xi_n)\in\mathbb{R}^{n-1},$$

$$P^0(x', 0, -i\vartheta) \neq 0,$$

where $P^0(x, \xi)$ is the principal part of $P(x, \xi)$:

$$P(x, \rho^{-1}\xi) = \rho^{-m}(P^0(x, \xi) + o(1)) \quad \text{as} \quad \rho\to 0, \quad P^0(x, \xi)\not\equiv 0 \quad \text{in} \quad \xi,$$

for each fixed $x\in\Omega_T^+$, and that

(ii) the number of the roots with positive imaginary part of the equation $P(x', 0, -i(\gamma_0+1)\vartheta', \lambda) = 0$ is equal to ℓ and $\ell\geqq 1$.

Let $\Gamma_{x'}(P)$ be the connected component of the set $\{\xi\in\mathbb{R}^n;\ P^0(x', 0, -i\xi) \neq 0\}$ containing ϑ for each fixed $x'\in\hat{\Omega}_T^+$ and put

$$\dot{\Gamma}_{x'}(P) = \{\xi'\in\mathbb{R}^{n-1};\ \xi\in\Gamma_{x'}(P) \quad \text{for some} \quad \xi_n\in\mathbb{R}\}.$$

In particular, we write

$$\Gamma(P) = \Gamma_{x'}(P), \quad \dot{\Gamma}(P) = \dot{\Gamma}_{x'}(P),$$

if $P(x, \xi)$ is independent of x. Then from the above assumptions it follows that

$$\deg_\lambda P(x', 0, \xi', \lambda) = \deg_\lambda P(x', 0, -i(\gamma_0+1)\mathscr{I}', \lambda)$$

for $x'\in\hat{\Omega}^+_T$ and $\xi'\in\mathbf{R}^{n-1}-i\gamma_0\mathscr{I}'-i\dot{\Gamma}_{x'}(P)$, and that the number of the roots with positive imaginary part of $P(x', 0, \xi', \lambda) = 0$ is equal to ℓ for $x'\in\hat{\Omega}^+_T$, $\operatorname{Im}\xi_1<-\gamma_0$ and $\xi'''=(\xi_2,\cdots,\xi_{n-1})\in\mathbf{R}^{n-2}$. Put

$$\mathscr{A}(x', \xi) = B(x', \xi)\,{}^t\mathrm{cof}\, L(x', 0, \xi),$$

$$P_+(x', 0, \xi) = \Pi_{j=1}^{\ell}(\xi_n-\lambda_j^+(x', \xi')),$$

for $x'\in\hat{\Omega}^+_T$, $\operatorname{Im}\xi_1<-\gamma_0$ and $\xi'''\in\mathbf{R}^{n-2}$, where the $\lambda_j^+(x', \xi')$ are the roots with positive imaginary part of $P(x', 0, \xi', \lambda) = 0$ and ${}^t\mathrm{cof}\, L$ denotes the transposed matrix of the cofactor matrix of L, i.e., ${}^t\mathrm{cof}\, L\cdot L = \det L\cdot I_N$. Here I_N denotes the $N\times N$ identity matrix. Moreover define the $\ell\times N$ matrix $\tilde{\mathscr{A}}(x', \xi) \equiv (a_{ij}(x', \xi))$ and the $\ell\times\ell N$ matrix $\mathscr{L}(x', \xi')$ by

$$\tilde{\mathscr{A}}(x', \xi) \equiv \mathscr{A}(x', \xi) \quad (\mathrm{mod}\ P_+(x', 0, \xi) \text{ as a polynomial of } \xi_n),$$

$$a_{ij}(x', \xi) = \textstyle\sum_{s=1}^{\ell} a_{ij}^{(s)}(x', \xi')\xi_n^{s-1},$$

$$\mathscr{L}(x', \xi') = \begin{pmatrix} a_{11}^{(1)} \cdots a_{11}^{(\ell)} a_{12}^{(1)} \cdots a_{1N}^{(\ell)} \\ \vdots \qquad\quad \vdots \quad \vdots \qquad\quad \vdots \\ a_{\ell 1}^{(1)} \cdots a_{\ell 1}^{(\ell)} a_{\ell 2}^{(1)} \cdots a_{\ell N}^{(\ell)} \end{pmatrix}.$$

Definition 2.3. For $x'\in\hat{\Omega}^+_T$, $\operatorname{Im}\xi_1<-\gamma_0$ and $\xi'''\in\mathbf{R}^{n-2}$ we define the Lopatinski determinant $\Delta(x', \xi')$ by

$$\Delta(x', \xi') = (\textstyle\sum_i |\Delta_i(x', \xi')|^2)^{1/2},$$

where the $\Delta_i(x', \xi')$ are the minor determinants of order ℓ of $\mathscr{L}(x', \xi')$.

Remark. $P_+(x', 0, \xi)$ can be defined for $x'\in\hat{\Omega}^+_T$, $\xi'\in\mathbf{R}^{n-1}-i\gamma_0\mathscr{I}'-i\dot{\Gamma}_{x'}(P)$ and $\xi_n\in\mathbf{C}$ ([37], [51]). Therefore $\Delta(x', \xi')$ can be also defined for $x'\in\hat{\Omega}^+_T$ and $\xi'\in\mathbf{R}^{n-1}-i\gamma_0\mathscr{I}'-i\dot{\Gamma}_{x'}(P)$.

Lemma 2.3 ([3], [44]). For $x^{0\prime} \varepsilon \hat{\Omega}_T^+$ and $\xi^{0\prime} \varepsilon R^{n-1} - i\gamma_0 \mathcal{J}' - i\dot{\Gamma}_{x^{0\prime}}(P)$ the following conditions are equivalent:

(i) The system

$$\begin{cases} L(x^{0\prime}, 0, \xi^{0\prime}, D_n)v(x_n) = 0, \quad x_n > 0, \\ B(x^{0\prime}, \xi^{0\prime}, D_n)v(x_n)|_{x_n=0} = g, \\ \exp[x_n \cdot \xi_n^0]v(x_n) \ \varepsilon\ L^2(\mathbb{R}_+) \end{cases}$$

has a (unique) solution $v(x_n)$ for any $g \varepsilon \mathbb{C}^\ell$, where ξ_n^0 is chosen so that $(-\mathrm{Im}\ \xi^{0\prime}, -\xi_n^0) \varepsilon \gamma_0 \mathcal{J} + \Gamma_{x^{0\prime}}(P)$.

(ii) The row vectors of $\mathcal{A}(x^{0\prime}, \xi^{0\prime}, \xi_n)$ are linearly independent modulo $P_+(x^{0\prime}, 0, \xi^{0\prime}, \xi_n)$.

(iii) rank $\mathcal{L}(x^{0\prime}, \xi^{0\prime}) = \ell$, i.e., $\Delta(x^{0\prime}, \xi^{0\prime}) \neq 0$.

Lemma 2.4. For each fixed $x' \varepsilon \hat{\Omega}_T^+$ and $\xi' \varepsilon R^{n-1} - i\gamma_0 \mathcal{J}' - i\dot{\Gamma}_{x'}(P)$ there exist ℓ column vectors of the matrix

$$({}^t\mathrm{cof}\ L(x', 0, \xi), P_{+1}\,{}^t\mathrm{cof}\ L, \cdots, P_{+\ell-1}\,{}^t\mathrm{cof}\ L)$$

which are linearly independent modulo $P_+(x', 0, \xi)$, where

$$P_+(x', 0, \xi) = \sum_{j=0}^{\ell} a_j(x', \xi')\xi_n^{\ell-j},$$

$$P_{+s}(x', 0, \xi) = \sum_{j=0}^{s} a_j(x', \xi')\xi_n^{s-j}, \quad 0 \leqq s \leqq \ell-1.$$

Moreover any $(\ell+1)$ column vectors of the above matrix are linearly dependent modulo $P_+(x', 0, \xi)$.

The above lemma can be proved by using the theory of elementary divisers.

3. A NECESSARY CONDITION FOR THE MIXED PROBLEM TO BE C^∞ WELL-POSED

In this section we assume that

(L-1) there exist integers $m_1, \cdots, m_N$, $n_1, \cdots, n_N$, such that $\deg_\xi L_{ij}(x, \xi) \leqq m_i + n_j$, and $P^0(x, \xi) = \det L^0(x, \xi)$, where

$L^0_{ij}(x, \xi)$ is the homogeneous part of degree m_i+n_j of $L_{ij}(x, \xi)$ and $L^0(x, \xi)=(L^0_{ij}(x, \xi))$. Here it is to be understood that the degree of $\ell(\xi)\equiv 0$ is arbitrary. Moreover assume that

(L-2) $P^0(x, -i\vartheta)\neq 0$ for $x\in\Omega^+_T$.

Then the Lax-Mizohata theorem yields us the following

Proposition 3.1 ([21], [27], [34]). Under the conditions (L-1) and (L-2), it is necessary for the mixed problem (2. 1) to be C^∞ well-posed in Ω^+_T that $P^0(x, \xi)$ is hyperbolic with respect to ϑ for each x in Ω^+_T, i.e.,

$$P^0(x, \xi) \neq 0 \text{ for } x\in\Omega^+_T, \text{ Im } \xi_1<0 \text{ and } \xi''\in R^{n-1}.$$

By Proposition 3.1 we may assume instead of the condition (L-2) that

(L-2)' $P^0(x, \xi)$ is hyperbolic with respect to ϑ for each $x\in\Omega^+_T$.

Moreover we impose the following conditions on the system $\{L, B\}$:

(L-3) $\deg_\lambda P^0(x', 0, -i\vartheta', \lambda)$ is independent of x' in $\hat{\Omega}^+_T$ and $\hat{\Omega}^+_T$ is connected, $0\in\hat{\Omega}^+_T$.

(L-4) The number of the roots with positive imaginary part of the equation $P^0(0, -i\vartheta', \lambda) = 0$ is equal to ℓ and $\ell\geqq 1$.

(B-1) There exist integers $b_1,\cdots,b_\ell$ such that

$$\deg_\xi B_{ij}(x', \xi) \leqq b_i+n_j, \quad \Delta^0(x', \xi')\neq 0 \text{ in } \xi',$$

where $B^0_{ij}(x', \xi)$ is the homogeneous part of degree b_i+n_j of $B_{ij}(x', \xi)$, $B^0(x', \xi)=(B^0_{ij}(x', \xi))$ and $\Delta^0(x', \xi')$ is the Lopatinski determinant of the system $\{L^0, B^0\}$.

(B-3) $\Delta^0(x', -i\vartheta')\neq 0$ for every $x'\in\hat{\Omega}^+_T$.

It follows from (L-2)', (L-3) and (L-4) that the number of the roots with positive imaginary part of $P^0(x', 0, -i\vartheta', \lambda)=0$ is equal to ℓ for $x'\in\hat{\Omega}^+_T$. Therefore we can define $\Delta^0(x', \xi')$.

Theorem 3.1 ([52]). Assume that the conditions (L-1)-(L-4), (B-1) and (B-2) are valid. Then it is necessary for the mixed problem (2. 1) to be C^∞ well-posed in Ω_T^+ that the system $\{L^0, B^0\}$ satisfies the Lopatinski condition, i.e.,

$$\Delta^0(x', \xi') \neq 0 \quad \text{for} \quad x' \varepsilon \hat{\Omega}_T^+, \quad \text{Im } \xi_1 < 0 \quad \text{and} \quad \xi'' \varepsilon \mathbb{R}^{n-2}.$$

The following proposition was proved by Mizohata [35] (see, also, [21], [24]).

Proposition 3.2. Assume that the condition (L-1) is valid and that the mixed problem (2. 1) is C^∞ well-posed in Ω_T^+ and has the finite propagation property with Γ. Then we have

$$P^0(x^0, \xi) \neq 0 \quad \text{for} \quad x^0 \varepsilon \Omega_T^+, \quad \Gamma(x^0) \cap \Omega_T^+ \subset\subset \Omega_T^+ \quad \text{and} \quad \xi \varepsilon \mathbb{R}^n - i(\text{int } \Gamma^*),$$

where $\Gamma^* = \{\xi \varepsilon \mathbb{R}^n;\ x\cdot\xi \geqq 0 \text{ for all } x \varepsilon \Gamma\}$ and int Γ^* denotes the interior of Γ^* in $\mathbb{R}^n$.

Theorem 3.2 ([52]). Assume that the conditions (L-1), (L-3), (L-4) and (B-1) are valid and that the mixed problem (2. 1) is C^∞ well-posed in Ω_T^+ and has the finite propagation property with Γ. Then we have

$$\Delta^0(x^{0\prime}, \xi') \neq 0 \quad \text{for} \quad x^{0\prime} \varepsilon \hat{\Omega}_T^+, \quad \Gamma(x^{0\prime}, 0) \cap \Omega_T^+ \subset\subset \Omega_T^+ \quad \text{and} \quad \xi' \varepsilon \mathbb{R}^{n-1} - i(\text{int } \dot{\Gamma}^*),$$

where $\dot{\Gamma}^* = \{\xi' \varepsilon \mathbb{R}^{n-1};\ \xi \varepsilon \Gamma^* \text{ for some } \xi_n \varepsilon \mathbb{R}\}$ and int $\dot{\Gamma}^*$ denotes the interior of $\dot{\Gamma}^*$ in $\mathbb{R}^{n-1}$.

Theorems 3.1 and 3.2 can be proved by improving the method in [21] and using Lemmas 2.1 and 2.2. For details we refer to [52]. We note that Chazarain [8] and Beals [6] gave sufficient conditions for the mixed problem to be well-posed in Gevrey classes.

4. C^∞ WELL-POSEDNESS (CONSTANT COEFFICIENT CASE)

From now on we assume that the coefficients of $L(x, \xi)$ and $B(x', \xi)$ are constant. So we write $L(\xi)=L(x, \xi)$, $B(\xi)=B(x', \xi)$, $P(\xi)=P(x, \xi), \cdots$, and so on. We shall consider the mixed problem

$$\left.\begin{aligned} &L(D)u(x) = f(x), \quad x_n>0, \\ &B(D)u(x)|_{x_n=0} = g(x'), \quad x'\in\mathbb{R}^{n-1}, \\ &\text{supp } u \subset \{x_1 \geqq 0\}, \end{aligned}\right\} \qquad (4.\ 1)$$

where $f\in C^\infty(\overline{R^n_+})$ with supp $f\subset\{x_1\geqq 0\}$ and $g\in C^\infty(R^{n-1})$ with supp $g\subset\{x_1\geqq 0\}$. Assume that

(A-1) $P^0(-i\vartheta)\neq 0$.

Then we have the following

Theorem 4.1. Under the condition (A-1) it is necessary for the mixed problem (4. 1) to be C^∞ well-posed that $P(\xi)$ is hyperbolic with respect to ϑ.

Proof. Let Γ be an open convex cone with its vertex at the origin in $\mathbb{R}^n$ such that $\vartheta\in\Gamma$, $\Gamma\subset\{\xi\in\mathbb{R}^n;\ P^0(\xi)\neq 0\}$ and $\overline{\Gamma}\subset\{\xi\in\mathbb{R}^n;\ \xi_1>0\}\cup\{0\}$. Put

$$\mathcal{C}\equiv \text{int } \Gamma^* = \{x\in\mathbb{R}^n;\ x\cdot\xi>0 \text{ for any } \xi\in\Gamma\}\ (=\emptyset).$$

We choose $x^0\in\mathbb{R}^n$ so that $x^0_1>0$ and $x_n>0$ for $x\in\mathcal{C}(x^0)\cap\{x_1\geqq 0\}$, where $\mathcal{C}(x^0)=\{x^0\}-\mathcal{C}$. Then it follows from the Holmgren uniqueness theorem that $u=0$ in $\mathcal{C}(x^0)$, provided that $u\in C^\infty(\overline{\mathbb{R}^n_+})$, supp u $\subset\{x_1\geqq 0\}$ and $Lu=0$ in $\mathcal{C}(x_0)$ (see Corollary 5.3.3 in [16]). Let U be a neighborhood of x^0 such that $\bar{U}$ is compact, $\mathcal{C}(x^0)$ $\cap\{x_1\geqq 0\}\subset\subset U$ and $x_n>0$ for $x\in U$. For any $u\in C_0^\infty(U)$ with supp $u\subset\{x_1\geqq 0\}$ there exists $f\in C_0^\infty(U)$ such that $f(x)=Lu(x)$ in $\mathcal{C}(x^0)$, supp $f\subset\{x_1\geqq 0\}$ and

$$|f|_{q,U} \leqq C_q|Lu|_{q,\ \mathcal{C}(x^0)\cap\{x_1\geqq 0\}}$$

(see Bierstone [7]). Let $v(x)\in C^\infty(\overline{R^n_+})$ be a solution of (4. 1) with $g=0$. Then we have $v(x)=u(x)$ in $\mathcal{C}(x^0)$. Thus it follows from Banach's closed graph theorem that for any integer $p\geqq 0$ there exist an integer $q\geqq 0$ and a constant $C_p>0$ such that

$$|u|_{p,\ \mathcal{C}(x^0)\cap\{x_1\geqq 0\}} \leqq C_p|Lu|_{q,\ \mathcal{C}(x^0)\cap\{x_1\geqq 0\}} \qquad (4.\ 2)$$

for every $u \in C_0^\infty(U)$ with $\operatorname{supp} u \subset \{x_1 \geqq 0\}$. Now assume that for any $\gamma > 0$ there is $\xi \in \mathbf{C}^n$ such that $\operatorname{Im} \xi_1 < -\gamma$, $\xi'' \in \mathbf{R}^{n-1}$ and $P(\xi) = 0$. Consider the system

$$P(\xi) = 0, \quad \xi'' \in \mathbf{R}^{n-1}, \ \phi = \operatorname{Im} \xi_1 < 0, \quad |\xi| = r, \ r > 0$$

and put $\phi(r) = \min \phi$. Then Seidenberg's lemma means that there exist $c > 0$ and $p > 0$ $(p \in \mathbf{Q})$ such that

$$\phi(r) = -c(r^p + o(1)) \quad \text{as} \quad r \to \infty.$$

Therefore there exist $\xi(r)$ for $r > r_0 >> 0$ such that $|\xi(r)| = r$, $\operatorname{Im} \xi_1(r) = \phi(r)$, $\xi''(r) \in \mathbf{R}^{n-1}$ and $P(\xi(r)) = 0$. Since $\det L(\xi(r)) = 0$, there exist $v(r) \in \mathbf{C}^N$ such that

$$L(\xi(r))v(r) = 0, \quad |v(r)| = 1.$$

Choose $\chi \in C_0^\infty(U)$ so that $\chi(x) = 1$ on U_1, where U_1 is a neighborhood of x^0, and put

$$u_r(x) = \exp[i(x - x^0) \cdot \xi(r)]\chi(x)v(r).$$

Then we have

$$|Lu_r|_{q,\, \mathcal{C}(x^0) \cap \{x_1 \geqq 0\}} \leqq C_k r^{-k}, \quad k = 0, 1, 2, \cdots,$$

$$Bu_r|_{x_n = 0} = 0,$$

$$|u_r|_{0,\, \mathcal{C}(x^0) \cap \{x_1 \geqq 0\}} \geqq 1,$$

since $\operatorname{Re} i(x - x^0) \cdot \xi(r) \leqq -c' r^p$, $c' > 0$, if $x \in (U \backslash U_1) \cap \mathcal{C}(x^0)$. This contradicts (4. 2). Q.E.D.

By Theorem 4.1 we can assume without loss of generality that

(A-1)' $P(\xi)$ is hyperbolic with respect to ϑ, i.e., there exists a positive constant γ_0 such that

$$P(\xi) \neq 0 \quad \text{for} \quad \operatorname{Im} \xi_1 < -\gamma_0 \quad \text{and} \quad \xi'' \in \mathbf{R}^{n-1},$$

$$P^0(-i\vartheta) \neq 0.$$

Theorem 4.2. Assume that (A-1)' is valid and that (4. 1) is C^∞ well-posed. Then the number ℓ' of the roots with positive imaginary part of $P(-i(\gamma_0 + 1)\vartheta', \lambda) = 0$ is equal to ℓ.

Proof. We can prove that $\ell' \leqq \ell$, using the same method as in the proof of Theorem 4.3 below. Thus let us prove that $\ell' \geqq \ell$. Assume that $\ell' < \ell$. Let $u(x)$ be a solution of (4. 1) with $f=0$ and let $\tilde{v}(x)$ be a solution of the system

$$\begin{cases} {}^t\mathrm{cof}\ L(D)\tilde{v}(x) = \tilde{u}(x) \quad \text{in} \quad \mathbb{R}^n, \\ \mathrm{supp}\ \tilde{v} \subset \{x_1 \geqq 0\}, \end{cases}$$

where $\tilde{u} \in C^\infty(\mathbb{R}^n)$ with $\mathrm{supp}\ \tilde{u} \subset \{x_1 \geqq 0\}$ is an extension of u (see [39]). Put $v = \tilde{v}|_{x_n \geqq 0}$. Then we have

$$\begin{cases} P(D)v(x) = 0, \quad x_n > 0 \\ B(D)\,{}^t\mathrm{cof}\ L(D)v(x)|_{x_n=0} = g(x'), \quad x' \in \mathbb{R}^{n-1}, \\ \mathrm{supp}\ v \subset \{x_1 \geqq 0\}. \end{cases}$$

Since the $a_j(\xi')$ are analytic in $\mathbb{R}^{n-1} - i\gamma_0 \vartheta' - i\dot{\Gamma}(P)$, we can define

$$A_j(x') = (2\pi)^{-n+1} \int_{\mathbb{R}^{n-1} - i\gamma\vartheta'} \exp[ix' \cdot \xi'] a_j(\xi') d\xi', \quad \gamma > \gamma_0,$$

and we have $\mathrm{supp}\ A_j \subset (\dot{\Gamma}(P))^*$ by the Paley-Wiener-Schwartz theorem, where $(\dot{\Gamma}(P))^* = \{x' \in \mathbb{R}^{n-1};\ x' \cdot \xi' \geqq 0 \ \text{for any} \ \xi' \in \dot{\Gamma}(P)\}$. Define

$$P_+(D)f(x) = \sum_{j=0}^{\ell'} A_j(x') \underset{x'}{*} (D_n^{\ell'-j} f(x))$$

for $f \in C^\infty(\overline{R^n_+})$ with $\mathrm{supp}\ f \subset \{x_1 \geqq 0\}$. Since $v(x)$ is uniquely determined by its Dirichlet data, we have

$$P_+(D)v(x) = 0, \quad x_n > 0.$$

From Lemma 2.4 it follows that

$$\mathrm{rank}\ \mathcal{L}(\xi') \leqq \ell' \quad \text{for} \quad \xi' \in \mathbb{R}^{n-1} - i\gamma_0 \vartheta' - i\dot{\Gamma}(P).$$

Thus there exists an integer ℓ_1 such that $0 \leqq \ell_1 \leqq \ell'$, rank $\mathcal{L}(\xi') \leqq \ell_1$ and there is a minor determinant of order ℓ_1 of $\mathcal{L}(\xi')$ which does not identically vanish. We can assume without loss of generality that

$$\det\,(a_{ij_\mu}^{(s_\mu)})_{\substack{\mu\to 1,\cdots,\ell_1\\ i\downarrow 1,\cdots,\ell_1}} \not\equiv 0 \quad \text{in}\quad \xi'.$$

Put

$$b_k(\xi')=(-1)^{k-1}\det\begin{pmatrix} a_{1j_\mu}^{(s_\mu)}\\ \vdots \\ \widehat{a_{kj_\mu}^{(s_\mu)}} \\ \vdots \\ a_{\ell_1+1j_\mu}^{(s_\mu)}\end{pmatrix}, \quad 1\leqq k\leqq \ell_1+1,$$

where the k-th row is omitted. Then we have

$$(b_1(\xi'),\cdots,b_{\ell_1+1},\overbrace{0,\cdots,0}^{\ell-\ell_1-1})\mathcal{L}(\xi') \equiv 0 \quad \text{in}\quad \xi'.$$

Therefore we have

$$(b_1(\xi'),\cdots,b_{\ell_1+1},0,\cdots,0)B(\xi)^t\mathrm{cof}\ L(\xi) \equiv 0 \quad (\mathrm{mod}\ P_+(\xi))$$

for every $\xi'\in\mathbb{R}^{n-1}-i\gamma_0\vartheta'-i\dot{\Gamma}(P)$. Since the $b_k(\xi')$ are analytic in $\mathbb{R}^{n-1}-i\gamma_0\vartheta'-i\dot{\Gamma}(P)$, we can define

$$B_k(x') = (2\pi)^{-n+1}\int_{\mathbb{R}^{n-1}-i\gamma\vartheta'}\exp[ix'\cdot\xi']b_k(\xi')d\xi', \quad \gamma>\gamma_0,$$

and we have $\mathrm{supp}\ B_k\subset(\dot{\Gamma}(P))^*$. Define

$$b_k(D')f(x) = B_k(x')\underset{x'}{*}f(x)$$

for $f\in C^\infty(\overline{R^n_+})$ with $\mathrm{supp}\ f\subset\{x_1\geqq 0\}$. Then we have

$$(b_1(D'),\cdots,b_{\ell_1+1},0,\cdots,0)B(D)^t\mathrm{cof}\ L(D)v(x) = 0, \quad x_n>0.$$

By continuity of convolution operators we have

$$(b_1,\cdots,b_{\ell_1+1},0,\cdots,0)(B^t\mathrm{cof}\ L\cdot v(x)|_{x_n=0}) = 0.$$

Therefore we have

$$b_{\ell_1+1}(D')g_{\ell_1+1}(x') = 0 \tag{4. 3}$$

for any $g_{\ell_1+1}(x') \varepsilon C^\infty(\mathbb{R}^{n-1})$ with $\operatorname{supp} g_{\ell_1+1} \subset \{x_1 \geqq 0\}$. On the other hand we have $b_{\ell_1+1}(\xi') \neq 0$ in ξ', which contradicts (4. 3). Q.E.D.

By Theorem 4.2 we may assume that

(A-2) the number of the roots with positive imaginary part of the equation $P(-i(\gamma_0+1)\vartheta', \lambda)=0$ is equal to ℓ.

The following results were obtained by Hersh [13], [14], [15] (see, also, [25]).

Theorem 4.3. Under the condition (A-1) it is necessary for the mixed problem (4. 1) to be C^∞ well-posed that the Lopatinski determinant $\Delta(\xi')$ of the system $\{L, B\}$ is hyperbolic with respect to ϑ', i.e., there exist a constant $\gamma_1(\geqq\gamma_0)$ and an open convex cone Σ with its vertex at the origin in $\mathbb{R}^{n-1}$ such that $\vartheta' \varepsilon \Sigma$ and

$$\Delta(\xi') \neq 0 \quad \text{for} \quad \xi' \varepsilon \mathbb{R}^{n-1} - i\gamma_1\vartheta' - i\Sigma.$$

Proof. We shall prove the theorem, applying the method used in [37]. Assume that there exists a sequence $\{\xi^{h}{}'\} \subset \mathbb{C}^{n-1}$ such that $\operatorname{Im} \xi_1^h \to -\infty$, $|\operatorname{Im} \xi^{h}{}'''| < |\operatorname{Im} \xi_1^h|/h$, $\log(1+|\xi^{h}{}'|)+1 < |\operatorname{Im} \xi_1^h|/h$ and $\Delta(\xi^{h}{}') = 0$. Put

$$(c_1^{(k)}(\xi),\cdots,c_N^{(k)}(\xi)) \equiv P_{+k-1}(\xi)\,{}^t\mathrm{cof}\, L(\xi) \pmod{P_+(\xi)},$$

$$c_j^{(k)}(\xi) = \textstyle\sum_{s=1}^{\ell} \xi_n^{s-1} c_j^{(sk)}(\xi'),$$

$$c_j^{(sk)}(\xi') = {}^t(c_{1j}^{(sk)},\cdots,c_{Nj}^{(sk)}),$$

$$\mathcal{C}(\xi') = \begin{pmatrix} c_{11}^{(11)} & \cdots & c_{1N}^{(11)} c_{11}^{(12)} & \cdots & c_{1N}^{(1\ell)} \\ \vdots & & & & \vdots \\ c_{N1}^{(11)} & & & & \vdots \\ c_{11}^{(21)} & & & & \vdots \\ \vdots & & & & \vdots \\ c_{N1}^{(\ell 1)} & \cdots & \cdots & \cdots & c_{NN}^{(\ell\ell)} \end{pmatrix},$$

$$\mathcal{D}(\xi') = (\sum_i |\mathcal{D}_i(\xi')|^2)^{1/2},$$

where the $\mathcal{D}_i(\xi')$ are the minor determinants of order ℓ of $\mathcal{C}(\xi')$. It follows from Lemma 2.4 that $\mathcal{D}(\xi')>0$ for $\xi'\in \mathbb{R}^{n-1}-i\gamma_0\mathcal{J}'-i\dot{\Gamma}(P)$. Seidenberg's lemma gives that there exist $c>0$ and $\delta\in\mathbb{Q}$ such that

$$|\mathcal{D}_i(\xi^{h,})| \geqq cr_h^\delta(1+o(1)) \quad \text{as } h\to\infty,$$

for some fixed i, where $r_h=|\xi^{h,}|$, if necessary, choosing a subsequence of $\{\xi^{h,}\}$. Write

$$\mathcal{D}_i(\xi') = \det(C_{i_\mu j_\nu}^{(s_\mu k_\nu)})_{\substack{\nu\to 1,\cdots,\ell \\ \mu\downarrow 1,\cdots,\ell}}.$$

Then $\{P_{+k_\nu-1}(\xi)C_{j_\nu}^{(1)}(\xi)\}_{\nu=1,\cdots,\ell}$ are linearly independent modulo $P_+(\xi)$ for $\xi'=\xi^{h,}$. Moreover there exists $\Delta_{i'}(\xi')$ such that

$$\Delta_{i'}(\xi') = \det((2\pi i)^{-1}\oint P_+(\xi)^{-1}B(\xi)(P_{+k_1-1}C_{j_1}^{(1)},\cdots,P_{+k_\ell-1}C_{j_\ell}^{(1)})d\xi_n).$$

Since $\Delta_{i'}(\xi^{h,})=0$, there exist $\{\alpha_{h\nu}\}_{1\leqq\nu\leqq\ell}\subset\mathbb{C}$, $h=1,2,\cdots$, such that

$$\sum_{\nu=1}^{\ell} |\alpha_{h\nu}|^2 = 1,$$

$$v_h(x_n) = \sum_{\nu=1}^{\ell} \alpha_{h\nu}(2\pi i)^{-1}\oint \exp[ix_n\cdot\xi_n] \times P_+(\xi^{h,},\xi_n)^{-1}P_{+k_\nu-1}(\xi^{h,},\xi_n)C_{j_\nu}^{(1)}(\xi^{h,},\xi_n)d\xi_n,$$

$$B(\xi^{h,},D_n)v_h(x_n)|_{x_n=0} = 0.$$

It is obvious that

$$L(\xi^{h,},D_n)v_h(x_n) = 0, \quad x_n>0.$$

Since

$$((P_{+\ell-s_\mu}(\xi^{h,}, D_n)v_h|_{x_n=0})_{i_\mu})_{\mu\downarrow 1,\cdots,\ell}$$

$$= (\sum_{\nu=1}^{\ell} \alpha_{h\nu}C_{i_\mu j_\nu}^{(s_\mu k_\nu)}(\xi^{h,}))_{\mu\downarrow 1,\cdots,\ell},$$

there exist $c>0$ and $a\in\mathbb{Q}$ such that

$$|((P_{+\ell-s_\mu}(\xi^{h\prime}, D_n)v_h|_{x_n=0})_{i_\mu})_{\mu\neq 1,\cdots,\ell}|$$
$$\geqq cr_h^a(1+o(1)) \quad \text{as} \quad h\to\infty.$$

Let $x^0=(1,0,\cdots,0)$ and choose $\phi(x_1)\in C^\infty(\mathbb{R})$ so that $\phi(x_1)=0$ if $x_1\leqq 0$, and $\phi(x)=1$ if $x_1\geqq 1/2$. Put

$$u_h(x) = \exp[i(x'-x^{0\prime})\cdot\xi^{h\prime}]\phi(x_1)v_h(x_n).$$

C^∞ well-posedness means that for any compact set K in $\overline{\mathbb{R}^n_+}$ and any integer $p\geqq 0$ there exist a compact set K' in $\overline{\mathbb{R}^n_+}$, $C_{p,K}>0$ and an integer $q\geqq 0$ such that

$$|u|_{p,K_t^-} \leqq C_{p,K}\{|Lu|_{q,K'^-_t} + |Bu|_{x_n=0}|_{q,\hat{K}'^-_t}\} \tag{4. 4}$$

for every $u\in C^\infty(\overline{\mathbb{R}^n_+})$ with $\operatorname{supp} u\subset\{x_1\geqq 0\}$ and every $t\in\mathbb{R}$. It is easy to see that

$$|Lu_h|_{q,K'^-_1} \leqq C_k(1+r_h)^{-k},$$

$$|Bu_h|_{x_n=0}|_{q,\hat{K}'^-_1} \leqq C_k(1+r_h)^{-k}, \quad k=1,2,\cdots,$$

$$|((P_{+\ell-s_\mu}(\xi^{h\prime}, D_n)u_h|_{x=x^0})_{i_\mu})_{\mu\neq 1,\cdots,\ell}|$$
$$\geqq cr_h^a(1+o(1)) \quad \text{as} \quad h\to\infty,$$

which contradict (4. 4). Thus there exists $\gamma>0$ such that

$$\Delta(\xi') \neq 0 \quad \text{for} \quad \operatorname{Im}\xi_1<0, \quad |\operatorname{Im}\xi'''|<|\operatorname{Im}\xi_1|/\gamma$$
$$\text{and} \quad \log(1+|\xi'|)+1<|\operatorname{Im}\xi_1|/\gamma,$$

if (4. 1) is C^∞ well-posed. Therefore the theorem follows immediately from Seidenberg's lemma. Q.E.D.

By Theorem 4.3 we may assume that

(A-3) $\Delta(\xi')$ is hyperbolic with respect to ϑ', i.e., there exist a constant $\gamma_1(\geqq\gamma_0)$ and an open convex cone Σ in $\dot{\Gamma}(P)$ with its vertex at the origin such that $\vartheta'\in\Sigma$ and

$$\Delta(\xi') \neq 0 \quad \text{for} \quad \xi'\in\mathbb{R}^{n-1}-i\gamma_1\vartheta'-i\Sigma.$$

Define the $\ell N\times\ell$ matrix $\mathcal{B}_i(\xi')$ by

$$\mathcal{L}(\xi')\mathcal{B}_i(\xi') = \Delta_i(\xi')I_\ell,$$

where $\ell(N-1)$ rows of $\mathcal{B}_i(\xi')$ are equal to zero and $\mathcal{B}_i(\xi')$ is analytic in $\mathbb{R}^{n-1}-i\gamma_0\vartheta'-i\dot{\Gamma}(P)$. By the condition (A-3) we may assume that $\Delta_{i_0}(\xi')\neq 0$ for some i_0. Put

$$(\alpha_s^{(jk)}(\xi')) \equiv \begin{pmatrix} \alpha_1^{(11)} & \cdots & \alpha_1^{(1\ell)} \\ \vdots & & \vdots \\ \alpha_\ell^{(11)} & \cdots & \alpha_\ell^{(1\ell)} \\ \alpha_1^{(21)} & \cdots & \alpha_1^{(2\ell)} \\ \vdots & & \vdots \\ \alpha_\ell^{(N1)} & \cdots & \alpha_\ell^{(N\ell)} \end{pmatrix} = \Delta_{i_0}(\xi')^{-1}\mathcal{B}_{i_0}(\xi')$$

and define the $N\times\ell$ matrix $\mathcal{N}(\xi)\equiv(\mathcal{N}_{jk}(\xi))$ by

$$\mathcal{N}_{jk}(\xi) = \sum_{s=1}^{\ell} \alpha_s^{(jk)}(\xi')P_{+\ell-s}(\xi).$$

Moreover define the polynomial $\mathcal{K}(\xi)$ of ξ_n of degree less than ℓ, whose coefficients are meromorphic in $\xi'\in\mathbb{R}^{n-1}-i\gamma_0\vartheta'-i\dot{\Gamma}(P)$, by

$$\mathcal{K}(\xi) \equiv i^{-1}\cdot{}^t\mathrm{cof}\, L(\xi)\cdot\mathcal{N}(\xi) \pmod{P_+(\xi)}. \tag{4. 5}$$

Then it is easy to see that

$$\hat{F}(\xi', x_n) = (2\pi)^{-1}\oint \exp[ix_n\cdot\xi_n]P_+(\xi)^{-1}\mathcal{K}(\xi)d\xi_n$$

satisfies the system

$$\left.\begin{aligned} &L(\xi', D_n)\hat{F}(\xi', x_n) = 0, \quad x_n>0, \\ &B\hat{F}(\xi', x_n)|_{x_n=0} = I_\ell, \end{aligned}\right\} \tag{4. 6}$$

if $\Delta_{i_0}(\xi')\neq 0$ and $\xi'\in\mathbb{R}^{n-1}-i\gamma_0\vartheta'-i\dot{\Gamma}(P)$. Put

$$\tilde{\alpha}_s^{(jk)}(\xi') = \Delta(\xi')^{-2}\sum_i \overline{\Delta_i(\xi')}\mathcal{B}_i(\xi'),$$

$$\tilde{\mathcal{N}}_{jk}(\xi) = \sum_{s=1}^{\ell} \tilde{\alpha}_s^{(jk)}(\xi')P_{+\ell-s}(\xi), \; \tilde{\mathcal{N}}(\xi)=(\tilde{\mathcal{N}}_{jk}(\xi)).$$

Then it follows from the uniqueness of the solutions of ordinary differential equations that

$$i^{-1}\cdot{}^t\mathrm{cof}\, L(\xi)\cdot\tilde{\mathcal{N}}(\xi) \equiv \mathcal{K}(\xi) \pmod{P_+(\xi)}.$$

Therefore $\mathcal{K}(\xi)$ is meromorphic in $(\mathbb{R}^{n-1}-i\gamma_0\vartheta'-i\dot{\Gamma}(P))\times\mathbb{C}$ and analytic if $\Delta(\xi')\neq 0$. Moreover we can define a distribution $F(x)\in C^\infty([0,\infty);\mathcal{D}'(\mathbb{R}^{n-1}))(\subset\mathcal{D}'(\mathbb{R}^n_+))$ by

$$F(x) = (2\pi)^{-n+1}\int_{\mathbb{R}^{n-1}-i\gamma\vartheta'}\exp[ix'\cdot\xi']\hat{F}(\xi', x_n)d\xi', \quad \gamma>\gamma_1,$$

which is called the Poisson kernel of the system $\{L, B\}$. In fact, $\hat{F}(\xi', x_n)$ is analytic in $\xi'\in\mathbb{R}^{n-1}-i\gamma_1\vartheta'-i\Sigma$ and, by Seidenberg's lemma, we have

$$|\hat{F}(\xi', x_n)| \leqq C_\gamma(1+|\xi'|)^M, \quad \xi'\in\mathbb{R}^{n-1}-i\gamma\vartheta', \quad \gamma>\gamma_1.$$

We can also define $\tilde{F}(x)\in\mathcal{D}'(\mathbb{R}^n)$ by

$$\tilde{F}(x) = (2\pi)^{-n}\int_{\mathbb{R}^n-i\gamma\vartheta}\exp[ix\cdot\xi]P_+(\xi)^{-1}\mathcal{K}(\xi)d\xi, \quad \gamma>\gamma_1.$$

In fact, $\mathcal{K}(\xi)/P_+(\xi)$ is analytic in $\xi\in\mathbb{R}^n-i\gamma_1\vartheta-i\Gamma_0$, and for any compact set K in $\gamma_1\vartheta+\Gamma_0$

$$|\mathcal{K}(\xi)/P_+(\xi)| \leqq C_K(1+|\xi|)^{a_K}, \quad \xi\in\mathbb{R}^n-i\Lambda_K,$$

where $\Gamma_0=\Sigma\times R\cap\Gamma(P_+)$, $\Lambda_K=\{t\xi;\ \xi\in K,\ t\geqq 1\}$, $\Gamma(P_+)$ denotes the connected component of $\{\xi\in\mathbb{R}^n;\ P^0_+(-i\xi)\neq 0\}$ containing ϑ and a_K and C_K are constants (see [37]). Applying the Paley-Wiener-Schwartz theorem, we have

$$\operatorname{supp}\tilde{F}\subset\Gamma_0^*\subset\{x\in\overline{\mathbb{R}^n_+};\ x_1\geqq 0\}.$$

For we have $\Gamma_0\supset\{(1,0,\cdots,0,\lambda)\in\mathbb{R}^n;\ \lambda\geqq 0\}$. It is easily seen that

$$F(x) = \tilde{F}(x)|_{x_n>0}.$$

From (4. 6) it follows that $F(x)$ satisfies the system

$$\begin{cases} L(D)F(x) = 0, & x_n>0,\\ B(D)F(x)|_{x_n=0} = \delta(x')I_\ell, & \\ \operatorname{supp}F\subset\Gamma_0^*\subset\{x_1\geqq 0\}. & \end{cases}$$

Theorem 4.4. Assume that (A-1)', (A-2) and (A-3) are valid. Then the mixed problem (4. 1) is C^∞ well-posed.

Proof. We shall first prove the existence of solutions of (4. 1). Let $\tilde{f}\varepsilon C^\infty(\mathbb{R}^n)$ be an extension of $f(x)$ such that $\operatorname{supp} \tilde{f}\subset\{x_1\geqq 0\}$. Let $U(x)\varepsilon C^\infty(\mathbb{R}^n)$ be a solution of the Cauchy problem

$$\begin{cases} L(D)U(x) = \tilde{f}(x), & x\varepsilon\mathbb{R}^n, \\ \operatorname{supp} U(x) \subset \{x_1\geqq 0\}. \end{cases}$$

Then it suffices to solve the mixed problem

$$\left.\begin{aligned} &L(D)v(x) = 0, \quad x_n>0, \\ &B(D)v(x)|_{x_n=0} = g(x')-BU(x)|_{x_n=0}, \\ &\operatorname{supp} v \subset\{x_1\geqq 0\}. \end{aligned}\right\} \tag{4. 7}$$

Put

$$v = F\underset{x'}{*}(g-BU|_{x_n=0}).$$

Since $\operatorname{supp} F\subset\Gamma_0^*$ and $\operatorname{supp} (g-BU|_{x_n=0})\subset\{x_1\geqq 0\}$, v is well-defined. Moreover v satisfies (4. 7) and

$$\operatorname{supp} v\subset \operatorname{supp} (g-BU|_{x_n=0})\times\{0\}+\Gamma_0^*, \quad v\varepsilon C^\infty(\overline{\mathbb{R}^n_+}).$$

Thus, putting $u(x)=v(x)+U(x)|_{x_n\geqq 0}$, $u(x)$ is a solution of (4. 1). Next we shall prove the uniqueness of solutions of (4. 1), applying the method used in Proposition 5.1 in [40]. Let $u(x)\varepsilon C^\infty(\overline{\mathbb{R}^n_+})$ be a solution of (4. 1) with $f=0$ and $g=0$. Choose $\phi\varepsilon C^\infty(\overline{\mathbb{R}^n_+})$ so that $\phi(x)=1$ if $|x|\leqq b$, $=0$ if $|x|\geqq b+1$, where $b>0$, and put

$$f(x) = L(D)(\phi u), \quad x_n\geqq 0,$$

$$g(x') = B(D)(\phi u)|_{x_n=0}.$$

Then we have

$$f\varepsilon C^\infty(\overline{\mathbb{R}^n_+}), \quad \operatorname{supp} f\subset\{x\varepsilon\overline{\mathbb{R}^n_+};\ x_1\geqq 0,\ b\leqq|x|\leqq b+1\},$$

$$g\varepsilon C^\infty(\mathbb{R}^{n-1}), \quad \operatorname{supp} g\subset\{x'\varepsilon\mathbb{R}^{n-1};\ x_1\geqq 0,\ b\leqq|x'|\leqq b+1\}.$$

Taking the Fourier-Laplace transforms with respect to x' we have

$$\left.\begin{aligned} &L(\xi', D_n)\widehat{\phi u}(\xi', x_n) = \hat{f}(\xi', x_n), \quad x_n \geqq 0, \\ &B(\xi', D_n)\widehat{\phi u}(\xi', x_n)|_{x_n=0} = \hat{g}(\xi'), \end{aligned}\right\} \tag{4. 8}$$

where $\hat{f}(\xi', x_n)$ is the Fourier-Laplace transform with respect to x' of $f(x)$. Conversely, $\widehat{\phi u}(\xi', x_n)$ is uniquely determined by (4. 8) for $\operatorname{Im} \xi_1 < -\gamma_1$ and $\xi'' \in \mathbb{R}^{n-2}$, which is a consequence of the uniqueness of the solutions of ordinary differential equations.

Thus we have

$$\operatorname{supp} \phi u \subset \{\operatorname{supp} f + \Gamma(P)^*\}$$
$$\cup [\{\operatorname{supp} g \times \{0\} \cup ((\operatorname{supp} f + \Gamma(P)^*) \cap \{x_n = 0\})\} + \Gamma_0^*],$$

which was proved in the first part of the proof. Therefore there exists $a > 0$ such that

$$\operatorname{supp} \phi u \cap \{|x| \leqq a\} = \emptyset.$$

Here we can choose a so that $a \to \infty$ as $b \to \infty$. Q.E.D.

5. THE LOPATINSKI DETERMINANT AND HYPERBOLIC FUNCTIONS

From now on we assume that (A-1)' and (A-2) are valid. Write

$$\sigma^k(\xi') = \sum_{j=1}^{\ell} \lambda_j^+(\xi')^k, \quad 1 \leqq k \leqq \ell.$$

Let $P(\xi)$ be written in the form

$$P(\xi) = \Pi_{j=1}^{p} p_j(\xi)^{\nu_j},$$

where the $p_j(\xi)$ are irreducible polynomials. We assume that

$\Pi_{j=p'+1}^{p} p_j(\xi)^{\nu_j} = 0$ does not have roots with positive imaginary part for $\operatorname{Im} \xi_1 < -\gamma_0$ and $\xi'' \in \mathbb{R}^{n-2}$. Put

$$\dot{\Gamma}_{\xi^{0\prime}} = \bigcap_{j=1}^{p'} \{\bigcap_{\xi_n^0 \in \mathbb{R}} \dot{\Gamma}((p_j)_{\xi^0}) \cap \dot{\Gamma}(((p_j)_{(0,1)})_{(\xi^{0\prime}, 0)})\}$$

for $\xi^{0\prime} \in \mathbb{R}^{n-1}$, where $(p_j)_{\xi^0}$ denotes the localization of $p_j(\xi)$ at ξ^0:

$$p_j(\nu^{-1}\xi^0 + \eta) = \nu^{m(\xi^0)}((p_j)_{\xi^0}(\eta) + o(1)) \quad \text{as} \quad \nu \downarrow 0, \quad (p_j)_{\xi^0}(\eta) \not\equiv 0$$

(see [5]). We note that $\dot{\Gamma}_0=\bigcap_{j=1}^{p'}\dot{\Gamma}(p_j)$ and we can replace $\dot{\Gamma}(P)$ by $\dot{\Gamma}_0$ in §4.

Lemma 5.1. Let $\xi^{0\prime}\in\mathbb{R}^{n-1}$ and M be a compact set in $\Gamma((P_{(0,1)})_{(\xi^{0\prime},0)})$. Then there exist a conic neighborhood Γ of $\xi^{0\prime}$ in $\mathbb{R}^{n-1}$ and positive constants C, d and t_0 such that

$$P(\xi-i(t|\xi'|\eta+\gamma_0\vartheta))\neq 0$$

for $\xi'\in\Gamma$, $|\xi'|>C$, $\xi_n\in\mathbb{R}$, $|\xi_n|>d|\xi'|$, $\eta\in M$ and $0<t\leqq t_0$.

This lemma can be proved by the same argument as in Lemma 2.10 in [51] if we put

$$f(\nu,r,\zeta',\lambda,s,t,\eta)$$
$$=P(\nu^{-1}r\xi^{0\prime}+r\zeta'-irt\eta'-i(s+\gamma_0)\vartheta',\nu^{-1/\rho}r\lambda-irt\eta_n).$$

So we omit the proof.

Lemma 5.2. Let $\xi^{0\prime}\in\mathbb{R}^{n-1}$. Then $\dot{\Gamma}_{\xi^{0\prime}}$ is an open convex cone.

Proof. It is obvious that $\dot{\Gamma}_{\xi^{0\prime}}$ is a convex cone (see [5]). Let us prove that $\dot{\Gamma}_{\xi^{0\prime}}$ is open. We can assume without loss of generalities that $P(\xi)$ is irreducible. Now assume that there exist $\eta^{k\prime}\notin\dot{\Gamma}_{\xi^{0\prime}}$, $k=1,2,\cdots$, such that $\eta^{k\prime}\to\eta^{0\prime}\in\dot{\Gamma}_{\xi^{0\prime}}$ as $k\to\infty$. Since $\dot{\Gamma}((P_{(0,1)})_{(\xi^{0\prime},0)})$ is open, we may assume that $\eta^{k\prime}\in\dot{\Gamma}((P_{(0,1)})_{(\xi^{0\prime},0)})$, $k=1,2,\cdots$. Thus there exist $\xi_n^k\in\mathbb{R}$, $k=1,2,\cdots$, such that $\eta^{k\prime}\notin\dot{\Gamma}(P_{\xi^k})$, where $\xi^k=(\xi^{0\prime},\xi_n^k)$. Assume that $\xi_n^{k_j}\to\xi_n^0$ as $j\to\infty$. Since $\eta^{0\prime}\in\dot{\Gamma}(P_{\xi^0})$ and $\dot{\Gamma}(P_{\xi^0})$ is open, it follows from inner semi-continuity of $\dot{\Gamma}(P_\xi)$ that $\eta^{k_j\prime}\in\dot{\Gamma}(P_{\xi^{k_j}})$ for sufficiently large j. So we have $|\xi_n^k|\to\infty$ as $k\to\infty$. Now we may assume that $P^0_{\xi^k}(-i\eta^{k\prime},0)=0$, since $(\eta^{k\prime},0)\notin\Gamma(P_{\xi^k})$ and $\vartheta\in\Gamma(P_{\xi^k})$. Noting that

$$t^{-m(\xi^k)}P^0(\xi^{0\prime}-it\eta^{k\prime},\xi_n^k)\to P^0_{\xi^k}(-i\eta^{k\prime},0)=0 \quad\text{as}\quad t\downarrow 0,$$

Rouche's theorem means that there exist $t_1>0$ and $\zeta(t)\in\mathbf{C}^n$, $0<t\leqq t_1$, such that $|\zeta(t)|\to 0$, as $t\downarrow 0$, and

$$P^0(\xi^{0\prime}-it(\eta^{k\prime}+\zeta(t)'),\ \xi_n^k-it\zeta_n(t)) = 0,\quad 0<t\leqq t_1.$$

On the other hand it follows from Lemma 5.1 that for any compact set M in $\Gamma((P_{(0,1)})_{(\xi^{0\prime},0)})$ there exist a neighborhood U of $\xi^{0\prime}$, $d>0$ and $t_0>0$ such that

$$P^0(\xi-it\eta)\neq 0 \quad\text{for}\quad \xi'\in U,\ \ \xi_n\in\mathbb{R},\ \ |\xi_n|>d,\ \ \eta\in M \ \text{and}\ \ 0<t\leqq t_0.$$

Since $\Gamma((P_{(0,1)})_{(\xi^{0\prime},0)}) = \dot{\Gamma}((P_{(0,1)})_{(\xi^{0\prime},0)})\times\mathbb{R}$, we can choose a compact set M in $\Gamma((P_{(0,1)})_{(\xi^{0\prime},0)})$ so that $(\eta^{k\prime},0)\in M$, $k=1,2,\cdots$. For sufficiently large k we have $|\xi_n^k| >d$, which leads us to a contradiction. Q.E.D.

Lemma 5.3 ([51]). Let $1\leqq k\leqq \ell$. For any compact set K in $\mathbb{R}^{n-1}-i\gamma_0\vartheta'-i\dot{\Gamma}_{\xi^{0\prime}}$, there exists $\nu_K>0$ such that $\sigma^k(\nu^{-1}\xi^{0\prime}+\eta')$ is well-defined for $\eta'\in K$ and $0<\nu\leqq\nu_k$ and

$$\nu^{s_k}\sigma^k(\nu^{-1}\xi^{0\prime}+\eta') = \sum_{j=0}^{\infty}\nu^{j/L}\sigma^k_{\xi^{0\prime},j}(\eta'),\quad \sigma^k_{\xi^{0\prime},0}(\eta')\neq 0,$$

whose convergence is uniform in $K\times\{\nu;\ |\nu|\leqq\nu_K\}$, where s_k is a rational number and L is a positive integer. Moreover the $\sigma^k_{\xi^{0\prime},j}$ are analytic in $\mathbb{R}^{n-1}-i\gamma_0\vartheta'-i\dot{\Gamma}_{\xi^{0\prime}}$.

Lemma 5.4 ([51]). Let $1\leqq k\leqq \ell$. For any compact set K in $\mathbb{R}^{n-1}-i\dot{\Gamma}_{\xi^{0\prime}}$, there exist ν_K and ρ_K (>0) such that

$\sigma^k(\nu^{-1}\rho^{-1}\xi^{0\prime}+\rho^{-1}\eta')$ is well-defined when $\rho_K^{-1}\eta'\in\mathbb{R}^{n-1}-i\gamma_0\vartheta'-i\dot{\Gamma}_{\xi^{0\prime}}$, $\alpha\eta'\in K$ for some $\alpha\in\mathbf{C}$ ($|\alpha|=1$), $0<\nu\leqq\nu_K$ and $0<\rho\leqq\rho_K$. Then we have

$$(\nu\rho)^{s_k}\sigma^k(\nu^{-1}\rho^{-1}\xi^{0\prime}+\rho^{-1}\eta') = \sum_{j=0}^{\infty}\sum_{i=0}^{\infty}\nu^{j/L}\rho^{i+q_{kj}}\sigma^{ki}_{\xi^{0\prime},j}(\eta'),$$

$$\sigma^{k0}_{\xi^{0\prime},j}(\eta')\neq 0 \quad\text{if}\quad \sigma^k_{\xi^{0\prime},j}(\eta')\neq 0,$$

whose convergence is uniform in $\{(\eta',\nu,\rho);\ \rho_K^{-1}\eta'\in\mathbb{R}^{n-1}-i\gamma_0\vartheta'-i\dot{\Gamma}_{\xi^{0\prime}}$, $\alpha\eta'\in K$ for some $\alpha\in\mathbf{C}$ ($|\alpha|=1$), $0\leqq\nu\leqq\nu_K$ and $\delta\leqq\rho\leqq\rho_K\}$ for any $\delta>0$, where the q_{kj} are rational numbers. Moreover the $\sigma^{kj}_{\xi^{0\prime},j}(\eta')$ are analytic in $\mathbb{R}^{n-1}-i\dot{\Gamma}_{\xi^{0\prime}}$ and homogeneous, and

$$\sigma^{k}_{\xi^{0\prime},j}(\rho^{-1}\xi') = \rho^{q_{kj}-j/L}\sum_{i=0}^{\infty}\rho^{i}\sigma^{ki}_{\xi^{0\prime},j}(\eta'),$$

whose convergence is uniform in $\{(\eta', \rho);\ \rho_K^{-1}\eta'\in\mathbb{R}^{n-1}-i\gamma_0\mathcal{J}'-i\dot{\Gamma}_{\xi^{0\prime}},$ $\alpha\eta'\in K$ for some $\alpha\in\mathbf{C}$ $(|\alpha|=1)$ and $\delta\leqq\rho\leqq\rho_K\}$ for any $\delta>0$. Here $q_{kj}-s_k$ is an integer.

Assume that $\Delta(\xi')\neq 0$ in ξ'. Then $\mathcal{K}(\xi)$ is well-defined by (4. 5) and meromorphic in $(\mathbb{R}^{n-1}-i\gamma_0\mathcal{J}'-i\dot{\Gamma}_0)\times\mathbf{C}$. For any $\xi^{0\prime}$ $\in\mathbb{R}^{n-1}-i\gamma_0\mathcal{J}'-i\dot{\Gamma}_0$ there exist a neighborhood U of $\xi^{0\prime}$ in $\mathbb{R}^{n-1}-i\gamma_0\mathcal{J}'-i\dot{\Gamma}_0$ and $\Delta_U(\xi')$ and $H_U(\xi)$ analytic in $U\times\mathbf{C}$ such that $\mathcal{K}(\xi)=\Delta_U(\xi')^{-1}H_U(\xi)$ for $\xi\in U\times\mathbf{C}$ and $\Delta_U(\xi')$ and the entries of $H_U(\xi)$ are relatively prime at each point in $U\times\mathbf{C}$. If $U\cap U'\neq\emptyset$, then $\Delta_U(\xi')/\Delta_{U'}(\xi')$ and $\Delta_{U'}(\xi')/\Delta_U(\xi')$ are analytic for ξ' $\in U\cap U'$, since the ring of germs of analytic functions is a unique factorization domain. Therefore there exists $\tilde{\Delta}(\xi')$ analytic in $\mathbb{R}^{n-1}-i\gamma_0\mathcal{J}'-i\dot{\Gamma}_0$ such that $\tilde{\Delta}(\xi')/\Delta_U(\xi')$ and $\Delta_U(\xi')/\tilde{\Delta}(\xi')$ are analytic in U for any U, since the second Cousin problem has a solution for the domain $(\mathbb{R}^{n-1}-i\gamma_0\mathcal{J}'-i\dot{\Gamma}_0)$. Putting $H(\xi)$ $=\tilde{\Delta}(\xi')\mathcal{K}(\xi)$, $\tilde{\Delta}(\xi')$ and the entries of $H(\xi)$ are relatively prime at each point in $(\mathbb{R}^{n-1}-i\gamma_0\mathcal{J}'-i\dot{\Gamma}_0)\times\mathbf{C}$, and

$$\mathcal{K}(\xi) = \tilde{\Delta}(\xi')^{-1}H(\xi).$$

If one choose $\tilde{\Delta}(\xi')$ and $H(\xi)$ which are relatively prime in $(\mathbb{R}^{n-1}-i\gamma_0\mathcal{J}'-i\dot{\Gamma}_0)\times\mathbf{C}$, then they are relatively prime at each point. We note that $\tilde{\Delta}(\xi')$ is only used to simplify the discussions below.

Lemma 5.5. Let $\xi'\in\mathbb{R}^{n-1}-i\gamma_0\mathcal{J}'-i\dot{\Gamma}_0$. Then, $\tilde{\Delta}(\xi')\neq 0$ if and only if $\tilde{\Delta}(\xi')\neq 0$.

Proof. If $\tilde{\Delta}(\xi')\neq 0$, then the system

$$\left.\begin{aligned} &L(\xi', D_n)v(x_n) = 0, \quad x_n>0, \\ &B(\xi', D_n)v(x_n)|_{x_n=0} = g, \\ &v(x_n) = \textstyle\sum_{j=1}^{\ell} \exp[i\lambda_j^+(\xi')x_n]C_j \end{aligned}\right\} \qquad (5.\ 1)$$

has a solution v for any $g\varepsilon\mathbf{C}^{\ell}$. In fact,

$$v(x_n) = (2\pi)^{-1}\oint \exp[ix_n\cdot\xi_n]P_+(\xi)^{-1}\mathcal{K}(\xi)gd\xi_n$$

is a solution of (5. 1). Thus it follows from Lemma 2.3 that $\Delta(\xi')\neq 0$. Conversely, if $\tilde{\Delta}(\xi')=0$, then there exists a sequence $\{\xi^k\}$ such that $\xi^{k}{}'\to\xi'$, as $k\to\infty$, and $\tilde{\Delta}(\xi^{k}{}')=0$ and $H(\xi^k)\neq 0$. Thus we have $\Delta(\xi^{k}{}')=0$ and, therefore, $\Delta(\xi')=0$. Q.E.D.

For each i with $\Delta_i(\xi')\neq 0$ we can write

$$\mathcal{K}(\xi) = \Delta_i(\xi')^{-1}H_i(\xi).$$

$\Delta_i(\xi')$ and $H_i(\xi)$ are polynomials of $\sigma^k(\xi')$, $1\leqq k\leqq\ell$, and ξ. Thus it follows from Lemma 5.3 that for $\xi^{0}{}'\varepsilon\mathbb{R}^{n-1}$ and any compact set K in $\mathbb{R}^{n-1}-i\gamma_0\mathscr{Y}'-i\dot{\Gamma}_{\xi^{0}{}'}$ there exists $\nu_K>0$ such that

$$\nu^{h_{i,\xi^{0}{}'}}\Delta_i(\nu^{-1}\xi^{0}{}'+\eta') = \textstyle\sum_{j=0}^{\infty} \nu^{j/L}\Delta_{i,\,\xi^{0}{}',j}(\eta'),$$

$$\nu^{k_{i,\xi^{0}{}'}}H_i(\nu^{-1}\xi^{0}{}'+\eta',\ \xi_n) = \textstyle\sum_{j=0}^{\infty} \nu^{j/L}H_{i,\xi^{0}{}',j}(\eta';\ \xi_n),$$

whose convergence is uniform in $K\times\{\nu;\ 0\leqq\nu\leqq\nu_K\}$. Put

$$\Delta_{i,\xi^{0}{}'}(\mu,\ \eta') = \textstyle\sum_{j=0}^{\infty} \mu^j\Delta_{i,\,\xi^{0}{}',j}(\eta'),$$

$$H_{i,\xi^{0}{}'}(\mu,\ \eta';\ \xi_n) = \textstyle\sum_{j=0}^{\infty} \mu^jH_{i,\xi^{0}{}',j}(\eta';\ \xi_n).$$

Then $\Delta_{i,\xi^{0}{}'}(\mu,\ \eta')$ and $H_{i,\xi^{0}{}'}(\mu,\ \eta';\ \xi_n)$ are analytic in $\{(\mu,\ \eta',\ \xi_n);\ \mu\varepsilon\mathbf{C},\ |\mu|\leqq\nu_K^{1/L},\ \eta'\varepsilon K \text{ and } \xi_n\varepsilon\mathbf{C}\}$. If K is a compact polydisc in $\mathbb{R}^{n-1}-i\gamma_0\mathscr{Y}'-i\dot{\Gamma}_{\xi^{0}{}'}$, Cousin's theorem implies that there exist $\delta_{\xi^{0}{}',K}(\mu,\ \eta')$ and $h_{\xi^{0}{}',K}(\mu,\ \eta';\ \xi_n)$ analytic in $\{(\mu,\ \eta',\ \xi_n);\ \mu\varepsilon\mathbf{C},\ |\mu|\leqq\nu_K^{1/L},\ \eta'\varepsilon K \text{ and } \xi_n\varepsilon\mathbf{C}\}$ such that

$$\Delta_{i,\xi^{0\prime}}(\mu, \eta')^{-1}H_{i,\xi^{0\prime}}(\mu, \eta'; \xi_n)$$
$$= \delta_{\xi^{0\prime},K}(\mu, \eta')^{-1}h_{\xi^{0\prime},K}(\mu, \eta'; \xi_n)$$

and $\delta_{\xi^{0\prime},K}(\mu, \eta')$ and the entries of $h_{\xi^{0\prime},K}(\mu, \eta'; \xi_n)$ are relatively prime at each point. If K is sufficiently small, then it is obvious that there exist $\delta_{\xi^{0\prime},K}(\mu, \eta')$ and $h_{\xi^{0\prime},K}(\mu, \eta'; \xi_n)$ which satisfy the above properties. In the discussion below, we need only the existence of $\delta_{\xi^{0\prime},K}(\mu, \eta')$ and $h_{\xi^{0\prime},K}(\mu, \eta'; \xi_n)$ for some K.

Lemma 5.6. Let $\eta' \varepsilon K$ and $0<\nu\leqq\nu_K$. Then, $\Delta(\nu^{-1}\xi^{0\prime}+\eta')\neq 0$ if and only if $\delta_{\xi^{0\prime},K}(\nu^{1/L}, \eta')=0$, where $\nu^{1/L}>0$.

Write

$$\delta_{\xi^{0\prime},K}(\mu, \eta') = \mu^{h_{\xi^{0\prime}}}\sum_{j=0}^{\infty} \mu^j \delta_{\xi^{0\prime},K,j}(\eta'), \tag{5. 2}$$
$$\delta_{\xi^{0\prime},K,0}(\eta')\neq 0,$$

where the $\delta_{\xi^{0\prime},K,j}(\eta')$ are analytic in K and $h_{\xi^{0\prime}}\geqq 0$. Then we have the following

Lemma 5.7. There exists $\Delta_{\xi^{0\prime}}(\eta')$ analytic in $\mathbb{R}^{n-1}-i\gamma_0\mathscr{I}'-i\dot{\Gamma}_{\xi^{0\prime}}$ such that $\Delta_{\xi^{0\prime}}(\eta')/\delta_{\xi^{0\prime},K,0}(\eta')$ and $\delta_{\xi^{0\prime},K,0}(\eta')/\Delta_{\xi^{0\prime}}(\eta')$ are analytic in K for any K.

Proof. Let K and K' be compact polydiscs in $\mathbb{R}^{n-1}-i\gamma_0\mathscr{I}'-i\dot{\Gamma}_{\xi^{0\prime}}$. If $\eta'\varepsilon \mathring{K}\cap\mathring{K}'$ and $|\mu|\leqq\min(\nu_K^{1/L}, \nu_{K'}^{1/L})$, then $\delta_{\xi^{0\prime},K}(\mu, \eta')^{-1}h_{\xi^{0\prime},K}(\mu, \eta'; \xi_n) = \delta_{\xi^{0\prime},K'}(\mu, \eta')^{-1}h_{\xi^{0\prime},K'}(\mu, \eta'; \xi_n)$. Therefore $\delta_{\xi^{0\prime},K}(\mu, \eta')/\delta_{\xi^{0\prime},K'}(\mu, \eta')$ and $\delta_{\xi^{0\prime},K'}(\mu, \eta')/\delta_{\xi^{0\prime},K}(\mu, \eta')$ are analytic for $\eta'\varepsilon \mathring{K}\cap\mathring{K}'$ and $|\mu| \leqq \min(\nu_K^{1/L}, \nu_{K'}^{1/L})$, since the ring of germs of analytic functions is a unique factorization domain. So $\delta_{\xi^{0\prime},K,0}(\eta')/\delta_{\xi^{0\prime},K',0}(\eta')$ and

$\delta_{\xi^{0\prime},K',0}(\eta')/\delta_{\xi^{0\prime},K,0}(\eta')$ are analytic in $\mathring{K}\cap\mathring{K}'$. Since the second Cousin problem has a solution for the domain $(\mathbb{R}^{n-1}-i\gamma_0\mathscr{I}'-i\dot{\Gamma}_{\xi^{0\prime}})$, the lemma easily follows. Q.E.D.

Definition 5.1. Let $\xi^{0\prime}\varepsilon\mathbb{R}^{n-1}$. $\Delta_{\xi^{0\prime}}(\xi')$ is called the localization of $\Delta(\xi')$ at $\xi^{0\prime}$.

We can similarly define the principal part $\Delta^0(\xi')$ of $\Delta(\xi')$, the principal part $(\Delta_{\xi^{0\prime}})^0(\xi')$ of $\Delta_{\xi^{0\prime}}(\xi')$ and the localization $(\Delta^0)_{\xi^{0\prime}}(\xi')$ of $\Delta^0(\xi')$ at $\xi^{0\prime}$. By Lemma 5.4, for $\xi^{0\prime}\varepsilon\mathbb{R}^{n-1}$ and any compact set K in $\mathbb{R}^{n-1}-i\dot{\Gamma}_{\xi^{0\prime}}$ there exist ν_K and $\rho_K(>0)$ such that

$$\nu^{h_{i,\xi^{0\prime}}}\rho^{q_{i,\xi^{0\prime}}}\Delta_i(\nu^{-1}\rho^{-1}\xi^{0\prime}+\rho^{-1}\eta')$$

$$=\sum_{j=0}^{\infty}\sum_{k=0}^{\infty}\nu^{j/L}\rho^{k+q_{i,\xi^{0\prime},j}}\Delta^k_{i,\xi^{0\prime},j}(\eta'), \qquad (5.\ 3)$$

$$\nu^{k_{i,\xi^{0\prime}}}\rho^{p_{i,\xi^{0\prime}}}H_i(\nu^{-1}\rho^{-1}\xi^{0\prime}+\rho^{-1}\eta';\ \xi_n)$$

$$=\sum_{j=0}^{\infty}\sum_{k=0}^{\infty}\nu^{j/L}\rho^{k+p_{i,\xi^{0\prime},j}}H^k_{i,\xi^{0\prime},j}(\eta';\ \xi_n), \qquad (5.\ 4)$$

whose convergence is uniform in $\{(\eta',\ \nu,\ \rho);\ \rho_K^{-1}\eta'\varepsilon\mathbb{R}^{n-1}-i\gamma_0\mathscr{I}'-i\dot{\Gamma}_{\xi^{0\prime}},\ \alpha\eta'\varepsilon K$ for some $\alpha\varepsilon\mathbb{C}(|\alpha|=1)$, $0\leqq\nu\leqq\nu_K$ and $0\leqq\rho\leqq\rho_K\}$, where the $q_{i,\xi^{0\prime},j}$ and $p_{i,\xi^{0\prime},j}$ are non-negative. Define $\Delta_{i,\xi^{0\prime}}(\mu,\ \rho,\ \eta')$ and $H_{i,\xi^{0\prime}}(\mu,\ \rho,\ \eta';\ \xi_n)$ by the right-hand sides of (5. 3) and (5. 4), respectively, which are analytic in $\{(\mu,\ \rho,\ \eta',\ \xi_n);\ \mu,\rho\varepsilon\mathbb{C},\ |\mu|\leqq\nu_K^{1/L},\ |\rho|\leqq\rho_K,\ \eta'\varepsilon K$ and $\xi_n\varepsilon\mathbb{C}\}$. If K is a compact polydisc in $\mathbb{R}^{n-1}-i\gamma_0\mathscr{I}'-i\dot{\Gamma}_{\xi^{0\prime}}$, Cousin's theorem implies that there exist $\delta_{\xi^{0\prime},K}(\mu,\ \rho,\ \eta')$ and $h_{\xi^{0\prime},K}(\mu,\ \rho,\ \eta';\ \xi_n)$ analytic in $\{(\mu,\ \rho,\ \eta',\ \xi_n);\ \mu,\rho\varepsilon\mathbb{C},\ |\mu|\leqq\nu_K^{1/L},\ |\rho|\leqq\rho_K,\ \eta'\varepsilon K$ and $\xi_n\varepsilon\mathbb{C}\}$ such that

$$\Delta_{i,\xi^{0\prime}}(\mu, \rho, \eta')^{-1}H_{i,\xi^{0\prime}}(\mu, \rho, \eta'; \xi_n)$$

$$= \delta_{\xi^{0\prime},K}(\mu, \rho, \eta')^{-1}h_{\xi^{0\prime},K}(\mu, \rho, \eta'; \xi_n),$$

and $\delta_{\xi^{0\prime},K}(\mu, \rho, \eta')$ and the entries of $h_{\xi^{0\prime},K}(\mu, \rho, \eta'; \xi_n)$ are relatively prime at each point.

Lemma 5.8. Assume that $0<\nu\leqq\nu_K$, $0<\rho\leqq\rho_K$, $\rho_K^{-1}\eta'\in\mathbb{R}^{n-1}-i\gamma_0\vartheta'-i\dot{\Gamma}_{\xi^{0\prime}}$ and $\alpha\eta'\in K$ for some $\alpha\in\mathbb{C}$ $(|\alpha|=1)$. Then, $\Delta(\nu^{-1}\rho^{-1}\xi^{0\prime}+\rho^{-1}\eta')\neq 0$ if and only if $\delta_{\xi^{0\prime},K}(\nu^{1/L}, \rho, \eta')\neq 0$, where $\nu^{1/L}>0$.

Write

$$\delta_{\xi^{0\prime},K}(\mu, \rho, \eta')$$
$$= \mu^{h_{\xi^{0\prime}}}\sum_{j=0}^{\infty}\sum_{k=0}^{\infty}\mu^j\rho^{k+q_{\xi^{0\prime},j}}\delta^k_{\xi^{0\prime},K,j}(\eta'), \tag{5. 5}$$

$$\delta^0_{\xi^{0\prime},K,0}(\eta')\neq 0, \quad \delta^0_{\xi^{0\prime},K,j}(\eta')\neq 0$$
$$\text{if } \sum_{k=0}^{\infty}\rho^k\delta^k_{\xi^{0\prime},K,j}(\eta')\neq 0, \quad j\geqq 1,$$

where the $\delta^k_{\xi^{0\prime},K,j}(\eta')$ are analytic in K. Define

$$q_{\xi^{0\prime}} = \min_j q_{\xi^{0\prime},j}, \quad \omega = \min\{j;\ q_{\xi^{0\prime}} = q_{\xi^{0\prime},j}\},$$

where $q_{\xi^{0\prime},j} = +\infty$ if $\sum_{k=0}^{\infty}\rho^k\delta^k_{\xi^{0\prime},K,j}(\eta')\equiv 0$.

Lemma 5.9. There exist $(\Delta_{\xi^{0\prime}})^0(\eta')$ and $(\Delta^0)_{\xi^{0\prime}}(\eta')$ analytic in $\mathbb{R}^{n-1}-i\dot{\Gamma}_{\xi^{0\prime}}$ such that $(\Delta_{\xi^{0\prime}})^0(\eta')/\delta^0_{\xi^{0\prime},K,0}(\eta')$, $\delta^0_{\xi^{0\prime},K,0}(\eta')/(\Delta_{\xi^{0\prime}})^0(\eta')$, $(\Delta^0)_{\xi^{0\prime}}(\eta')/\delta^0_{\xi^{0\prime},K,\omega}(\eta')$ and $\delta^0_{\xi^{0\prime},K,\omega}(\eta')/(\Delta^0)_{\xi^{0\prime}}(\eta')$ are analytic in K for any K.

Definition 5.2. Let $\xi^{0\prime}\varepsilon\mathbb{R}^{n-1}$. $(\Delta_{\xi^{0\prime}})^{0}(\xi')$ is called the principal part of $\Delta_{\xi^{0\prime}}(\xi')$. We define the principal part $\Delta^{0}(\xi')$ of $\Delta(\xi')$ by

$$\Delta^{0}(\xi') = (\Delta_{\xi^{0\prime}})^{0}(\xi') \quad \text{for} \quad \xi^{0\prime} = 0.$$

$(\Delta^{0})_{\xi^{0\prime}}(\xi')$ is called the localization of $\Delta^{0}(\xi')$ at $\xi^{0\prime}$.

Theorem 5.1. Let Σ be an open connected cone in $\dot{\Gamma}_0$ with its vertex at the origin such that $\vartheta'\varepsilon\Sigma$. Then the condition

$$\Delta(\xi') \neq 0 \quad \text{for} \quad \xi'\varepsilon\mathbb{R}^{n-1}-i\gamma_1\vartheta'-i\Sigma \tag{5. 6}$$

is equivalent to the following conditions:

$$\Delta(\xi') \neq 0 \quad \text{for} \quad \text{Im}\ \xi_1<-\gamma_1 \quad \text{and} \quad \xi'''\varepsilon\mathbb{R}^{n-2}. \tag{5. 7}$$

$$\Delta^{0}(-i\xi') \neq 0 \quad \text{for} \quad \xi'\varepsilon\Sigma. \tag{5. 8}$$

Proof. Assume that (5. 7) is valid, and that there exists $\xi^{0\prime}\varepsilon\Sigma$ such that $\Delta^{0}(-i\xi^{0\prime})=0$. Let K be a compact polydisc in $\mathbb{R}^{n-1}-i\Sigma$ and a complex neighborhood of $\xi^{0\prime}$. By Lemma 5.9 we have $\delta^{0}_{0,K,0}(-i\xi^{0\prime})=0$. Since $\delta^{0}_{0,K,0}(\xi')\not\equiv 0$, there exist $\varepsilon,\delta>0$ and $\eta^{0\prime}\varepsilon\mathbb{C}^{n-1}$ such that $-i\xi^{0\prime}+s\eta^{0\prime}\varepsilon K$ for $|s|\leqq\delta$, and

$$|\delta^{0}_{0,K,0}(-i\xi^{0\prime}+s\eta^{0\prime})| > \varepsilon \quad \text{for} \quad |s|=\delta.$$

On the other hand we have

$$\delta_{0,K}(0,\ \rho,\ -i\xi^{0\prime}+s\eta^{0\prime}) = \rho^{q_{0,0}}(\delta^{0}_{0,K,0}(-i\xi^{0\prime}+s\eta^{0\prime})+o(1)) \quad \text{as} \quad \rho\to 0.$$

Thus, by Rouche's theorem, there exist $\rho_0>0$ and $s(\rho)\varepsilon\mathbb{C}$, $0<\rho\leqq\rho_0$, such that $|s(\rho)|<\delta$ and $\delta_{0,K}(0,\ \rho,\ -i\xi^{0\prime}+s(\rho)\eta^{0\prime})=0$, $0<\rho\leqq\rho_0$. By Lemma 5.8 we have

$$\Delta(\rho^{-1}(-i\xi^{0\prime}+s(\rho)\eta^{0\prime})) = 0, \quad 0<\rho\leqq\rho_0.$$

If ρ_0 is sufficiently small, then $\rho^{-1}(-i\xi^{0\prime}+s(\rho)\eta^{0\prime})\varepsilon\mathbb{R}^{n-1}-i\gamma_1\vartheta'-i\Sigma$ for $0<\rho\leqq\rho_0$. Therefore (5. 6) is not valid. Next assume that the conditions (5. 7) and (5. 8) are valid. Let

$\xi^{0\prime}\in\mathbb{R}^{n-1}$ and $\eta^{0\prime}\in\Sigma$, and let $\eta'(\theta)$, $0\leqq\theta\leqq 1$, be a continuous curve in Σ such that $\eta'(0)=\eta^{0\prime}$ and $\eta'(1)=\vartheta'$. Put

$$f(s, t, \theta) = \tilde{\Delta}(\xi^{0\prime}-is\eta'(\theta)-it\vartheta'-i\gamma_1\vartheta'),$$

where Re s, Re $t\geqq 0$, Re $(s+t)>0$ and $0\leqq\theta\leqq 1$. Then (5. 7) implies that

$$f(s, t, \theta) \neq 0 \quad \text{for} \quad \mathrm{Re}\ s=0 \quad \text{and} \quad \mathrm{Re}\ t>0. \tag{5. 9}$$

Let t be fixed, Re $t>0$. Denote by $K(\theta;d)$ the closed polydisc with center $-i\eta'(\theta)$ and radius d. Then there exist $\theta_1,\cdots,\theta_r$ $\in[0, 1]$ such that $\{-i\eta'(\theta);\ 0\leqq\theta\leqq 1\}\subset\bigcup_{j=1}^{r} K(\theta_j;d/2)$, where d is chosen so that $K(\theta;d)\subset\mathbb{R}^{n-1}-i\dot{\Gamma}_0$. There exists $\rho_0>0$ such that $-i\eta'(\theta)+s^{-1}(\xi^{0\prime}-it\vartheta'-i\gamma_1\vartheta')\in K(\theta_j;d)$ if $-i\eta'(\theta)\in K(\theta_j;d/2)$ and $|s|^{-1}\leqq\rho_0$. Moreover we have

$$\begin{aligned}&\delta_{0,K}(0, s^{-1}, -i\eta'(\theta)+s^{-1}(\xi^{0\prime}-it\vartheta'-i\gamma_1\vartheta'))\\ &= s^{-q_{0,0}}(\delta^0_{0,K,0}(-i\eta'(\theta)+s^{-1}(\xi^{0\prime}-it\vartheta'-i\gamma_1\vartheta'))+o(1))\\ &\qquad\qquad \text{as} \quad |s|\to\infty,\end{aligned} \tag{5.10}$$

where $-i\eta'(\theta)\in K(\theta_j;d/2)$ and $K=K(\theta_j;d)$. Here (5.10) is valid uniformly for $-i\eta'(\theta)\in K(\theta_j;d/2)$. Therefore we have

$$\delta_{0,K}(0, s^{-1}, -i\eta'(\theta)+s^{-1}(\xi^{0\prime}-it\vartheta'-i\gamma_1\vartheta')) \neq 0, \quad |s|^{-1}\leqq\rho_0,$$

if necessary, taking ρ_0 sufficiently small, Thus we have

$$f(s, t, \theta) \neq 0 \quad \text{for} \quad |s|^{-1}\leqq\rho_0 \quad \text{and} \quad 0\leqq\theta\leqq 1. \tag{5.11}$$

Now assume that $f(s_0, t_0, 0)=0$, where Re s_0, Re $t_0\geqq 0$ and Re $(s_0+t_0)>0$. Since $f(s, t_0, 0) \not\equiv 0$ in s, we may assume that Re $t_0>0$. Then there exists a continuous function $s(\theta)$ defined on $[0, 1]$ such that

$$f(s(\theta), t_0, \theta) = 0 \quad \text{and} \quad s(0) = s_0. \tag{5.12}$$

In fact, by (5. 9) and (5.11) we have Re $s(\theta)>0$ and $|s(\theta)|<\rho_0^{-1}$. So we can define $s(\theta)$ on $[0, 1]$. From (5.12) it follows that

$$\Delta(\xi^{0\prime}-i(s(1)+t_0+\gamma_1)\vartheta') = 0,$$

which is a contradiction to (5. 7). Therefore we have

$$\Delta(\xi^{0\prime}-is\eta^{0\prime}-it\vartheta'-i\gamma_1\vartheta') \neq 0 \tag{5.13}$$

for Re s, Re $t \geqq 0$ and Re $(s+t)>0$. This proves the theorem.

Q.E.D.

Definition 5.3. Assume that $\Delta^0(-i\vartheta')\neq 0$. Then we denote by $\Gamma(\Delta)$ the connected component of the set $\{\xi'\in\dot{\Gamma}_0;\ \Delta^0(-i\xi')\neq 0\}$ containing ϑ'.

Corollary 5.1. If (A-3) is valid, then

$$\Delta(\xi') \neq 0 \quad \text{for} \quad \xi'\in\mathbb{R}^{n-1}-i\gamma_1\vartheta'-i\Gamma(\Delta). \tag{5.14}$$

Corollary 5.2. If (A-3) is valid, then $\Gamma(\Delta)$ is an open convex cone in $\mathbb{R}^{n-1}$ with its vertex at the origin.

Proof. From (5.13) and Theorem 5.1 it follows that $\Gamma(\Delta)$ is star-shaped with respect to ϑ'. For each fixed $\eta^{0\prime}\in\Gamma(\Delta)$ there exists $\gamma_1'\in\mathbb{R}$ such that $\Delta(\xi')\neq 0$ if $\operatorname{Im}\xi'\in\{\lambda\eta^{0\prime};\ \lambda<-\gamma_1'\}$, since $\Gamma(\Delta)$ is an open cone. Therefore $\Delta(\xi')$ is hyperbolic with respect to $\eta^{0\prime}$, which implies that $\Gamma(\Delta)$ is star-shaped with respect to $\eta^{0\prime}$. Thus $\Gamma(\Delta)$ is convex. Q.E.D.

Corollary 5.3. If (A-3) is valid, then $\Delta^0(\xi')\neq 0$ for $\xi'\in\mathbb{R}^{n-1}-i\Gamma(\Delta)$.

Proof. We can prove this corollary by the same argument as in the proof of Theorem 5.1. Q.E.D.

From now on we assume that (A-3) is valid, i.e., the mixed problem (4. 1) is C^∞ well-posed.

Lemma 5.10. Let $\xi^{0\prime}\in\mathbb{R}^{n-1}$. Then

$$\Delta_{\xi^{0\prime}}(\eta') \neq 0 \quad \text{for} \quad \eta'\in\mathbb{R}^{n-1}-i\gamma_1\vartheta'-i\Gamma(\Delta).$$

Proof. Assume that there exists $\eta^{0\prime}\in\mathbb{R}^{n-1}-i\gamma_1\vartheta'-i\Gamma(\Delta)$ such that $\Delta_{\xi^{0\prime}}(\eta^{0\prime})=0$. Let K be a compact polydisc and a complex neighborhood of $\eta^{0\prime}$ in $\mathbb{R}^{n-1}-i\gamma_1\vartheta'-i\Gamma(\Delta)$. Then we have $\delta_{\xi^{0\prime},K,0}(\eta^{0\prime})=0$. Since $\delta_{\xi^{0\prime},K,0}(\eta')\neq 0$, there exist $\varepsilon,\delta>0$ and

$\eta^{1\prime}\in\mathbf{C}^{n-1}$ such that

$$|\delta_{\xi^{0\prime},K,0}(\eta^{0\prime}+s\eta^{1\prime})|>\varepsilon \text{ for } |s|=\delta,$$

and $\eta^{0\prime}+s\eta^{1\prime}\in K$ for $|s|\leq\delta$. By (5. 2) and Rouche's theorem there exist $\nu_0>0$ and $s(\nu)$, $0<\nu\leqq\nu_0$, such that

$$\delta_{\xi^{0\prime},K}(\nu^{1/L},\ \eta^{0\prime}+s(\nu)\eta^{1\prime}) = 0 \text{ and } |s(\nu)|<\delta,\ \ 0<\nu\leqq\nu_0,$$

which contradicts (5.14). Q.E.D.

Lemma 5.11. Let $\xi^{0\prime}\in\mathbb{R}^{n-1}$ and Σ be an open connected cone in $\dot{\Gamma}_{\xi^{0\prime}}$ with its vertex at the origin such that $\vartheta'\in\Sigma$. Then the condition

$$\Delta_{\xi^{0\prime}}(\xi')\neq 0 \text{ for } \xi'\in\mathbb{R}^{n-1}-i\gamma_1\vartheta'-i\Sigma \tag{5.15}$$

is equivalent to the following conditions:

$$\Delta_{\xi^{0\prime}}(\xi')\neq 0 \text{ for } \operatorname{Im}\xi_1<-\gamma_1 \text{ and } \xi''\in\mathbb{R}^{n-2}. \tag{5.16}$$

$$(\Delta_{\xi^{0\prime}})^0(\xi')\neq 0 \text{ for } \xi'\in-i\Sigma(\text{or } \xi'\in\mathbb{R}^{n-1}-i\Sigma). \tag{5.17}$$

Proof. The proof is essentially the same as the proof of Theorem 5.1. Assume that there exists $\eta^{0\prime}\in\mathbb{R}^{n-1}-i\Sigma$ such that $(\Delta_{\xi^{0\prime}})^0(\eta^{0\prime})=0$. Let K be a complex neighborhood (compact polydisc) of $\eta^{0\prime}$ in $\mathbb{R}^{n-1}-i\Sigma$. Then there exist $\rho_0>0$, $s(\rho)$, $0<\rho\leqq\rho_0$ and $\eta^{1\prime}\in\mathbf{C}^{n-1}$ such that $\eta^{0\prime}+s(\rho)\eta^{1\prime}\in\mathbb{R}^{n-1}-i\gamma_1\vartheta'-i\Sigma$ for $0<\rho\leqq\rho_0$ and

$$\mu^{-h_{\xi^{0\prime}}}\delta_{\xi^{0\prime},K}(\mu,\ \rho,\ \eta^{0\prime}+s(\rho)\eta^{1\prime})|_{\mu=0} = 0,\ \ 0<\rho\leqq\rho_0. \tag{5.18}$$

Put $K'=\rho_0^{-1}K$. Then we have

$$\delta_{\xi^{0\prime},K',0}(\rho_0^{-1}\xi') = v(\xi')(\mu^{-h_{\xi^{0\prime}}}\delta_{\xi^{0\prime},K}(\mu,\ \rho_0,\ \xi'))|_{\mu=0},\quad \xi'\in K, \tag{5.19}$$

where $v(\xi')\neq 0$ in K. In fact, we have

$$\delta_{\xi^{0\prime},K'}(\nu^{1/L}, \rho^{-1}\xi')^{-1}h_{\xi^{0\prime},K'}(\nu^{1/L}, \rho^{-1}\xi'; \xi_n)$$

$$= \delta_{\xi^{0\prime},K}(\nu^{1/L}, \rho, \xi')^{-1}h_{\xi^{0\prime},K}(\nu^{1/L}, \rho, \xi'; \xi_n)$$

for $0<\nu\leqq\min(\nu_K, \nu_{K'})$, $\xi'\varepsilon K$ and $\rho^{-1}\xi'\varepsilon K'$. Thus there exist analytic functions $u(\mu, \xi')$ and $v(\mu, \rho, \xi')$ such that

$$\delta_{\xi^{0\prime},K}(\mu, \rho_0, \xi') = u(\mu, \xi')\delta_{\xi^{0\prime},K'}(\mu, \rho_0^{-1}\xi'),$$

$$\delta_{\xi^{0\prime},K'}(\mu, \rho^{-1}\xi') = v(\mu, \rho, \xi')\delta_{\xi^{0\prime},K}(\mu, \rho, \xi'). \tag{5.20}$$

(5.19) easily follows from this. By (5.18) and (5.19) we have

$$\Delta_{\xi^{0\prime}}(\rho_0^{-1}(\eta^{0\prime}+s(\rho_0)\eta^{1\prime})) = 0.$$

Thus (5.15) is not valid. Next assume that the conditions (5.16) and (5.17) are valid. Let $\xi^{1\prime}\varepsilon\mathbb{R}^{n-1}$ and $\eta^{0\prime}\varepsilon\Sigma$ and let $\eta'(\theta)$, $0\leqq\theta\leqq1$, be a continuous curve in Σ such that $\eta'(0)=\eta^{0\prime}$ and $\eta'(1)=\vartheta'$. Put

$$f(s, t, \theta) = \Delta_{\xi^{0\prime}}(\xi^{1\prime}-is\eta'(\theta)-it\vartheta'-i\gamma_1\vartheta'),$$

where $\mathrm{Re}\ s$, $\mathrm{Re}\ t\geqq0$, $\mathrm{Re}\ (s+t)>0$ and $0\leqq\theta\leqq1$.

Then (5.16) implies that

$$f(s, t, \theta) \neq 0 \quad \text{for} \quad \mathrm{Re}\ s=0 \quad \text{and} \quad \mathrm{Re}\ t>0.$$

Since

$$\mu^{-h_{\xi^{0\prime}}}\delta_{\xi^{0\prime},K}(\mu, \rho, \xi')|_{\mu=0}$$

$$= \rho^{q_{\xi^{0\prime},0}}(\delta^0_{\xi^{0\prime},K,0}(\xi')+o(1)) \quad \text{as} \quad \rho\to0,$$

by the same argument as in the proof of Theorem 5.1 and (5.20), we have

$$f(s, t, \theta) \neq 0 \quad \text{for} \quad |s|^{-1}\leqq\rho_0 \quad \text{and} \quad 0\leqq\theta\leqq1,$$

where ρ_0 is small enough and t is fixed, $\mathrm{Re}\, t>0$. The proof easily follows from this. Q.E.D.

Definition 5.3. Let $\xi^{0\prime}\in\mathbb{R}^{n-1}$. We denote by $\Gamma(\Delta_{\xi^{0\prime}})$ the connected component of the set $\{\xi'\in\dot{\Gamma}_{\xi^{0\prime}};\ (\Delta_{\xi^{0\prime}})^0(-i\xi')\neq 0\}$ containing ϑ'.

Remark. By Lemmas 5.10 and 5.11 we have $(\Delta_{\xi^{0\prime}})^0(-i\vartheta')\neq 0$.

Corollary 5.4. Let $\xi^{0\prime}\in\mathbb{R}^{n-1}$. Then,

$\Delta_{\xi^{0\prime}}(\xi')\neq 0$ for $\xi'\in\mathbb{R}^{n-1}-i\gamma_1\vartheta'-i\Gamma(\Delta_{\xi^{0\prime}})$.

Corollary 5.5. Let $\xi^{0\prime}\in\mathbb{R}^{n-1}$. Then $\Gamma(\Delta_{\xi^{0\prime}})$ is an open convex cone in $\mathbb{R}^{n-1}$ with its vertex at the origin and $\Gamma(\Delta_{\xi^{0\prime}})\supset\Gamma(\Delta)$.

Corollary 5.6. Let $\xi^{0\prime}\in\mathbb{R}^{n-1}$. Then,

$(\Delta_{\xi^{0\prime}})^0(\xi')\neq 0$ for $\xi'\in\mathbb{R}^{n-1}-i\Gamma(\Delta_{\xi^{0\prime}})$.

Lemma 5.12. Let $\xi^{0\prime}\in\mathbb{R}^{n-1}$. Then,

$(\Delta^0)_{\xi^{0\prime}}(\xi')\neq 0$ for $\xi'\in\mathbb{R}^{n-1}-i\Gamma(\Delta)$.

Lemma 5.13. Let $\xi^{0\prime}\in\mathbb{R}^{n-1}$ and Σ be an open connected cone in $\dot{\Gamma}_{\xi^{0\prime}}$ with its vertex at the origin such that $\vartheta'\in\Sigma$. Then the condition

$(\Delta^0)_{\xi^{0\prime}}(\xi')\neq 0$ for $\xi'\in\mathbb{R}^{n-1}-i\Sigma$

is equivalent to the following conditions:

(i) $(\Delta^0)_{\xi^{0\prime}}(\xi')\neq 0$ for $\mathrm{Im}\,\xi_1<0$ and $\xi''\in\mathbb{R}^{n-2}$.

(ii) $(\Delta^0)_{\xi^{0\prime}}(\xi')\neq 0$ for $\xi'\in-i\Sigma$.

Definition 5.4. Let $\xi^{0\prime}\in\mathbb{R}^{n-1}$. We denote by $\Gamma((\Delta^0)_{\xi^{0\prime}})$ the connected component of the set $\{\xi'\in\dot{\Gamma}_{\xi^{0\prime}};\ (\Delta^0)_{\xi^{0\prime}}(-i\xi')\neq 0\}$ containing ϑ'.

Corollary 5.7. Let $\xi^{0\prime}\in\mathbb{R}^{n-1}$. Then,

$(\Delta^0)_{\xi^0{}'}(\xi') \neq 0$ for $\xi' \in \mathbb{R}^{n-1} - i\Gamma((\Delta^0)_{\xi^0{}'})$.

Corollary 5.8. Let $\xi^0{}' \in \mathbb{R}^{n-1}$. Then $\Gamma((\Delta^0)_{\xi^0{}'})$ is an open convex cone in $\mathbb{R}^{n-1}$ with its vertex at the origin and $\Gamma((\Delta^0)_{\xi^0{}'}) \supset \Gamma(\Delta)$.

Lemma 5.14 ([51]). Let $\xi^0{}' \in \mathbb{R}^{n-1}$ and let M be a compact set in $\dot{\Gamma}_{\xi^0{}'}$. Then there exist a conic neighborhood Γ_1 of $\xi^0{}'$ in $\mathbb{R}^{n-1}$ and C, $t_0 > 0$ such that $P_+(\xi)$ is analytic in $\xi \in \Lambda \times \mathbb{C}$, where $\Lambda = \{\xi' - it|\xi'|\eta' - i\gamma_0 \vartheta';\ \xi' \in \Gamma_1,\ |\xi'| \geqq C,\ \eta' \in M \text{ and } 0 < t \leqq t_0\}$.

Lemma 5.15. Let $\xi^0{}' \in \mathbb{R}^{n-1}$. For any compact set M in $\Gamma((\Delta^0)_{\xi^0{}'})$, there exist a conic neighborhood Γ_1 of $\xi^0{}'$ in $\mathbb{R}^{n-1}$ and C, $t_0 > 0$ such that

$$\Delta(\xi' - it|\xi'|\eta' - i\gamma_1 \vartheta') \neq 0$$

if $\xi' \in \Gamma_1$, $|\xi'| \geqq C$, $\eta' \in M$ and $0 < t \leqq t_0$.

Proof. Put

$$f(\nu, \rho, \xi', s, t, \eta') = \Delta(\nu^{-1}\rho^{-1}\xi^0{}' + \rho^{-1}\xi' - it\rho^{-1}\eta' - is\vartheta' - i\gamma_1 \vartheta'),$$

where $0 < \nu \leqq \nu_0$, $0 < \rho \leqq \rho_0$, $\xi' \in \mathbb{R}^{n-1}$, $|\xi'| \leqq \varepsilon$, Re $s \geqq 0$, Re $t \geqq 0$, Re $(t+s) > 0$, $|s| \leqq s_0$, $|t| \leqq t_0$ and $\eta' \in M$. Here positive numbers ν_0, ρ_0, ε, s_0 and t_0 are chosen so that Lemma 5.14 is applicable and $f(\nu, \rho, \xi', s, t, \eta')$ is well-defined. Applying the same argument as in the proof of Lemma 3.7 in [49], we can prove the lemma. In fact, let K be a compact set in $\mathbb{R}^{n-1} - i\Gamma((\Delta^0)_{\xi^0{}'})$ such that

$$\bar{t}|t|^{-1}(\xi' - it\eta' - i\rho s \vartheta' - i\rho\gamma_1 \vartheta') \in K$$

when $0 < \rho \leqq \rho_0$, $|\xi'| \leqq \varepsilon$, $|s| \leqq s_0$, $|t| = t_0$ and $\eta' \in M$, if necessary, modifying ρ_0, ε and s_0. Then $f(\nu_0, \rho, \xi', t, s, \eta') = 0$ if and only if $\delta_{\xi^0{}', K}(\nu_0^{1/L}, \rho, \xi' - it\eta' - i\rho(s+\gamma_1)\vartheta') = 0$. By (5. 5) we have

$$\delta_{\xi^0{}',K}(\nu_0^{1/L}, \rho, \xi'-it\eta'-i\rho(s+\gamma_1)\vartheta')$$

$$= \nu_0^{h_{\xi^0{}'}/L}\{\textstyle\sum_{j=0}^{\omega} \nu_0^{j/L}\rho^{q_{\xi^0{}',j}}\delta^0_{\xi^0{}',K,j}(\eta')$$

$$+ \textstyle\sum_{j=0}^{\omega} O(\nu_0^{j/L}\rho^{q_{\xi^0{}',j}+1}) + O(\nu_0^{(\omega+1)/L}\rho^{q_{\xi^0{}'}})\}.$$

Thus we can apply the same argument as in [49]. Q.E.D.

Lemma 5.16 ([49]). Let $\xi^0{}'\in\mathbb{R}^{n-1}$. Then we have $\Gamma(\Delta_{\xi^0{}'}) \supset\Gamma((\Delta^0)_{\xi^0{}'})$.

The following theorem can be proved by the same methods as in [51].

Theorem 5.2. Let $\xi^0{}'\in\mathbb{R}^{n-1}$ and let M be a compact set in $\Gamma((\Delta^0)_{\xi^0{}'})$. Then there exists a neighborhood U of $\xi^0{}'$ in $\mathbb{R}^{n-1}$ such that

$$M \subset \Gamma((\Delta^0)_{\xi'}) \quad \text{for} \quad \xi'\in U.$$

6. PROPAGATION OF SINGULARITIES

In this section we assume that (A-1) is valid and that the mixed problem (4. 1) is C^∞ well-posed. Let $\xi^0\in\mathbb{R}^n$ and put

$$\Gamma_{\xi^0} = \Gamma(P_{+\xi^0}) \cap (\Gamma(\Delta_{\xi^0{}'})\times\mathbb{R}),$$

$$\Gamma^0_{\xi^0} = \Gamma(P_{+\xi^0}) \cap (\Gamma((\Delta^0)_{\xi^0{}'})\times\mathbb{R}),$$

where $\Gamma(P_{+\xi})$ denotes the connected component of the set $\{\eta\in\dot{\Gamma}_{\xi'}\times\mathbb{R};\ P^0_{+\xi}(-i\eta)\neq 0\}$ containing ϑ (see [51]). From the results in §5 it follows that for any compact set K in $\mathbb{R}^n-i\gamma_1\vartheta$ there exist $\nu_k>0$, $p_0\in\mathbb{Q}$ and $L\in\mathbb{N}$ such that

$$\nu^{-p_0}P_+(\nu^{-1}\xi^0+\xi)^{-1}\mathcal{K}(\nu^{-1}\xi^0+\xi) = \textstyle\sum_{j=0}^{\infty} \nu^{j/L}\hat{F}_{\xi^0,j}(\xi),$$

whose convergence is uniform for $\xi\in K$ and $0\leqq\nu\leqq\nu_K$ (see [48]). Then we have the following

Theorem 6.1 ([51]). Let $\xi^0\in\mathbb{R}^n$. Then we have

$$\nu^{-p_0}\exp[-i\nu^{-1}x\cdot\xi^0]\tilde{F}(x)$$

$$\sim\sum_{j=0}^{\infty}\nu^{j/L}\tilde{F}_{\xi^0,j}(x) \quad \text{in} \quad \mathcal{D}'(R^n) \quad \text{as} \quad \nu\to 0,$$

where

$$\tilde{F}_{\xi^0,j}(x) = (2\pi)^{-n}\int_{\mathbb{R}^n-i\gamma\vartheta}\exp[ix\cdot\xi]\hat{F}_{\xi^0,j}(\xi)d\xi, \quad \gamma>\gamma_1.$$

Moreover

$$\bigcup_{\xi\in\mathbb{R}^n\setminus\{0\}}\bigcup_{j=0}^{\infty}\operatorname{supp} F_{\xi,j}\times\{\xi\}\subset WF(F)$$

$$\subset WF_A(F)\subset\bigcup_{\xi\in\mathbb{R}^n\setminus\{0\}}(\Gamma_\xi^0)^*\times\{\xi\},$$

$$\overline{\text{ch}}\,[\bigcup_{j=0}^{\infty}\operatorname{supp} F_{\xi,j}]\subset(\Gamma_\xi)^*\subset(\Gamma_\xi^0)^*,$$

where $F_{\xi,j}(x)=\tilde{F}_{\xi,j}(x)|_{x_n>0}$ and $(\Gamma_\xi)^*=\{x\in\mathbb{R}_+^n;\ x\cdot\eta\geqq 0$ for any $\eta\in\Gamma_\xi\}$.

Theorem 6.2 ([51]) $\bigcup_{\xi\in\mathbb{R}^n\setminus\{0\}}(\Gamma_\xi^0)^*\times\{\xi\}$ is closed in $T^*\mathbb{R}_+^n\setminus 0$.

These theorems can be proved by the same argument as in [51], using the results in §5. Next we shall study singularities near the boundary of the Poisson kernel F(x).

Definition 6.1 ([10]). Let $u(\cdot, x_n)\in C^\infty([0,\infty);\ \mathcal{D}'(\mathbb{R}^{n-1}))$. We say that a point $(x^{0\prime}, \xi^{0\prime})$ in $T^*\mathbb{R}^{n-1}\setminus 0$ is not in the set $WF_0(u)$ if there exist $\phi\in C_0^\infty(\mathbb{R}^{n-1})$, a conic neighborhood Γ of $\xi^{0\prime}$ and $\varepsilon>0$ such that $\phi(x^0)\neq 0$ and

$$|\mathcal{F}_{x'}[\phi(x')D_n^j u(x)](\xi')|\leqq C_{jk}(1+|\xi'|)^{-k}$$

when $\xi'\in\Gamma$, $x_n\in[0, \varepsilon)$ and $j,k=0,1,2,\cdots$.

Theorem 6.3. We have

$$WF_0(F)\subset\bigcup_{\xi\in\mathbb{R}^{n-1}\setminus\{0\}}\Gamma((\Delta^0)_{\xi'})^*\times\{\xi'\}.$$

Proof. First let us prove that for any $\xi_n\in\mathbb{R}$

$$\Gamma((\Delta^0)_{\xi'})^* = \{x'\in\mathbb{R}^{n-1};\ x'\cdot\eta'\geqq 0 \text{ for any } \eta\in\Gamma_\xi^0\}. \tag{6. 1}$$

Put $\dot{\Gamma}^0_\xi=\{\eta'\in\mathbb{R}^{n-1};\ \eta\in\Gamma^0_\xi$ for some $\eta_n\in\mathbb{R}\}$. Then we have

$$(\dot{\Gamma}^0_\xi)^* = \{x'\in\mathbb{R}^{n-1};\ x'\cdot\eta'\geqq 0 \text{ for any } \eta\in\Gamma^0_\xi\},$$

$$\dot{\Gamma}^0_\xi = \Gamma((\Delta^0)_{\xi'}) \cap \dot{\Gamma}(P_{+\xi}).$$

On the other hand we have $\dot{\Gamma}(P_{+\xi})=\dot{\Gamma}_{\xi'}$, which implies (6. 1). In fact, by Rouche's theorem we have $P^0_{+\xi}(-i\vartheta', -i\lambda)\neq 0$ for $\lambda\geqq 0$. $P^0_{+\xi}(\zeta)$ is a polynomial of ζ_n and its coefficient $a(\zeta')$ of the highest power of ζ_n does not vanish for $\zeta'\in -i\dot{\Gamma}_{\xi'}$. Therefore we have $\dot{\Gamma}(P_{+\xi})\supset\dot{\Gamma}_{\xi'}$. Let $(x^{0\prime}, \xi^{0\prime})\in T^*\mathbb{R}^{n-1}\setminus 0$, $|\xi^{0\prime}|=1$. Now assume that $(x^{0\prime}, 0)\notin(\Gamma^0_{(\xi^{0\prime},\xi_n)})^*$ for any $\xi_n\in\mathbb{R}$, i.e., $x^{0\prime}\notin\Gamma((\Delta^0)_{\xi^{0\prime}})^*$. Thus there exist $\eta(\xi_n)\in\Gamma^0_{(\xi^{0\prime},\xi_n)}$, $\xi_n\in\mathbb{R}$, such that $x^{0\prime}\cdot\eta(\xi_n)'<0$. By Lemma 5.1 there exist $d>0$ and $\eta^0\in\mathbb{R}^n$ such that $\eta^0=(\eta^{0\prime}, 0)$, $x^{0\prime}\cdot\eta^{0\prime}<0$ and $\eta^0\in\Gamma^0_{(\xi^{0\prime},\xi_n)}$ for any $\xi_n\in\mathbb{R}$ with $|\xi_n|>d$. For it is obvious that there exists $\eta^0\in\Gamma((\Delta^0)_{\xi^{0\prime}})\times\mathbb{R}$ $\subset\bigcap_{j=1}^{p'}\Gamma(((p_j)_{(0,1)})_{(\xi^{0\prime},0)})$ such that $\eta^0=(\eta^{0\prime}, 0)$ and $x^{0\prime}\cdot\eta^{0\prime}<0$. Let M be the line segment joining ϑ and η^0. By Rouche's theorem, we have

$$M \subset \bigcap_{j=1}^{p'} \Gamma((p_j)_{(\xi^{0\prime},\xi_n)}) \quad \text{if} \quad |\xi_n|>d,$$

applying Lemma 5.1 with $\xi'=\lambda\xi^{0\prime}$ and $t=\lambda^{-1}$. Thus we have $\eta^0\in\Gamma^0_{(\xi^{0\prime},\xi_n)}$. Moreover, by Lemma 5.1, there exist a conic neighborhood Γ of $\xi^{0\prime}$ in $\mathbb{R}^{n-1}$ and C, d, $t_0>0$ such that

$$P_+(\xi-i(t|\xi'|\eta^0+\gamma_2\vartheta)) \neq 0 \tag{6. 2}$$

for $\xi'\in\Gamma$, $|\xi'|\geqq C$, $\xi_n\in\mathbb{R}$, $|\xi_n|>d|\xi'|$ and $0\leqq t\leqq t_0$, where $\gamma_2=\gamma_1+1$. From Theorem 5.2 it follows that there exist cones $\Gamma(\xi_n)$, $\xi_n\in\mathbb{R}$, such that

$$\Gamma(\xi_n) = \{\zeta\in\mathbb{R}^n;\ \zeta'\in\gamma(\xi_n) \text{ and } b_1(\xi_n)<\zeta_n/|\zeta'|<b_2(\xi_n)\}$$

and $\eta(\xi_n)\in\Gamma^0_\zeta$ for $\zeta\in\Gamma(\xi_n)$ and $\xi_n\in\mathbb{R}$, where $\gamma(\xi_n)$ is a conic neighborhood of $\xi^{0\prime}$ in $\mathbb{R}^{n-1}$ and $\xi_n\in(b_1(\xi_n), b_2(\xi_n))$. Then there exist $\xi_n^j \in \mathbb{R}$, $1 \leqq j \leqq s$, such that $|\xi_n^j| \leqq d$ and $\bigcup_{j=1}^s I_j \supset [-d,d]$, where $I_j = (b_1(\xi_n^j), b_2(\xi_n^j))$. Choose $\psi_j \in C^\infty(\mathbb{R})$, $0 \leqq j \leqq s$, so that $0 \leqq \psi_j \leqq 1$ for $0 \leqq j \leqq s$, supp $\psi_j \subset I_j$ for $1 \leqq j \leqq s$, supp $\psi_0 \subset (-\infty,-d) \cup (d,\infty)$ and $\sum_{j=0}^s \psi_j \equiv 1$. Put

$$\tilde{\eta}(\xi_n) = \psi_0(\xi_n)\eta^0 + \sum_{j=1}^s \psi_j(\xi_n)\eta(\xi_n^j).$$

Then we have $\tilde{\eta}(\xi_n) \in \Gamma^0_{(\xi^{0\prime},\xi_n)}$ for $\xi_n \in \mathbb{R}$. By Lemma 5.15, (6. 2) and Seidenberg's lemma there exist a conic neighborhood Γ of $\xi^{0\prime}$ in $\mathbb{R}^{n-1}$, positive constants C, t_0 and δ, and $a \in \mathbb{Q}$ such that

$$|\Delta(\zeta' - i(t|\zeta'|\tilde{\eta}(\zeta_n)' + \gamma_2 \vartheta'))P_+(\zeta - i(t|\zeta'|\tilde{\eta}(\zeta_n) + \gamma_2 \vartheta))|$$
$$\geqq \delta|\zeta'|^a, \quad \zeta' \in \Gamma, \quad |\zeta'| \geqq C, \quad \zeta_n \in \mathbb{R} \text{ and } 0 \leqq t \leqq t_0.$$

By Seidenberg's lemma, there exist $b \in \mathbb{Q}$, $b \geqq 1$, and $C' > 0$ such that

$$P_+(\zeta' - i(t|\zeta'|\tilde{\eta}(\zeta_n)' + \gamma_2 \vartheta'), \lambda) \neq 0$$

if $\zeta' \in \Gamma$, $|\zeta'| \geqq C$, $\zeta_n \in \mathbb{R}$, $0 \leqq t \leqq t_0$ and $|\lambda| \geqq C'|\zeta'|^b$. In fact, $\{\tilde{\eta}(\xi_n);\ \xi_n \in \mathbb{R}\} \subset\subset \bigcap_{j=1}^{p'} \Gamma(((p_j)_{(0,1)})_{(\xi^{0\prime},0)})$. We may assume that

$$P_+(\zeta' - i\gamma_2 \vartheta', \lambda) \neq 0 \quad \text{for } \zeta' \in \mathbb{R}^{n-1} \text{ and } |\lambda| \geqq C'|\zeta'|^b,$$

and that $C' \geqq 2d$ and $C \geqq 1$. Put

$$\begin{aligned} \mathscr{C} = \{\zeta \in \mathbb{C}^n;\ \zeta' \in \mathbb{R}^{n-1} \text{ and } &(i)\ \ \zeta_n/|\zeta'|^b \in [-C', C'] \\ &\text{or } \zeta_n = C'|\zeta'|^b e^{i\theta},\ 0 \leqq \theta \leqq \pi, \text{ if } |\zeta'| \geqq C, \\ &(ii)\ \ \zeta_n \in [-C^b C', C^b C'] \text{ or } \zeta_n = C^b C' e^{i\theta}, \\ &0 \leqq \theta \leqq \pi, \text{ if } |\zeta'| < C\}. \end{aligned}$$

Let $\Gamma_1 (\subset\subset \Gamma)$ be a conic neighborhood of $\xi^{0\prime}$ in $\mathbb{R}^{n-1}$ and choose $\phi \in C^\infty(\mathbb{R}^{n-1})$ so that ϕ is positively homogeneous of degree zero in $|\xi'| \geqq C+1$ and $\phi(\xi') = 1$ for $\xi' \in \Gamma_1 \cap \{|\xi'| \geqq C+1\}$, $0 \leqq \phi \leqq 1$, supp $\phi \subset \Gamma$ and $\phi(\xi') = 0$ for $|\xi'| \leqq C$. Then we have

$$|\Delta(\zeta' - i(tv(\zeta)' + \gamma_2 \vartheta'))P_+(\zeta - i(tv(\zeta) + \gamma_2 \vartheta))|$$
$$\geqq \delta|\zeta'|^a \quad \text{for } \zeta \in \mathscr{C} \text{ and } 0 \leqq t \leqq t_0,$$

where

$$v(\zeta) = \begin{cases} \phi(\zeta')|\zeta'|\tilde{\eta}(\zeta_n) & \text{if } \zeta\in\mathbb{R}^n, \\ \phi(\zeta)|\zeta'|\eta^0 & \text{if } \operatorname{Im}\zeta_n>0. \end{cases}$$

Moreover there exist a neighborhood U of $x^{0\prime}$ in $\mathbb{R}^{n-1}$ and $\varepsilon>0$ such that $x\cdot v(\zeta)\leqq 0$ for $\zeta\in\mathscr{C}$ and $x\in U\times[0,\varepsilon]$. For $\chi\in C_0^\infty(U)$ we have

$$\mathscr{F}_{x'}[\chi(x')D_n^jF(x)](\xi')$$
$$= (2\pi)^{-n}\int_{\mathscr{C}-i\gamma_2\vartheta} \hat{\chi}(\xi'-\zeta')\exp[ix_n\cdot\zeta_n]\zeta_n^jP_+(\zeta)^{-1}\mathscr{K}(\zeta)d\zeta.$$

It is easy to see that

$$|\hat{\chi}(\xi'-\zeta'+i(tv(\zeta)'+\gamma_2\vartheta'))\exp[ix_n\cdot(\zeta_n-itv_n(\zeta))]|$$
$$\leqq C_k(1+|\xi'-\zeta'+i(tv(\zeta)'+\gamma_2\vartheta')|)^{-k}$$

for $\xi'\in\mathbb{R}^{n-1}$, $\zeta\in\mathscr{C}$, $x_n\in[0,\varepsilon]$, $0\leqq t\leqq t_0$ and $k=0,1,2,\cdots$. Write

$$\mathscr{C}_t = \{\zeta-itv(\zeta);\ \zeta\in\mathscr{C}\}.$$

Then, by Stokes' formula, we have

$$\mathscr{F}_{x'}[\chi(x')D_n^jF(x)](\xi')$$
$$= (2\pi)^{-n}\int_{\mathscr{C}_{t_0}-i\gamma_2\vartheta} \hat{\chi}(\xi'-\zeta')\exp[ix_n\cdot\zeta_n]\zeta_n^jP_+(\zeta)^{-1}\mathscr{K}(\zeta)d\zeta.$$

Since $|\xi'-\zeta'+i(t_0v(\zeta)'+\gamma_2\vartheta')|\geqq c(|\xi'|+|\zeta'|)$, $c>0$, for $\xi'\in\Gamma_1$ and $\zeta\in\mathscr{C}$, we have

$$|\mathscr{F}_{x'}[\chi(x')D_n^jF(x)](\xi')| \leqq C_{jk}(1+|\xi'|)^{-k}$$

for $\xi'\in\Gamma_1$, $x_n\in[0,\varepsilon]$ and $j,k=0,1,2,\cdots$, which proves $(x^{0\prime}, \xi^{0\prime})\notin WF_0(F)$. Q.E.D.

Put

$$\hat{F}^{(j)}(\xi') = (2\pi)^{-1}\oint \xi_n^jP_+(\xi)^{-1}\mathscr{K}(\xi)d\xi_n,\quad j=0,1,2,\cdots.$$

Then the $\hat{F}^{(j)}(\xi')$ are rational functions of $\sigma^k(\xi')$, $1\leqq k\leqq \ell$, and ξ'. Let $\xi^{0\prime}\in\mathbb{R}^{n-1}$ and let K be a compact set in $\mathbb{R}^{n-1}-i\gamma_1\vartheta'-i\Gamma(\Delta_{\xi^{0\prime}})$. By Lemma 5.3 and Corollary 5.4, there exists $\nu_K>0$ such that

$$\nu^{-s_j}\hat{F}^{(j)}(\nu^{-1}\xi^{0\prime}+\xi') = \sum_{k=0}^\infty \nu^{k/L}\hat{F}^{(j)}_{\xi^{0\prime},k}(\xi')$$

for $\xi'\varepsilon K$ and $0<\nu\leqq\nu_K$, whose convergence is uniform. The $\hat{F}^{(j)}_{\xi^{0},k}(\xi')$ are analytic in $\mathbb{R}^{n-1}-i\gamma_1\mathcal{J}'-i\Gamma(\Delta_{\xi^{0}'})$ and we can apply the Paley-Wiener-Schwartz theorem. Write

$$F^{(j)}_{\xi^{0},k}(x')=\mathcal{F}^{-1}[\hat{F}^{(j)}_{\xi^{0},k}(\xi')](x').$$

Then we have the following

Theorem 6.4. Let $\xi^{0}{}'\varepsilon\mathbb{R}^{n-1}$. Then,

$$\nu^{-s_j}\exp[-i\nu^{-1}x'\cdot\xi^{0}{}'](D_n^jF(x))|_{x_n=0}$$

$$\sim\sum_{k=0}^{\infty}\nu^{k/L}F^{(j)}_{\xi^{0}{}',k}(x') \quad \text{in} \quad \mathcal{D}'(R^{n-1}) \quad \text{as} \quad \nu\to 0.$$

Moreover we have

$$\bigcup_{\xi'\varepsilon\mathbb{R}^{n-1}\setminus\{0\}}\bigcup_{j,k=0}^{\infty}\operatorname{supp}F^{(j)}_{\xi',k}\times\{\xi'\}\subset WF_0(F),$$

$$\overline{\mathrm{ch}}\,[\bigcup_{j,k=0}^{\infty}\operatorname{supp}F^{(j)}_{\xi^{0}{}',k}]\subset\Gamma(\Delta_{\xi^{0}{}'})^*\subset\Gamma((\Delta^0)_{\xi^{0}{}'})^*.$$

Remark. By Theorem 6.1, we also have $(x^{0}{}', \xi^{0}{}')\notin WF_0(F)$ if there exist $\{x^j\}\subset\mathbb{R}^n_+$ and $\{\xi^j\}\subset\mathbb{R}^n$ such that $x^{j}{}'\to x^{0}{}'$, $x^j_n\to 0$ and $\xi^{j}{}'\to\xi^{0}{}'$ as $j\to\infty$, and $x^j\varepsilon\bigcup_{k=0}^{\infty}\operatorname{supp}F_{\xi^j,k}$.

Let $x^{0}{}'\varepsilon\mathbb{R}^{n-1}$ and let $U_j(j=1,2)$ be neighborhoods of $x^{0}{}'$, $U_1\supset\supset U_2$. Then there exists $\{\phi_k\}\subset C_0^{\infty}(\mathbb{R}^{n-1})$ such that $\operatorname{supp}\phi_k\subset U_1$, $\phi_k(x')=1$ on U_2 and

$$|D'^{\alpha'}\phi_k(x')|\leqq C(Ck)^{|\alpha'|} \quad \text{for} \quad |\alpha'|\leqq k$$

(see [18]).

Definition 6.2. Let $u(\cdot, x_n)\varepsilon C^{\infty}([0,\infty);\mathcal{D}'(R^{n-1}))$. We say that a point $x^{0}{}'\varepsilon\mathbb{R}^{n-1}$ is not in $\operatorname{sing\,supp}_{A0}u$ if there exist $\{\phi_k\}(\subset C_0^{\infty}(\mathbb{R}^{n-1}))$ of the above type and M, C, $\varepsilon>0$ such that

$$|\mathcal{F}_{x'}[\phi_k(x')D_n^ju(x)](\xi')|\leqq C^{k+1}(1+|\xi'|)^M(k+|\xi'|)^j$$

for $x_n\varepsilon[0,\varepsilon)$, $0\leqq j\leqq k$ and $k=1,2,\cdots$. Moreover we say that a point $(x^{0}{}', \xi^{0}{}')\varepsilon T^*(\complement\operatorname{sing\,supp}_{A0}u)\setminus 0$ is not in $WF_{A0}(u)$ if there exist $\{\phi_k\}$ of the above type, a conic neighborhood Γ of $\xi^{0\prime}$ in $\mathbb{R}^{n-1}$ and C, $\varepsilon>0$ such that

$$|\mathcal{F}_{x'}[\phi_k(x')D_n^j u(x)](\xi')| \leqq C(Ck)^k(k+|\xi'|)^{j-k}$$

for $\xi'\in\Gamma$, $x_n\in[0,\varepsilon)$, $0\leqq j\leqq k$ and $k=1,2,\cdots$, where $U_1 \subset\subset$ sing supp_{A0} u.

Lemma 6.1. Let $u\in C^\infty([0,\infty);\mathcal{D}'(\mathbb{R}^{n-1}))$. Then,

$$\pi(WF_{A0}(u)) \cup \text{sing supp}_{A0}\, u$$
$$= \mathbb{R}^{n-1}\setminus\{x'\in\mathbb{R}^{n-1};\ \text{there exist a neighborhood } U \text{ of } x' \text{ and } \varepsilon>0 \text{ such that } u \text{ is analytic in } U\times[0,\varepsilon)\},$$

where π: $T^*(\complement\text{sing supp}_{A0}\, u)\setminus 0\to \complement\text{sing supp}_{A0}\, u$ is the natural projection.

Lemma 6.2. Let $u\in C^\infty([0,\infty);\mathcal{D}'(\mathbb{R}^{n-1}))$. Then,

$$WF_0(u) \subset \text{sing supp}_{A0}\, u\times(\mathbb{R}^{n-1}\setminus\{0\}) \cup WF_{A0}(u).$$

Assume that

(A-4) $\Pi_{j=1}^{p'}\ p_j(0,\ 1) \neq 0.$

Then we have the following

Theorem 6.5. Assume that (A-4) is valid. Then,

$$\text{sing supp}_{A0}\, F(x) = \emptyset.$$

By the same argument as in the proof of Theorem 6.3 we have the following

Theorem 6.6. Assume that (A-4) is valid. Then,

$$WF_{A0}(F) \subset \bigcup_{\xi'\in\mathbb{R}^{n-1}\setminus\{0\}} \Gamma((\Delta^0)_{\xi'})^*\times\{\xi'\}.$$

Define

$$Fg(x)=F(x)\underset{x'}{*}g(x')$$

for $g\in\mathcal{D}'(\mathbb{R}^{n-1})$ with $\mathrm{supp}\, g\subset\{x_1\geqq 0\}$. Put

$$C = \{((x,\ \xi),\ (y',\ \eta'))\in T^*\mathbb{R}^n_+\times T^*\mathbb{R}^{n-1}\setminus 0;\ \xi'=\eta' \text{ and } (x'-y',\ x_n)\in(\Gamma^0_\xi)^*\},$$

$$C_0 = \{((x', \xi'), (y', \eta')) \varepsilon T^*\mathbb{R}^{n-1} \times T^*\mathbb{R}^{n-1} \setminus 0;\ \xi' = \eta' \text{ and } x'-y' \varepsilon \Gamma(((\Delta^0)_{\xi'})^*\}.$$

Then the following theorem is a consequence of Theorems 6.1, 6.3, 6.5 and 6.6.

Theorem 6.7. Let $g \varepsilon \mathcal{D}'(\mathbb{R}^{n-1})$ with $\operatorname{supp} g \subset \{x_1 \geq 0\}$.

Then,

$$WF(Fg) \subset C \circ WF(g),\quad WF_A(Fg) \subset C \circ WF_A(g),$$

$$WF_0(Fg) \subset C_0 \circ WF(g).$$

Moreover, if (A-4) is valid, then we have

$$\text{sing supp}_{A0}\, Fg = \emptyset,\quad WF_{A0}(Fg) \subset C_0 \circ WF_A(g).$$

Remark. If $u \varepsilon C^\infty([0,\infty); \mathcal{D}'(\mathbb{R}^{n-1}))$ satisfies the system (4. 1) with $f=0$ and $g=0$, then $u \equiv 0$ (see the proof of Theorem 4. 4). Therefore $Fg(x)$ is a unique solution in $C^\infty([0,\infty); \mathcal{D}'(\mathbb{R}^{n-1}))$ of the system (4. 1) with $f=0$ and $g \varepsilon \mathcal{D}'(\mathbb{R}^{n-1})$.

REFERENCES

1. Agemi, R., and Shirota, T.: On necessary and sufficient conditions for L^2-well-posedness of mixed problems for hyperbolic equations, J. Fac. Sci. Hokkaido Univ., 21 (1970), pp 133-151.
2. ______________________: On necessary and sufficient conditions for L^2-well-posedness of mixed problems for hyperbolic equations, II, J. Fac. Sci. Hokkaido Univ., 22 (1972), PP 137-149.
3. Agmon, S., Douglis, A., and Nirenberg, L.: Estimates near the boundary for solutions of elliptic partial differential equations satisfying general boundary conditions II, Comm. Pure Appl. Math., 17 (1964), PP 35-92.
4. Andersson, K. G., and Melrose, R. B.: The propagation of singularities along gliding rays, Invent. Math., 41 (1977), pp 197-232.
5. Atiyah, M. F., Bott, R. and Gårding, L.: Lacunas for hyperbolic differential operators with constant coefficients, I, Acta Math., 124 (1970), pp 109-189.
6. Beals, R.: Mixed boundary value problem for nonstrict hyperbolic equations, Bull. Amer. Math. Soc., 78 (1972), pp 520-521.
7. Bierstone, E.: Extension of Whitney fields from subanalytic set, Invent. Math., 46 (1978), pp 277-300.

8. Chazarain, J.: Problèmes de Cauchy abstraits et application à quelques problèmes mixtes, J. Functional Anal., 7 (1971), pp 386-446.
9. ____________: Construction de la paramétrix du problème mixte hyperbolique pour l'équation des ondes, C. R. Acad. Sci. Paris, 276 (1973), pp 1213-1215.
10. ____________: Reflection of C^∞ singularities for a class of operators with multiple characteristics, Publ. RIMS, Kyoto Univ., 12 Suppl. (1977), pp 39-52.
11. Duff, G. F. D.: On wave fronts, and boundary waves, Comm. Pure Appl. Math., 17 (1964), pp 189-225.
12. Eskin, G.: A parametrix for interior mixed problems for strictly hyperbolic equations, J. Anal. Math., 32 (1977), pp 17-62.
13. Hersh, R.: Mixed problem in several variables, J. Math. Mech., 12 (1963), pp 317-334.
14. _________: Boundary conditions for equations of evolution, Arch. Rational Mech. Anal., 16 (1964), pp 243-264.
15. _________: On surface waves with finite and infinite speed of propagation, Arch. Rational Mech. Anal., 19 (1965), pp 308-316.
16. Hörmander, L.: "Linear Partial Differential Operators," Springer-Verlag, Berlin, Göttingen, Heidelberg, 1963.
17. ____________: "On the singularities of solutions of partial differential equations," International Conference of Functional Analysis and Related Topics, Tokyo, 1969.
18. ____________: Uniqueness theorems and wave front set for solutions of linear differential equations with analytic coefficients, Comm. Pure Appl. Math., 24 (1971), pp 671-703.
19. Ikawa, M.: Mixed problem for the wave equation with an oblique derivative boundary condition, Osaka J. Math., 7 (1970), pp 495-525.
20. ________: Sur les problèmes mixtes pour l'équation des ondes, Publ. RIMS, Kyoto Univ., 10 (1975), pp 669-690.
21. Ivrii, V. Ja., and Petkov, V. M.: Necessary conditions for the Cauchy problem for non-strictly hyperbolic equations to be well-posed, Uspehi Mat. Nauk, 29 (1974), pp 3-70. (Russian; English translation in Russian Math. Surveys.)
22. Kajitani, K.: Sur la condition nécessaire du problème mixte bien posé pour les systémes hyperboliques à coefficients variables, Publ. RIMS, Kyoto Univ., 9 (1974), pp 261-284.
23 ____________: A necessary condition for the well posed hyperbolic mixed problem with variable coefficients, J. Math. Kyoto Univ., 14 (1974), pp 231-242.
24. ____________: On the $\mathcal{E}$-well posed evolution equations, Comm. in Partial Differential Equations, 4 (1979), pp 595-608.
25. Kasahara, K.: On weak well posedness of mixed problem for hyperbolic systems, Publ. RIMS, Kyoto Univ., 6 (1971), pp 503-514.
26. Kreiss, H.-O.: Initial boundary value problems for hyperbolic systems, Comm. Pure Appl. Math., 23 (1970), pp 277-298.

27. Lax, P. D.: Asymptotic solutions of oscillatory initial value problems, Duke Math. J., 24 (1957), pp 624-646.
28. Lejeune-Rifaut, E.: "Solutions élémentaires des problèmes aux limites pour des opérateurs matriciels de dérivation hyperbolique posés dans un demi-espace," Dissertation, Liège, 1978.
29. Matsumura, M.: Localization theorem in hyperbolic mixed problems, Proc. Japan Acad., 47 (1971), pp 115-119.
30. Melrose, R. B.: Microlocal parametrices for diffractive boundary value problems, Duke Math. J., 42 (1975), pp 605-635.
31. Melrose, R. B., and Sjöstrand, J.: Singularities of boundary problems I, Comm. Pure Appl. Math., 31 (1978), pp 593-617.
32. Miyatake, S.: Mixed problem for hyperbolic equation of second order, J. Math. Kyoto Univ., 13 (1973), pp 435-487.
33. __________: Mixed problems for hyperbolic equations of second order with first order complex boundary operators, Japanese J. Math., 1 (1975), pp 111-158.
34. Mizohata, S.: Some remarks on the Cauchy problem, J. Math. Kyoto Univ., 1 (1961), pp 109-127.
35. __________: On evolution equations with finite propagation speed, Israel J. Math., 13 (1972), pp 173-187.
36. Nirenberg, L.: "Lectures on Linear Differential Equations," Regional Conference Series in Math., 20 (1973), Amer. Math. Soc., Providence, Rhode Island.
37. Sakamoto, R.: $\mathcal{E}$-well posedness for hyperbolic mixed problems with constant coefficients, J. Math. Kyoto Univ. 14 (1974), pp 93-118.
38. __________: L^2-well-posedness for hyperbolic mixed problems, Publ. RIMS, Kyoto Univ., 8 (1972/73), pp 265-293.
39. Seeley, R. T.: Extension of C^∞ functions defined in a half space, Proc. Amer. Math. Soc., 15 (1964), pp 625-626.
40. Shibata, Y.: A characterization of the hyperbolic mixed problems in a quarter space for differential operators with constant coefficients, Publ. RIMS, Kyoto Univ., 15 (1979), pp 357-399.
41. __________: $\mathcal{E}$-well posedness of mixed initial-boundary value problem with constant coefficients in a quarter space, to appear.
42. Shirota, T.: On the propagation speed of hyperbolic operator with mixed boundary conditions, J. Fac. Sci. Hokkaido Univ., 22 (1972), pp 25-31.
43. Sjöstrand, J.: Propagation of analytic singularities for second order Dirichlet problems, Comm. in Partial Differential Equations, 5 (1980), pp 41-94.
44. Solonnikov, V. A.: On general boundary problems for systems which are elliptic in the sense of A. Douglis and L. Nirenberg. I, Izv. Akad. Nauk, 28 (1964), pp 665-706. (Russian; English translation in Amer. Math. Soc. Transl., 56 (1966).)
45. Taylor, M. S.: Grazing rays and reflection of singularities of solutions to wave equations, Comm. Pure Appl. Math., 29 (1976), pp 1-38.

46. Tsuji, M.: Fundamental solutions of hyperbolic mixed problems with constant coefficients, Proc. Japan Acad., 51 (1975), pp 369-373.
47. Wakabayashi, S.: Singularities of the Riemann functions of hyperbolic mixed problems in a quarter-space, Proc, Japan Acad., 50 (1974), pp 821-825.
48. ______________: Singularities of the Riemann functions of hyperbolic mixed problems in a quarter-space, Publ. RIMS, Kyoto Univ., 11 (1976), pp 417-440.
49. ______________: Analytic wave front sets of the Riemann functions of hyperbolic mixed problems in a quarter-space, Publ. RIMS, Kyoto Univ., 11 (1976), pp 785-807.
50. ______________: Microlocal parametrices for hyperbolic mixed problems in the case where boundary waves appear, Publ. RIMS, Kyoto Univ., 14 (1978), pp 283-307.
51. ______________: Propagation of singularities of the fundamental solutions of hyperbolic mixed problems, Publ. RIMS, Kyoto Univ., 15 (1979), pp 653-678.
52. ______________: A necessary condition for the mixed problem to be C^∞ well-posed, to appear in Comm. in Partial Differential Equations.

INDEX

A

B

C

G

H

I

K

L

M